THE ROCKS ARE SMILING AT US

by

Bevan Warris

Fifty Years in the Australian Oil Industry

ISBN No: 978-0-646-81794-1

Cover Picture by: Syncline outcropping in the lower parking lot of Calico Ghost Town, San Bernardino County, California, USA.

Printed by: iPrintPlus

CONTENTS

EARLY DAYS

I went to university to be a chemist. My mother bought me a chemistry set when I was 12 years old and I just loved it. I bought a chemistry book and learned the theory behind all the reactions. By the time I went to high school, I already knew the first year's chemistry curriculum.

I was born in Murwillumbah, NSW in 1943 but moved to Sydney when I was very young. I was the only child of a single mum **(Fig 1)**. My childhood was interesting to say the least because we moved about a lot as my mum went from job to job. Whilst we were poor, there were some advantages. I learned to cook at an early age as I would help get dinner ready when my mum came home from work.

Life was simpler then. No TV. No computers. But we made our own fun. I remember living in St Marys in the early 1950's and attending the local primary school. It was just great as there was lots of bush and our gang would spend lots of time there playing war games, catching spiders and tadpoles, and fishing for yabbies. Sometimes there was too much bush and I remember one summer coming home from school early as bushfires were threatening our homes. Some things never change.

The highlights of the week were the Saturday afternoon pictures. My mum would give me a shilling (provided I'd tidied my room and dried the dishes) and the pictures cost 11 pence so I had a penny left over to buy a chocolate bar. The pictures always started with a cartoon followed by a serial which generally consisted of 12 episodes, one each week. This was followed by another cartoon and then the main attraction, either a Western, Tarzan movie or a war movie. Adults were not allowed in and we could make as much noise as we liked – which we did! It was just great.

I must have been relatively smart in primary school because I qualified to go to North Sydney Boys High School **(Fig 2)**. This was a terrific school and I reckon I got the best education and start in life from the teachers at this school. I learned Latin, French and German; ancient and modern history; maths, physics and chemistry; and all aspects of English. The school motto is "Vincit qui se vincit" which in Latin roughly means "He who conquers must first conquer himself" but the older boys preferred "Wink at those who wink at you" as a better translation!

Figure 1
Three year old Bevan gives the thumbs up

Figure 2
Thank you North Sydney Boys High

Our teachers were mostly people who took up teaching in later life and they had their own life experiences to impart to us. Our German teacher had a million mark note from the great inflation period of the 1920's – it would then only buy a loaf of bread. Our Latin teacher made us translate Caesar's Gallic Wars from Latin to English so we learned some history (even if it was somewhat biased) on the way to learning Latin. Our physics teacher, in our last year, told us that as most of us would be going to University, we should get into training and learn the necessary disciplines. He told us that we didn't have to attend his class or submit our homework anymore because that's how we were going to be treated at University. It turned out to be very practical advice and I have never forgotten it.

As it turned out, I was good at maths and science. Although I also loved ancient history and German, I was hopeless at these subjects so it was a scientific career for me. And of course I just loved chemistry. Apart from getting good marks, my chemistry was also useful in other areas. The Russians had launched Sputnik I in October, 1957 and I don't know what the adults thought about it but to a 13 year old boy, it was absolutely stupendous! The space race was on - and we were not to be outdone. We had a club who designed and built model rockets. I joined the club to help provide the rocket fuel. We experimented with a number of chemicals but zinc dust and sulphur was the best. In those days you could buy powdered sulphur from any chemist shop and, with a bit of fast talking, zinc dust could be bought from Selbey's, the local industrial chemical supplier. Zinc dust and sulphur was a very flexible mixture. In certain proportions, it would burn slowly generating voluminous clouds of choking white smoke. In a different proportion, it would burn very quickly – perfect for rockets. The rockets were made out of copper piping. Some of the kid's dads had access to welding gear and they made up the fins. I would add the zinc dust and sulphur and we would take the rockets down to the Tunk's Park and set them off pointing towards the harbour. How we didn't kill someone (including ourselves) is still a mystery to me. We did come close once. After watching a war film on the Allied bombing of German cities during WW2, I decided to experiment with phosphorous. What I needed was the right oxidising agent. After some research, I learned that Condie's Crystals was the go. These could be bought in any chemist shop until their sale was banned because kids kept dropping them into fountains where they made the water a bright fluorescent green. Little buggers! Again the phosphorous could be bought from Selbey's. The first time we tried it out it blew the rocket to smithereens and showered us all with blobs of molten phosphorous. It turned out that phosphorous was really nasty stuff which continued to burn after the explosion! It was unanimously decided by the club that it was never to be used again. So we returned to good old faithful – zinc dust and sulphur.

Now zinc dust and sulphur was not just good for rocket fuel. It had other uses. On Monday mornings there was an assembly to go through the weekend sporting events. This was a tedious affair which everyone hated. So I was commandeered to make smoke bombs to disrupt these assemblies and bring sanity and order back into our school lives. The zinc dust and sulphur was prefect because in the right mixture (not the rocket fuel one) it would burn slowly generating voluminous clouds of choking white smoke, just perfect for disrupting assemblies. I would make the smoke bombs at home and bring them to school. One of the gang would drop it into a rubbish bin and another would, at the appropriate time, set fire to the rubbish bin. The great white clouds of choking white smoke were very effective in disrupting assemblies. However, we weren't invincible and these escapades resulted in many visits to the headmaster and more "six of the best" canings than I could count. The cunning bastards always did it on your left hand so that you still had to do your homework.

SYDNEY UNIVERSITY

I passed my matriculation and made it to Sydney University in 1961. Although I was in the top class at North Sydney Boys High, there were lots of boys brighter than me in this class so I never thought I was very smart. When I got my matriculation results, I was very surprised how well I did. I was the dumbest kid in the top class in one of the best schools in the state. Thank you again North Sydney Boys High for giving me such a great start in my academic life.

My high matriculation score also won me a Commonwealth Scholarship to fund my university fees. This scheme, envisaged by Ben Chifley, was implemented by the Menzies government in 1951 and boosted university enrolments from 12,000 in 1938 to 96,000 in 1966. Without it, I would not have been able to go to university – at least not initially. It did have a kicker to it – if you failed a year, you lost it. So I studied very hard at university because I knew I wouldn't get a second chance. I guess I was a bit of a nerd at university.

The way the first year science course was set up in those days, you had to do four units: maths, physics, chemistry and something else out of biology, zoology, psychology or geology. The first three sounded really boring so I decided on geology. Pure maths was the toughest – there were 1,000 students in the lecture theatre and by the end of the hour, the lecturer had filled twelve blackboards so you had to write quickly. And if you didn't know anything about calculus, you were doomed. The drop out rate was horrendous, and of those who made it to the year end exams, 50% failed.

In April, 1961, the Russians put the first man, Yuri Gagarin, into space and his spacecraft, Vostok completed one full orbit of the earth. The Russians were certainly off to a flying start to the space race.

I found geology interesting except for crystallography which was crushingly boring. But the best thing about geology was the field trips. Our first field trip was to Lithgow, a coal mining town on the other side of the Blue Mountains. We wandered about the bush, chipping on outcrops and mapping the various sedimentary deposits – but we were a motley crew including the lecturers **(Fig 3)**!

Figure 3
Come on guys – show a bit of enthusiasm

We were staying in the local pub and on the first night, after dinner we all descended to the bar. Because my birthday was in December, I was still only 17 and was a bit nervous about drinking in such a public place. But I was quickly told by Tony Gates not to worry as we were in the country – "she'll be right" he assured me. In Lithgow in those days, 10 o'clock closing was just a rumour. So we were all drinking away when the cops marched into the pub at midnight. So there I was, 17 years old, standing in a pub, illegally drinking at midnight – I thought I was doomed. But it was OK – the cops were also the SP bookies and were just there to collect the bets for the Saturday races. Tony Gates was right – "she was right", and we all carried on drinking.

The next night, the local bodgies descended on the pub to check out these "poofta students". One of these bodgies had some stones which he reckoned were diamonds but his mates all scoffed at him. He showed them to us and, low and behold, there were a couple of very small diamonds. Nothing valuable but he felt pretty good. This unfortunately didn't change the mood of the rest of the bodgies and a fight quickly broke out and spread rapidly to the whole bar. Being 5' 6" and eight stone (50 kilos), I made a hasty retreat – everybody in the room was bigger than me! The fight didn't last long but the damage to the pub was considerable and we were firmly told to leave and find other accommodation. The lecturers were furious but I thought that if this was Geology 101, then I'd definitely be back next year for Geology II.

Starting in August, 1961, the communists built the infamous Berlin Wall around the East German part of city and completely enclosed it from West Berlin. To me this was by far the greatest indictment of the communist system of government. Throughout the last 5,000 years of written human history, civilisations had often built walls around their cities to keep out barbarians, invaders and other types of riff-raff. But this was the first time that a civilisation had built a wall around a city to keep its own citizens in!

After the exciting field trip to Lithgow, in second year university, I enrolled in Geology II as well as Chemistry II and Maths II. Big mistake!!!! The work load was enormous. To make matters worse, October, 1962 was not a good time for doing a lot of study. The Americans and the Russians seemed hell bent on blowing each other to smithereens and taking the rest of the world with them. Most people today don't realise how close the world came to a nuclear war during those two tense weeks of the Cuban Missile Crisis. I myself thought it was a bloody miracle that we didn't. We spent most of the time in the local pub drinking and playing pool – not much point in going to lectures if Western civilisation was about to come to a screaming end. And living in Australia, it wouldn't have been a quick death by being instantly raised to a temperature of 10,000 degrees but a slow lingering death by radiation poisoning. When people talk about the Cuban Missile Crisis, I always remember that line in a song "If the Russians love their children too". As it turned out "the Russians did love their children too" and the Americans and Russians didn't blow the world up so we all went back to lectures. But there was always a lingering doubt that, somewhere out there in the vast wastelands of Siberia and Nevada, there was a maniac in charge of one of those missile silos. They even made a film about it – Dr Strangelove – a classic, very funny film if you didn't think too deeply about it. But we all lived life to the full as none of us expected to live past 30.

I almost failed second year – scraped through with three passes. The worst exam was Applied Maths. I calmly walked into the exam room, sat down and the supervisor announced at the appropriate time, that we could start. I opened the paper and it said "answer all seven questions". So far so good!

Then as I studied the questions, my heart sank. I didn't have a clue. I answered one question – it is permanently seared into my memory. It was "you have a board inclined at 30° and a billiard ball with a co-efficient of friction of x with the board, you spin it backwards with an angular momentum of y, how far up the board will the billiard ball travel". That was the easiest question! For the rest of the questions, I just wrote down some formulas so the marker would know that I actually did attend some of the lectures. Then it was over. Fortunately, everyone else was equally as ignorant as me and I just managed to scrape through with a pass (which must have been a very low percentage).

Even the Geology exams were hard. During the year, our lecturer, Dr Gordon Packham, gave a throw away line that he didn't have time to discuss lacustrine sedimentation and we should read up on it ourselves. Of course, guess what was in the exam? "Discuss lacustrine sedimentation". I couldn't even spell "lacustrine", let alone write anything meaningful about it. However, I still managed to pass Geology 2. I also passed Chemistry but found that it no longer interested me and decided that a career in Chemistry was not for me. So I majored in Geology for my Third Year at University.

Before I started Third Year, I got some work experience with the Bureau of Mineral Resources (BMR) in Canberra. The BMR was essentially the Commonwealth Geological Survey and they were busy mapping the remote areas of Australia. My work experience was only about three weeks but it was a lot of fun. We stayed in a student hostel where there were students from all over the world under the Columbo Plan. This plan, introduced in 1951, extended Australian university education to students in our neighbouring countries across Southeast Asia. So we were definitely a mixed multicultural bag. At the BMR, most of the geology students were simply colouring and folding maps. However, my supervisor was Dave Forman and he had just returned from the Musgrave Ranges on the South Australian and Northern Territory border. So he got me describing all the thin sections of the rock samples he collected. I was definitely full bottle on the petrology of metamorphic rocks when I finished at the BMR and started Third Year Geology.

One weekend, we all got together and went to Yass where there was a good outcrop of Silurian **(see Appendix 1 Geological Timescale)** limestones full of fossils. It was truly spectacular and I remember picking up a perfect trilobite (long extinct). I held it in my hand and suddenly realised that this wasn't just a rock but was actually once an animal that swam around in a warm shallow sea on the very spot where I was standing over 400 million years ago! And I was the first living creature in 400 million years to see its remains! That was it – I was hooked. So I started Geology III. Again it was the field trips that broke up the tedium of attending lectures **(Fig 4)**.

Figure 4
I'm the little fellow on the right

One of field trips was visiting the Broken Hill mine and studying the surrounding geology. The mine visit was fascinating. They decked us out in hard hats and lamps etc, and we assembled in the cage to take us down into the mine. The bloke on the brake thought he'd have some fun and gave us the "rapid descent tour". I remember my stomach floating just below my eyeballs! Once we got down, we went along one of tunnels to study the geology. It was very impressive. The sheer size of the tunnels – you could run a train along them – which of course is exactly what they did! I remember one section where they walls and the roof of the tunnel all sparkled. It was all pure galena (the lead ore) – the tunnel was cut through solid galena. The mine itself was very hot, very dusty and very noisy. As we reached the surface into the daylight (I never really appreciated sunlight until that moment) and watched the next shift of miners descend into the mine, I thought to myself "whatever they pay those blokes, it isn't enough".

After that, it was the surface geology – very complicated but our lecturer Dr Hobbs seemed to know what was going on. One amusing situation presented itself when we ended up in a town called Cockburn on the South Australian border. In the men's toilet there was an interesting sign which informed me that "You are now pissing on South Australia".

In November, 1963, John Kennedy was assassinated. It was a momentous event which shocked the entire world. I heard it on the radio just after I'd finished cleaning windows for a local car yard. The job started at 4am, finished at 9am and I was driving down to Harborne Beach to go surfing. I think that everyone from my generation knows exactly where they were when he was assassinated.

Third Year was easier than Second Year (especially without any more nuclear war crises) and I passed, even with some credits and distinctions. So I decided to do an Honours Degree. As I was interested in sedimentary geology not metamorphic, my supervisor Gordon Packham suggested an area around the town of Orange as my mapping project. This was an area of Silurian deep water sediments with areas of shallow water limestone deposits along the ancient island arc systems. But first, I had to get set up. I contacted the local station owner and he allowed me to stay in the shearer's quarters. This was great and infinitely better than camping – it gets bloody cold in Orange during the winter. Now in my opinion, the greatest invention of the 20th century was the "frypan". This afforded me a certain amount of luxury on my field trips. I could get fresh supplies in Orange and ate very well. One evening, I even managed to cook a roast dinner – the station owner's wife was really impressed. The field work involved a lot of walking. The outcrop was poor and could only be accessed along the various creeks in the area **(Fig 5)**.

Figure 5
Not much outcrop here

In those days, there was no GPS.

Locating yourself was by air photos of the area and trying to recognise the roads, creeks and trees. You marked the outcrops on the air photos and wrote copious notes in your field book. Back at Uni, you put the air photos together and made a map. It wasn't high tech by today's standards but it worked – very well. The greatest danger was snakes and wasps. The snakes didn't bother me as I quickly learned that they were more frightened of me than I was of them. Still, you had to be careful. It was not good field practice to step on one because if you got bitten, you were on your own and most likely a long way from your vehicle. But the worst hazard was the wasps. These were the most aggressive beasties I have ever encountered. These things resented *Homo sapiens* and firmly believed that insects actually ruled the world – not humans. All humans were to be attacked (and presumably exterminated) on sight and they certainly made my life a misery. If they saw me they would immediately attack, even if I was nowhere near them or their nest. I carried a blue bag with me and every time I got bitten, I would wet it in the creek and rub it on the bite to ease the pain. For those of you who don't know what a blue bag is, these date back to the dark ages when people washed their clothes in copper boilers before the invention of washing machines! I even remember turning the wringer as my mum fed the clothes through to get the water out. But the blue bags were just great for insect bites.

Despite all these hazards, I completed my field work, wrote up my thesis and passed with first class honours. It was December, 1964 and I spent my 21st birthday at the famous Marble Bar at the old Adams Hotel. I was joined by a couple of my school mates, Ted Scribner and John Shedrick and we spent the evening drinking beer and listening to Kenny Ball and his Jazzmen. It was a great evening. The Adams Hotel was eventually sold but the historic Marble Bar was saved and reassembled in the new Hilton Hotel in 1973.

I also decided to do something about my puny physique. I went to the gym and started pushing weights. One of the other patrons asked what I was doing benchpressing weights heavier than me and I told him that I wanted to put on weight. He advised me to carry on doing what I was doing but in addition, drink a pint of milk a day. It worked a treat and I was soon up to 11 stone (70 kilos).

I received by Bachelor of Science with First Class Honours on 14 April, 1965 **(Fig 6)**.

Figure 6
With mum after the ceremomy

FIRST WELL-SITE JOB

I then decided to do a PhD. But before I started my PhD, I had the opportunity to do some well site work to supplement my meager finances. One of our geology class of 1963, Mike Wiltshire was already out working and suggested I contact Jim Cundell of Cundell, Myers & Associates who were looking for well site geologists. So I rang Jim Cundell and after a short conversation ended up on a flight to Melbourne and an overnight stay in a first class hotel (this Geology degree was looking better every day). I received a very quick lesson in well site geology from Jim and then I was on my way to Casterton-1, an exploration well drilling in the Otway Basin in western Victoria. It was April, 1965 and this was my first well-site job – the beginning of my career in the oil industry. When I arrived at the well site, I was amazed at the sheer size of everything - the derrick, tanks, motors and the noise. Now despite all this equipment, drilling an oil well is actually quite simple **(Fig 7)**.

The drill bit is connected to the drill-pipe and rotated to drill the hole. Drilling mud is pumped down the inside of the drill-pipe, through the drill bit and up the outside of the drill-pipe. The drill bit has a larger diameter than the drill-pipe to create the annulus. The drilling mud has several uses. Firstly, it cools the drill bit as it grinds its way through the rocks. Secondly, it brings the rock chips (called cuttings) up to surface to clean the hole and allows the well site geologist to examine the cuttings under a microscope and determine what rocks the well is drilling through. Thirdly, the mud lines the well bore and keeps it from collapsing, although under some conditions, this is not always successful. Fourthly, and most importantly, the mud provides weight and pressure at the bottom of the well which prevents the oil and gas, if discovered, from blowing out under its own pressure. Again, if everyone on the well is not on the ball, this may not be successful. Blowouts are the worst part of the oil industry and although extremely rare, can cause death, destruction and environmental pollution.

The other well site geologist was an American who gave me a run down on what they'd drilled through and informed me that I had the day shift - and then left. So there I was looking after this well without any on-site training. I had to bluff my way through and convince everyone on the rig that I

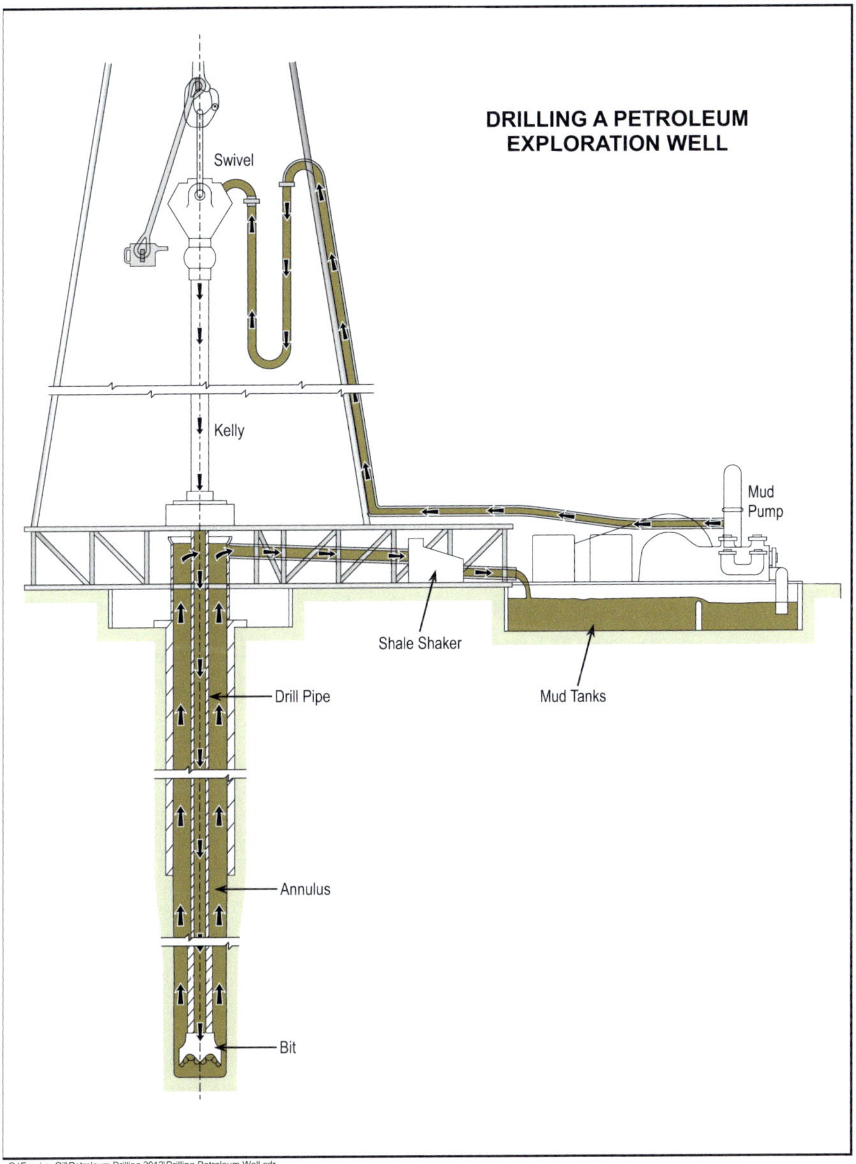

Figure 7
Drilling an Exploration Well

knew what I was doing. Fortunately it was a routine well and I didn't have to make any major decisions. As it turned out, the driller knew as much about what we were drilling through as I did. I learned to respect these men who could tell from the feel of the drill bit and the torque on the drill pipe what lithologies they were drilling through.

On the rig, you were supposed to wear a hard hat on duty. This rule was very lax and only about half the roughnecks wore one. As it was bloody hot, I thought I wouldn't wear one either. At the time, operations were pulling out of the hole to change the drilling bit. This involved the rig motors lifting the drill pipe up to a connection, the roughnecks putting the slips under the connection to prevent the drill pipe falling down the hole, and then unscrewing the pipe with tongs. Once the connection was broken, one tong was held in place and the pipe was unscrewed by pulling on a chain wrapped round the pipe. Unfortunately one time, the chain broke, unwrapped itself off the pipe and went flying through the air at the speed of sound, fortunately missing the roughnecks who had hit the deck at twice the speed of sound. I was standing in the dog house and saw the whole thing. It occurred on me that drilling oil wells was a seriously dangerous occupation and after that, I always wore my hard hat and steel capped boots on a well site – no matter how hot it was!

On the way back from the well site after one of my shifts, I ran out of petrol just as I entered the town. The gauge read full so I couldn't figure it out - nobody had told me that the vehicle had two tanks and all I had to do was flip a switch to change over tanks. It was 7-00pm so I decided to hitch a ride back to the motel and call for help from there. As I was standing by the vehicle, a car pulled up and this bloke got out, flashed his cop badge at me and started asking lots of questions about who I was, who's vehicle it was etc etc. Good grief I thought – he thinks I've stolen it! Telling him I was a geologist from Sydney working on an oil drilling rig outside of town didn't impress him. I told him I was staying at the motel and they could vouch for me. So he called the motel and they told him I was "a nice young man, clean cut etc". Of course, at that time I was in rig gear and boots covered in mud, unshaven and looking very scruffy. Not convinced, he decided to drive me to the motel so they could identify me. On the way, he got a call – there was a fight outside the local hotel. So away we went and sure enough the fight was in full flight – about six blokes going for it and the cops trying to break it up which they eventually did by beating the crap out of them. Those were the "good old days" when there was no such thing as "police brutality"! Those that hadn't run off, they threw in the back of a paddy wagon which drove off. After all this excitement, we finally got to the motel where I was recognised and finally released from custody. After that I organised to get the vehicle the next day and had a very good night's sleep.

Finally the operator decided that they had tested all the objectives and ordered that the well be plugged and abandoned. At the end of drilling a well, it has to be electric logged prior to plugging and abandonment. This consists of dropping a series of logging tools down the hole on the end of a thick wire connected to a spool inside a truck. All the tools are pulled up the hole, measuring the various properties of the rocks (radioactivity, density, resistivity, sonic velocity). These are ingenious tools – sandstones are non-radioactive while shales are so the gamma ray tool distinguishes your reservoirs and seals. Oil and fresh water are resistive while salt water is conductive so the resistivity tool distinguishes oil from water but you must know the salinity of the formation water – fresh water is resistive and can look like oil. Tight sandstones are denser than porous sandstones so the various tools can identify sands from shales, oil from water and the quality of the reservoirs. Of course "Mother Nature's a bitch" and the rocks are always more complicated than this, so the interpretation of electric logs is a specialist business all on its own.

But the best part of this story was the logging truck. Amazingly, the electric logging business in those days was dominated by a French company called Schlumberger – and this in the midst of an oil business controlled by the Americans. Now Schlumberger's trucks are dark blue and have more electrical gear inside than Houston space mission control. And they were designed by the Marquis de Sade. Because rig time is expensive, the logging must be carried out continuously until completed. This can take up to 36-48 hours so the trick is to keep the operator awake. This is done in two ways – firstly, if the operator falls asleep and his head falls forward, it hits a buzzer which wakes him up. Secondly, if he falls asleep and his head falls backward, the chair collapses and he falls off – which of course wakes him up. The only consolation to this job was that they were paid enormous amounts of money.

DOCTOR OF PHILOSOPHY DEGREE

After my exciting introduction to the world of well site geology, I returned to boring old university to start my PhD. My supervisor was again Gordon Packham who offered me a number of projects to choose from. The most intriguing was around Milparinka, an area north of Broken Hill where some Cambrian trilobites had been found but the geology had never been mapped in detail. So I decided to take on this project.

Only one problem – it was about 1,500 kilometres from Sydney and petrol costs were going to eat up my meager resources. The distance from Sydney to Milparinka is about the same as from Berlin to Moscow – and I hadn't left the State of New South Wales. The most direct route was from Sydney to Dubbo, then Nyngan, onto Cobar, Wilcannia and Broken Hill. From there you travelled up the road north to Tibooburra. In those days, I drove a Vauxhall Viva – OK for the city but not so good for the bush. The road from Narromine to Nyngan was straight but only one lane, and was known by the locals as "the crystal highway". I soon found out why. The road was littered with broken windscreens caused by stones flying up as cars passed each other with only one set of wheels on the bitumen. Fortunately, I made it up the crystal highway without breaking anything but my car got lots of chips in the paint work. But in 1966, the bitumen stopped at Cobar. It was tough on my car, rattling across the dirt, and very corrugated road from Cobar to Broken Hill. The half way mark was Wilcannia, a most welcome stop. But no – it was a Wednesday and everything was closed – even the pubs. By this stage of the journey my tongue was covered in red dust and I was in desperate need of a beer before I continued the next bone-rattling 200 kilometres to Broken Hill. As I sat in my car pondering my predicament, I noticed a steady trickle of blokes disappearing up this alley. So I followed one and found my way into the back of one of the pubs. Never did a beer taste so good! As a town, Wilcannia did not have much going for it. The area was in drought and the mighty Darling River we were taught about in school – well it was just a river of sand. As I drove out of the town, I thought to myself that I should try to be good for the rest of my life or when I died, the Devil would send me to Wilcannia for all eternity!

Eventually, I arrived at Broken Hill. Now here was a town. There was a pub on every corner and more new cars than you would see in Sydney. Despite the drought, there was a mining boom on and the miners were paid a "lead bonus" on the profits the companies made. Broken Hill was a very rich town. I stayed with Trevor Jones (another of the class of '63) and his wife Marina and headed off the next day to Milparinka. More dirt roads but now the country was in serious drought. There were dead sheep everywhere and trucks were bringing the barely live ones south to greener pastures. I found out later that the area had less than an inch of rain during the entire year of 1965.

When I got to Milparinka, I got instruction to get to the property of Mount Arrowsmith. Again, the property owner Bill Evans put me up in the shearer's quarters but alas, no electricity. The frypan would remain clean and I had to live on sandwiches, cheese and cold meats. Every now and then, Bill and his wife would take pity on me and invite me in for a decent meal. This was a real experience for me as I was wondering why any sane person would choose to live on such a barren, god forsaken landscape. But the house was surrounded by large trees and beautiful gardens, and I discovered that the family lacked for nothing. They had a huge wood stove which not only cooked their food but heated the house as well (it gets very cold in the desert in winter). Bill informed me that the area received good rains (up to12 inches) during the 1950's and he made a lot of money during those years when wool prices were high due to the Korean War. He then bought a pub in Broken Hill and the profits from this pub allowed him to survive the droughts – very smart was Bill.

So off I went mapping the desert in my Vauxhall Viva. There wasn't much outcrop so it was very easy to find my way around on the air photos. The geology was mostly Proterozoic (600 million years ago) sediments with inliers of Early Palaeozoic. The Proterozoic was amazing being mostly shale but with units of thick sandstone and thin limestones. The limestones were barely half a metre thick but extended for miles **(Fig 8)**. I called this sequence the Euriowie Formation. It must have been a very large, shallow shelf with occasional changes in sea level to deposit the limestones. The limestones also contained the odd stromatolite called *Collenia frequens*. These are an ancient form of marine algae which lived in the shallow sea which covered this part of NSW some 600-700 million years ago. They don't look much in outcrop – just some wavy structures and crinkles in the limestone but in thin section you can see the internal structure of the algae **(Fig 9)**.

Figure 8
These limestones are 700 million years old

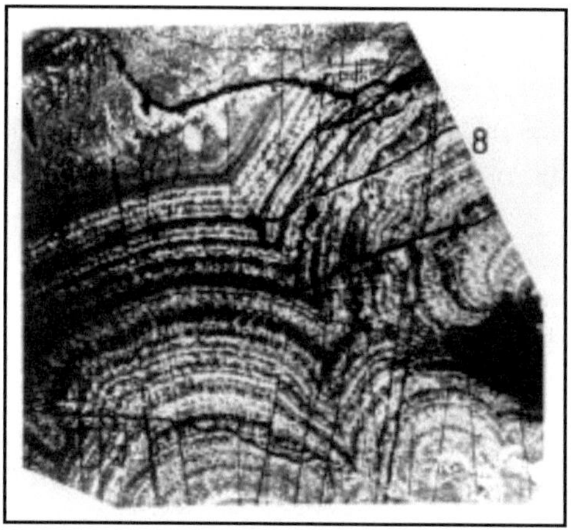

Figure 9
Algae structure of a stromatolite

Although primitive forms of life, stromatolites are very resilient and can still be found today living in Shark Bay, Western Australia. Stromatolites, being a form of marine plant, take in carbon dioxide and using the sun's energy via the process of photosynthesis, break the carbon dioxide into carbon which allows them to grow and oxygen which is expelled into the atmosphere. We should therefore give a vote of thanks to these creatures as they evolved some two billion (yes billion – this is not a typo) years ago. Very slowly over the next billion and a half years, they were responsible for slowly converting our then atmosphere of nitrogen and carbon dioxide into the nitrogen and oxygen atmosphere that allowed more complex life to evolve on this planet of ours some 500 million years ago. It then took another 500 million years for *Homo sapiens* to evolve!

Of course it didn't take me long to get the Viva bogged in sand. This was quite a disaster as it very hot and I was over 10 miles from the Arrowsmith homestead. But I had the air photos and knew if I followed a particular outcrop of sandstone, I would get back there. Incredibly stupid but I didn't know that I should have stayed in the car. And even if I had, it could have been days before anyone noticed me missing. So off I trudged along the outcrop but it was nightfall by the time I made it and I could see the lights of the homestead to guide me the last few miles. But I had a taste of what dying of thirst might feel like. By the time I got to the homestead, my tongue was so swollen in my mouth that I could barely speak.

After I retrieved the Viva, I returned to Sydney. I got an early start and was making good time and decided to keep driving. Between Bathurst and Lithgow, the road has lots of bends -not good for driving tired and at night. I must have dozed off for a second, hit the shoulder, skidded and drove sideways off the road and down the embankment. It's amazing what goes through your mind at moments like this. I remember as if it was yesterday – "What have you done you stupid bastard - you're going to die!" The car rolled down the embankment and eventually crashed against a tree. Fortunately, seat belts had just been introduced and this amazing device undoubtedly saved my life. The car was upside down and I was suspended inside. I saw the red ignition light and not particularly wanting the petrol tank to catch fire, I turned off the ignition. The windscreen had popped out and I was struggling to get out. I eventually realised that the seat belt was still done up so I released it and crawled out through the windscreen. I had a scratch on my knee. I walked up the embankment to the road and the next car that appeared was an ambulance. My guardian angel was definitely looking after me that night! I got a lift back to Bathurst, stayed overnight and next day arranged to retrieve the car. Incredibly, it was still drivable – a little breezy without a windscreen but I got it back to Bathurst for repairing. I then hitched a ride back to Sydney. All in all, it was a very expensive exercise and I vowed that for the sake of the cost of a motel room, I would never drive tired again.

When I got back to Sydney, I went to the Geology Department and when Gordon Packham saw me, he said "Oh, you're back". Like this was unusual! Did PhD students often disappear in the desert, never to be heard of again? Looking back, I now realise that "Oh, you're back" was the sum total of Sydney University's health and safety manual for 1965.

Before I went back, I knew I had to get my hands on a 4W drive vehicle if I was going to do any serious mapping on my PhD area. To the rescue came John Blumer who arranged for me to borrow a landrover from Aquitaine

Petroleum, a French company who was exploring for oil in the Amadeus and Bonaparte basins in the Northern Territory and northern Western Australia. The exploration targets in these basins were the same age as those on my PhD area so the company lent me the vehicle in return for a copy of my PhD thesis. This was a good deal for both of us as I got a 4W drive vehicle free and the company got a free report on another area to fill in their overall geology of the Early Palaeozoic of Australia. The vehicle I got was a long wheel base Landrover. It was just perfect. I could drive anywhere and camp overnight as I could sleep comfortably in the back of the vehicle **(Fig 10)**. This was great as my PhD area covered some 25,000 square kilometres, virtually all of it accessible only to 4W drive vehicles. One part of my area was interesting morphologically. It consisted of deep eroded trenches which were too steep to traverse even with a Landrover. I had to detour round each one before I found a suitable crossing and getting to the various outcrops took forever.

I should also comment on the wildlife or lack thereof. Because of the drought, there were virtually no kangaroos or emus. The only species which survived this god forsaken land was the goats. They were completely wild and came in assorted colours – black, white, grey, spotted, striped, an amazing variety. Their survival ability was improved by their ability to climb trees – yes climb trees – I wouldn't believe it either except I saw them with my own eyes. And I never smoked marijuana – I think I

Figure 10
Home away from home

missed the 60's. The goats climbed the trees to get at the juicy leaves that the sheep couldn't reach. The sheep died - the goats survived. It was an excellent example of how the process of evolution works on this little blue planet of ours.

The sheep also had other problems.While mapping I came across a dam with dozens of dead sheep around it. As the dam was half full of water, I was rather mystified. On my return to Mount Arrowsmith, I asked Bill what had happened. He told me that during a drought, the sheep are too afraid to move too far from the water in the dam. So they eat all the grass around the dam and eventually starve to death **(Fig 11)**.

Figure 11
Northwestern NSW was not a nice place in 1965.

Initially my mapping was concentrated on the Cambrian outcrops at Gnalta and Pincally homesteads **(Fig 12)**. This photo also gives you an idea of how desolate my PhD area was in 1965. Despite this, it was a palaeontologist's dream as there were plenty of trilobites to collect. The trilobite species were mostly identifiable from similar specimens collected from outcrops in the Northern Territory and described by Dr Opic at the BMR. My PhD area was the easternmost part of the Australian continent during the Cambrian, 500 million years ago. At that time, everything to the east of Gnalta was the deep oceanic crust of the Pacific Ocean. Most of NSW, Queensland, Victoria and Tasmania simply did not exist.

Figure 12
That's me on the right. Note the fault on the left

During one of my field mapping expeditions around Mootwingee, I was mapping the Late Devonian outcrops. These were made up mostly of sandstones and conglomerates which were hard and resistive and formed major hills in the area. They were deposited in rivers and flood plains which were the main environments in the area after the sea levels fell and the Cambrian and Ordovician seas had long retreated to the east. In one of the outcrops there was a large cave with aboriginal paintings including one of a snake which extended the full width of the cave. I named these rocks the Snake Cave Formation after the site. I had never seen Aboriginal rock art before and these were excellent examples being totally free of graffiti normally associated with any historical site. I tried to get a photo of the snake but the light wasn't right. I mentioned my find to the Nivens' at Acacia Downs and they groaned in disappointment. They swore me to secrecy as they had known of the site for years but had kept it secret to prevent possible future vandalism. I never divulged the location to anyone but later learned that the site was declared a national park in 1982 and is today a major tourist site (properly supervised of course).

The real gem in my PhD area was the Ordovician outcrop at Mount Arrowsmith. This was a syncline which formed a small amphitheater in the desert. The amphitheater ridge was formed by basal quartzites overlain by marine shales which eroded out to form the amphitheater floor. These shales had thin interbeds of limestones which were just chock a block full of fossils. There were brachiopods (extinct sea shells), pelecypods (normal sea shells), trilobites and nautiloids (extinct squid-like creatures but with a skeleton). All these fossils were new species and even new genera and I thought that I had found a whole list of extinct animals that were totally unknown to the rest of the world. However, after discussions with Joyce Gilbert-Tomlinson at the BMR in Canberra, I discovered that most of these fossils also occurred in the Amadeus Basin in the Northern Territory. My prized new genus of trilobite not only occurred in the Amadeus Basin but had also been described from China. Joyce referred me to a paper by Kobyashi who described it as *Asaphus illarensis*. Kobyashi described many other fossils from China and his was the only record in English at the time of my PhD of these fossils from China.

Another bonus from the Mount Arrowsmith limestones was that if you dissolved them in monochloroacetic acid, they were also full of microfossils called conodonts. These are extinct creatures which look like microscopic teeth. They aren't teeth but are the skeletal remains of an unknown life form which, because its soft body is never fossilised in its complete form. It would be like finding the individual bones of a human being but never seeing an in-tact skeleton. Because they swim in the ocean and drift with the currents, they are very widespread and are very good for age dating limestones. The sea in which the Mount Arrowsmith conodonts lived can be traced through the Amadeus Basin in the Northern Territory, the Canning Basin in Western Australia, through China, across Russia and into the Baltic State of Estonia. Even taking into account that, 450 million years ago, Western Australia was joined to Indochina, the extent of that Ordovician sea was quite impressive.

After a couple of trips back and forth via Cobar, I needed a change of scenery. On my next trip, I decided to go southwest to Cowra, Hay, then to Mildura and north to Broken Hill. It was longer but I enjoyed the change. During the drive between Hay and Balrandal, the land is so flat that you could see the trucks and cars coming for miles away. But Mildura was a pretty town with the Murray River flowing past and it was nice to take in the green countryside before reaching the red brown, desolate wastelands of northwestern NSW. In 1965, the road north to Broken Hill was dirt and in the drought, very dusty. Not long after leaving Mildura, I ended up behind a road train travelling very slowing up the road to Broken Hill. The road train was trailing a steady stream of brown dust obscuring visibility for some distance behind it - I couldn't see a damn thing. Eventually, there was a sufficiently curved part of the road that I could see no stream of brown dust coming in the other direction. I decided to overtake the road train and as I passed through and out of the dust stream, I thought that I was in the clear. But no - up ahead was a fence with a cattle grid across the road – and guess what – it was only one car wide. I wasn't stupid enough to take on a road train in my gutless Landrover so I jammed on the brakes, let the road train cross the cattle grid and stopped by the road in a cloud of dust. I reckon that road train driver is still laughing and it seemed like hours before the dust cleared and I could resume my journey.

In my desire to vary my boring trips between Sydney and Broken Hill, I decided to return to Sydney across "the "top end" via Milparinka, Bourke, Brewarrina and down to Sydney. Bad decision! The road from Milparinka to Bourke – well, it was a figment of the New South Wales map makers' imagination. I've seen sheep tracks that were more easily followed. The "road" crossed many sand dunes and these weren't easy to navigate. You needed to have a good run up, pick up as much speed as possible and hope like hell you had enough momentum to reach the top of the dune – after which there was an exciting drop down the other much steeper side. Sometimes it took me a couple of attempts to get across. In places, I was tempted to get off the road and onto the salt lakes to avoid these dunes. But after my many disasters in the bush, I was learning to be more careful before plunging headlong into the unknown. I decided to test the strength of the salt lakes before driving on them. To my horror, I discovered that even my puny weight was enough to break the salt crust and my boots descended into a disgusting stinking, black, clayey ooze. So I went back to plan A and my relentless battle with the dunes.

I eventually arrived in Bourke. With respect to the inhabitants of Bourke and Brewarrina, my impressions of these places were coloured by the dreadful drought that covered most of northern and western New South Wales in 1965 and 1966. Like most of the towns I traveled through, the countryside was covered with red-brown sand, thick dust, sparse vegetation, dead sheep, millions of flies and endless days with brilliant blue skies. They did have one redeeming feature for a weary geologist – the loveliest old-fashioned pubs where a man could wash down the dust and flies out of his throat. And after my struggle across the dunes, it took more than one middie.

After my discovery that I had the same Ordovician fossils as those in the Amadeus Basin, I needed to see the Camro-Ordovician section in that basin. So I managed to plead my way onto the next BMR field trip to the basin. All I had to do was get to Alice Springs and I could join up with the field party. This PhD thesis was threatening to bankrupt me before I even started working. The flight from Adelaide to Alice Springs was in a turbo-prop DC-8 which cruised at an altitude of 8,000 feet (~2400 metres). I had an excellent aerial view of the Flinders Ranges, the Leigh Creek open cut coal mine, the eastern edge of Lake Eyre and the southern outcrops of the Amadeus Basin. I was surprised by Alice Springs which was a pretty town with the McDonald Ranges as a picture postcard backdrop.

I caught up with Peter Cook who was leading the party. It turned out that there were only three of us with the third member being a French geologist from the Institute Francais de Petrole. It was the middle of winter and I was amazed at how cold it was at night. We slept out in the open with a piece of tarpaulin covering our swags. In the morning, there was a thin crust of ice on the tarpaulin so it was thick socks and two jumpers around the morning fire cooking breakfast. But by 10-00am, it was 25°C and we were in shorts and T-shirts. The fascinating thing about the rocks in the Amadeus Basin was the colours. I had seen paintings by Albert Namatjira and always thought that his colours were artificial. But in fact, he painted what he saw – Proterozoic grey, Cambrian blue, Ordovician yellow, Mereenie Sandstone orange and Devonian Pertnjara Formation dark red-purple. You can look at his paintings and, as well as admiring their beauty, name every geological formation on them!

Our field trip started at Ellery Creek to the west of Alice Springs. We started at the McDonald Ranges at the northern end of the creek where the Late Proterozoic Heavitree Quartzite (about 900 million years old) overlies the Archean basement. The creek then winds its way south through 600 million years of geological history **(Fig 13).**

Figure 13
Ellery Creek cutting through the Proterozoic

The overlying Bitter Springs Formation was mostly shale with interbeds of dark grey limestones which smelt of oil when broken open with a hammer. Despite its incredible age (800 million years old), this unit contains good source rocks for the generation of petroleum and this at a time when there was only primitive life on the planet. Oil is formed by the burial and heating of dead animals and plants preserved in fine grained sediments deposited under reducing environments. We call these source rocks. Given that complex life on the planet did not thrive for another 300 million years, the organisms that formed these source rocks were simple stromatolites, algae and bacteria.

The next unit was the Areyonga Formation of tillites formed under glacial conditions. This was the first known major ice age on the planet. These glacial conditions extended from my PhD area west to the Amadeus Basin in the Northern Territory and the Officer Basin in South and Western Australia.

Without boring you with every rock outcrop along Ellery Creek, we eventually arrived at the Horn Valley Siltstone which was the same age as my Mount Arrowsmith section. I collected samples of the limestones so I could process them for conodonts when I returned to Sydney and compare them with the conodonts in the limestones from Mount Arrowsmith.

Life in the field in the Amadeus Basin was magnificent. The geology was spectacular, the outcrops excellent, the unbelievable colours of the rocks, the pure white trunks and shiny green leaves of the ghost gums and the endless warm days with brilliant blue skies. I was in geology heaven **(Fig 14)**.

Figure 14
Spectacular cliffs of Mereenie Sandstone

And so was the Frenchman – almost. He couldn't believe Australia's secret weapon – the flies. They were thick, particularly at lunch and dinner. So much so that the Frenchman took a picture of our lunch plates covered with the little buggers – ah, that's after we'd finished eating of course. No doubt he dined out many times on that picture back in Paris – excuse the pun!

But the most interesting part of my Amadeus field trip was a visit to Gosse's Bluff. This was a perfectly round outcrop of vertically dipping Mereenie and Pernjara sandstones about 4 kilometres across in the middle of a sandy plain **(Fig 15)**.

Figure 15
Circular impact crater at Gosses Bluff

Figure 16
Shatter cones formed by the impact

This was originally interpreted to be a salt dome and was drilled as an oil exploration prospect in the early 1960's. Peter Cook interpreted it as an impact crater. Something certainly hit the earth with a thump. There was a piece of Mereenie and Pernjara sandstones the size of a five story apartment block lying upside down outside the edge of the crater. However, there were no iron-nickel fragments around the crater. Peter's theory was that the crater was caused by a comet impact. The Gosse's Bluff-1 exploration well had recovered some carbon monoxide whilst drilling through the sediments. Carbon monoxide is not common in nature on this planet but does occur in comets in a frozen state. However comets are mostly made up of frozen gases and have a low density and are unlikely to have caused the huge crator at Gosse's Bluff. It was probably caused by a carbonaceous chondritic asteroid with low metal content. The sandstones forming the edge of the crater were shattered into cones **(Fig 16)** with the apex of the shatter cones pointing towards the centre of the crater. Whatever it was, it must have been a spectacular impact!

When I returned to Sydney, I processed the Horn Valley limestones and - low and behold – they were full of conodonts. All the same species were present as at Mount Arrowsmith. I was also intrigued by the fact that my conodonts were black but the Horn Valley conodonts were a light golden brown colour. I found out later that the colour of conodonts become darker as they are heated up by increasing depth of burial. Their colour is now used to determine the thermal maturation of source rocks for evaluating the petroleum potential of an area.

In the 1960's, the petroleum potential of Australia was totally unknown. In 1947, an Australian oil company called Ampol Petroleum Limited acquired the first leases in Western Australia. Caltex joined Ampol in 1951 forming a company called Western Australian Petroleum Pty Ltd (WAPET). WAPET drilled Rough Range-1 in 1953 and this well was Australia's first oil discovery, flowing 550 barrels oil per day. Unfortunately, this discovery turned out to be small and non-commercial. Commercial gas was discovered at Roma in 1960 and the first commercial oil field was discovered by Union Oil at Moonie in the Surat Basin, Queensland in 1961. In 1964, WAPET discovered both the major Barrow Oil Field in the Carnarvon Basin and the Dongara Gas Field in the Perth Basin both in Western Australia, and Delhi Petroleum discovered the Gidgealpa Gas Field in the Cooper Basin, South Australia. The Mereenie Oil Field and

Palm Valley Gas Field were discovered by Magellan Petroleum in the Amadeus Basin, Northern Territory during 1964-65. So in 1966, there was an increasing interest in Australia as a potential petroleum province. I was contacted by Jim Heppert of American Overseas Petroleum who had petroleum exploration permits in the area, and we spent a number of weeks scouting and mapping the area. I was very appreciative of this trip as Jim was a very experienced field geologist and I learned a lot from him about field work and assessing petroleum potential.

After completing my field mapping of the area, I began to write up my thesis. This was a time consuming process. There were no computers or word processing programs in those days. The thesis had to be typed, corrected and then typed onto roneo sheets for printing on a machine invented in the 19th century. The conodonts had to be photographed at X50 on the microscope, then each one cut out and pasted onto a black sheet of A4 paper, numbered in white ink and then re-photographed for inclusion in the thesis. Most of the figures and maps, I drafted myself and each field photo had to individually pasted into each copy of the thesis.

In February, 1966, decimal currency was introduced in Australia with ten shillings becoming one dollar. This was a good plan as the new one cent piece was only slightly different in value to the old penny.

I submitted my thesis "The Palaeozoic Stratigraphy and Palaeontology of Northwestern New South Wales" in 1967. It took some time to have it accepted by the two independent reviewers as the international reviewer disputed my age dating of the Mount Arrowsmith conodonts. My age dating was Lower Ordovician based on correlations with similar faunas in the Amadeus and Canning basins. However, some of the conodonts were very similar to known species in the Middle Ordovician in North America. These either had a longer age range than previously thought or were actually different species despite their similarities. I did not obtain my Doctor of Philosophy degree until June 1969.

In June, 1967, the Arabs and the Israelis went to war yet again. The Israelis won but Egypt closed the Suez Canal throwing world trade into chaos. The Middle East was becoming progressively unstable and with 75% of the world's oil reserves located in this area, this was to have a major impact on the oil industry. The Suez Canal didn't reopen until 1975 causing international trade to go the long way around the Cape of Good Hope.

ESSO EXPLORATION

By this stage I had decided on a career in the oil industry. The oil business is quite different to the mining industry. In mining, exploration is relatively cheap but proving up and developing a mine is very expensive. In the oil industry, exploration is very expensive and oil is mostly very hard to find. However oil is the only mineral that comes out of the ground by itself so it is therefore very profitable. Onshore, oil wells can start production while the field is still being developed. Offshore it is not so easy and the reserves have to be proved up before the development can commence.

On the advice of Lewis G. Weeks, BHP had acquired leases in the offshore Gippsland Basin in 1960. They farmed out their permits to Esso in 1964. During 1965-66, Esso had discovered the large Barracouta and Marlin gas fields in the offshore Gippsland Basin and was recruiting. Jim Haffenbrack, the Chief Geologist visited Sydney University so I went for an interview. A week later, I was offered a job. Just like that – it was a different world in 1966. So I started work with Esso and finished off my thesis at night and on weekends. People used to ask my mother what I was doing my PhD thesis on and she would reply "the dining room table!" But I will never forget getting my first pay cheque – it felt so good as long as I didn't think about all the money I owed!

Esso Australia was the Australian subsidiary of Standard Oil New Jersey and was at that time the largest oil company in the world. It was part of what was popularly known as the "Seven Sisters". This group consisted of Esso, Mobil, Chevron, Texaco, Gulf Oil, Shell and BP. These seven companies controlled most of the oil industry from exploration, development, production, transport, refining and distribution to the petrol tank in your car. I was working for one of the largest and most powerful companies in the world.

This is a good spot to explain the origins of the Seven Sisters with acknowledgements to "The Prize" by Daniel Yergin. Standard Oil was founded in 1870 by John D. Rockefeller and became very successful by not only producing oil but also transporting and refining it and selling the refined products. The company became so large that by 1904, it controlled

90% of all oil refining in the USA. In 1909, the US Department of Justice sued Standard Oil under Sherman Anti Trust Act 1890 for sustaining a monopoly. In 1911, the US Supreme Court declared Standard Oil an "unreasonable" monopoly (is there such a thing as a reasonable monopoly?) and broke it up into 33 separate companies. Over the years many of the 33 companies got taken over and the successful survivors were Standard Oil New Jersey (Exxon), Standard Oil New York (Mobil), Standard Oil California (Chevron), Standard Oil Ohio (Sohio) and Standard Oil Indiana (Amoco). Amoco and Sohio were taken over by BP and BP together with Shell, Exxon, Mobil, Chevron, Gulf Oil and Texaco, became so successful that they ended up controlling most of the world's oil production, transportation, refining and distribution again. In the 1950's, they became known as the Seven Sisters, a term originally used by Enrico Mattei, head of the then Italian state oil company Eni.

Esso had their offices in the British Tobacco Building at Circular Quay. I discovered that I was one of many new recruits most of whom made great contributions to the Australian oil industry over the next four decades. I lived with my mum in a flat in Cremorne so it was very convenient – a quick, relaxing ferry ride from Cremorne wharf to Circular Quay. No traffic jams and no stress. Coming home on a Friday night was a different matter. At the Quay, there was a pub called the "First and the Last". It opened at 6am for the ships arriving at the docks and closed at 6pm. It reminded me of the good old days of 6 o'clock closing (just kidding). After we got thrown out of there, we ambled on down to the "Paragon" for a few more beers before stagging across the road for the ferry ride home.

I was put in Esso's Basin Study Group and my first assignment was to review the Sydney Basin. Graduating from Sydney University, I knew a lot about the geology of the basin so I had a head start. What I had to work out was the petroleum potential and whether there were any prospects for drilling. There were plenty of structures in the basin but the Permian sandstone reservoirs were tight. I developed a theory that better reservoirs might be developed where the sandstones were deposited in high energy environments. Such an area was on the Muswellbrook anticline in the Hunter Valley so I organised a field trip to confirm the mapping. Hey – this oil business was turning out just great. What other business would pay you to tour around the Hunter Valley!

After my field trip I had to fill in my expense account. Now I hired a car and stayed in motels so most of the expenses were straight forward. Lunches were the problem. These consisted of driving into the nearest pub and ordering a couple of beers and a steak sandwich – a couple of dollars. My boss was an American called Dan Edwards and when he saw my expense account he called me into his office and told me the facts of life in the oil business. "What's this" he said – "a couple of dollars for lunch, don't be ridiculous, you'll spoil the system" So he told me to fill it out again but with $20 for each lunch. I was really beginning to love this oil business.

My prospect was eventually drilled – Jerry Plains-1 in 1969-70 after I had left Esso. My theory was right – it was a high energy environment but the sandstones were still as tight as an iron bar and the well was a dry hole.

Just after I joined Esso, they made two giant oil discoveries, Kingfish and Halibut in 1967. As it turned out, Esso almost missed the Kingfish Oil Field as Kingfish-1 was drilled only 500 metres from the edge of the field. Subsequent seismic analysis of the overlying sediments indicated high velocity limestones in erosional channels. The seismic method gives maps in time but if there are velocity variations in the overlying sediments, these time maps will not give an accurate picture of the real structure in depth. In fact, the geophysicists drew their maps in pencil so that after each well was drilled, they could adjust the velocities, rub out the relevant pencil lines and draw in the revised depth contours. Revised depth maps were reconstructed from this new velocity information from Kingfish-1 and after additional appraisal drilling, the Kingfish discovery turned out to be a two billion barrel oil field.

Halibut-1 was drilled in 1967 and the decision to develop this field was made on the results of this one well. This was very brave decision at the time due to the velocity variations in the overlying sediments and the difficulty of depth conversion. Fortunately it turned out OK and subsequent development drilling confirmed the structure.

My next assignment was to evaluate the Cooper Basin. The first gas fields, Gidgealpa and Moomba, had been discovered by the then Operator and 50% partner with Santos, an American company called Delhi Petroleum. Now Delhi and Santos controlled the entire basin and the undiscovered oil and gas reserves could have been (and eventually turned out to be) immense.

Delhi was for sale in the US and Esso was evaluating the opportunity to buy the company. My supervisor was an American geophysicist Chuck Hogan. He did the seismic mapping and I did the well evaluations and correlations – we made a great team. The first thing that became obvious was that there were lots and lots of structures in the basin so things looked promising. The Permian reservoirs were not good but had sufficient permeability to give good gas flows at the high pressures. The difficult part was that some of the structures were gas bearing but lots were dry holes. Why? As well as mapping the various geological horizons, Chuck produced an isopach map between the "C" horizon in the Early Cretaceous and the "P" horizon at the top Permian. It turned out that all the successful wells had a "thin" over the top whereas the dry holes did not. What this meant was that the dry holes were drilled on recently formed structures which were not present when the petroleum was generated and migrated through the area. Since we had mapped lots of structures with "thins" on the top, the basin looked very prospective for additional discoveries. In addition, the wells indicated that the Permian reservoirs had more saline formation water towards the west. Therefore, there was a possibility of more oil prone source rocks to the west and therefore the possibility of finding oil in the basin in addition to the gas. Years later, the major 80 million barrel Tirrawarra Oil Field was found in the western part of the basin. This had absolutely nothing to do with my theory but as we say in the oil industry, I'd rather drill a discovery for the wrong geological theory than drill a dry hole with the right one.

Based on what we had worked up on the basin, we gave a strong recommendation for Esso to buy Delhi. The price in 1968 was about $US25 million. But in the end, Esso did not buy Delhi. Naturally we were very disappointed as we were convinced that this basin would turn out to be a major petroleum province, which it eventually did! However, Esso's Exploration Manager, Don Langston came into our office and explained that Esso didn't buy Delhi because it had other corporate priorities and complimented us on an excellent report. This really impressed me - that management would actually descend into the basement to congratulate the grunts. By this stage, I had realised that Esso was a great company to work for.

Working for Esso got even better as they had this great system of training. Rather than send the "golden haired boys" to the US for training, Esso sent the trainers to us. They had a huge research establishment in Houston,

Texas and employed the brightest scientists. Every couple of months, we had these experts travel round the world and drop into the Sydney office. We had experts on sandstones, carbonates, source rocks, structural geology, seismic interpretation, and plate tectonics. What Esso didn't know about geology and finding oil, wasn't worth knowing.

The two main courses that interested me were plate tectonics and seismic interpretation. During my university education, I was taught that there was no such thing as plate tectonics. Now I had read Van Wagonner's theories of continental drift and was a believer. But it was taboo in academic circles and I had to make sure that I never mentioned the concept when doing my exams. Esso's course on plate tectonics was a whole new world. It explained how continental drift actually took place. And all this before Menard's 1969 paper was published in Scientific American.

The course on seismic interpretation was even more impressive. Esso tried to convince us that seismic reflections followed time lines and not lithological boundaries but this did not seem logical to us. It took them a long time to get this concept into our thick skulls but eventually the evidence was so overwhelming that we were eventually convinced. Esso had been studying this for years and had sea level curves showing sea level falls (prograding seismic down-lap events) and sea level rises (seismic onlapping events). Fifteen years later, these concepts were expanded and became known as sequence stratigraphy, one of the most powerful exploration tools in the oil industry.

As well as all these courses, there was a wealth of experience in the Sydney office. The Americans and Canadians were all experienced geologists and geophysicists and had worked all over the world. So we all benefited from working with them and soaking up their experiences. One of the best stories was from a geophysicist who had worked in Libya. His party was scouting an area for a seismic survey. In this desert area, dynamite was the energy source. They laid out the first line and set off the dynamite charge. The whole desert erupted – the dynamite had set off every land mine within the immediate radius. They had actually been driving around for weeks totally oblivious of the danger. So from then on, nobody went anywhere without the area being cleared of mines.

My next task was to prepare a series of palaeogeography maps of Australia for the basin study group. This was a lot of fun for me as I already knew a lot about the Palaeozoic basins. The difficult part was what to do at the Australian coastline. Obviously many of the basins extended from Australia onto the next adjoining continent but exactly what continent was still being hotly debated.

The other concept that was being developed at this time was Risk Analysis. In those days, you developed your prospects and submitted them along with everyone else, to the senior management. They decided which ones were the best and they were drilled in order of merit, the number being dependent on the exploration budget. This was all very subjective. I was asked to research how these prospects could be better risked and put into some sort of rational order for drilling. I found a recently published book called "Decisions under Uncertainty" by C. Jackson Grayson. In his book, he outlined the concepts of decision trees, probability theory, net present values and expected monetary values. This was all new to me but I found it all very fascinating and we development a crude (by today's standards) risk analysis structure to rate our various prospects. In 1975, Newendorp published his book "Decision Analysis for Petroleum Exploration" and risk analysis became part of the everyday oil industry.

The Esso Sydney office was also very social. We used to have these fishing trips on the weekends. Very few fish were caught but lots of beer was drunk. The Americans were all mad gamblers and enjoyed playing "craps". This was OK on the way out when the betting was sensible, but and the return trip - well I hadn't seen so much money thrown onto the deck of a fishing boat. But the best social event was Andy Whittle's buck's party. This was held in the evening on one of the Sydney Harbour ferries. It was a great night, lots of booze was drunk and of course there was the obligatory stripper. Her pimp had a big bulge in his pocket (and it wasn't because he was pleased to see us) so we tried to behave ourselves. But we were all very pissed by the end of the night. I got dropped off at Cremorne wharf and stagged up the hill toward my place. But I woke up next morning lying on someone's front lawn. So I continued my journey and slept until midday. I woke up with a terrible hangover and headed on up to the Oakes for a heart starter. I was nursing a beer when this bloke walked into the bar wearing a tuxedo jacket with black bow tie, red shorts and thongs. The bartender knew him and asked how his friend's wedding had been. He crowed at the top of his voice "Bloody beauty". The bar erupted in laughter and I almost began to feel better. Eventually some of the riff-raff from Whittle's party gradually dragged themselves into the Oakes – they were a pathetic, sorry sight - and they had the hide to tell me that I looked terrible!

The Vietnam War had been slowly escalating during the mid 1960's. During February, 1968, the Viet Cong launched the Tet offensive in South Vietnam. Although initially successful, US and South Vietnamese forces regained the offensive and inflicted heavy losses on the Viet Cong. However, the Tet Offensive undermined public opinion in the USA and Australia with increasing protests and demonstrations against the war.

In 1968 there was a strong stock market and some of us were trading shares. Esso had just farmed into a permit where a company called Oil Search had a significant interest. They were about to drill a well called Ini-1 on a Miocene reef in the Gulf of Papua. I remember the geophysicist in charge of the seismic acquisition commenting "normally we budget seismic acquisition in terms of $'s per line mile but in PNG we budget seismic acquisition in $'s per line foot"! Even in those days, seismic acquisition in PNG was very expensive, as it still is today. But before the drilling, most of the office were buying Oil Search shares (insider trading wasn't illegal in those days – I think). I bought some at 40 cents and sold them at 80 cents but they continued up to $1-20. After that I stood aside but many others in the office continued to trade believing that Esso would get the results of the well before the market. The well was drilled during November-December, 1967 but by the time the results of the well reached the office (the well was not oil bearing) the market in Oil Search shares had collapsed. Somehow the information got from a rig stuck out in the middle of the Gulf of Papua to the market before it got to Esso, the Operator of the well! Well, it was total disaster in the office. One of the engineers had lost so much money he simply took off – left the company – left his wife and family and was never heard from again. One of the more experienced share traders immediately began to short sell the stock (short selling was illegal then but he either didn't know or didn't care) and ended up making about $500 after turning over in excess of $500,000 worth of shares. For me, it was a great lesson in the benefits and pitfalls of share trading. I had sold too early but was one of the few people in the office to make any money.

In January, 1969, an environmental disaster occurred in the offshore waters of California. The Santa Barbara oil spill was caused by a blowout on the Union Oil Platform A on the Dos Cuadros Oil Field. Over a 10 day period, 80,000 barrels of oil was spilt into the ocean and a considerable portion of this oil washed up on the beaches of California. This was very undesirable publicity for the oil industry but later studies showed that the environmental damage was not as serious as predicted. This was probably due to the fact that the area of the blowout is an area of active natural oil seeps and the environment had evolved to cope with some oil pollution. In any event, petroleum exploration is still banned offshore California and many other offshore parts of the USA.

Part of our duties included wellsite work on the offshore rigs in Bass Strait. We spent about two weeks offshore every couple of months. I remember my first trip to Flounder-2 in early 1969. I flew to Melbourne, then onto Bairnsdale and then helicopter to the rig. I arrived in Bairnsdale about midday and joined the rig crew waiting for the helicopter. I noticed that the helicopter was in pieces all over the tarmac and these two blokes were screwing it back together. I thought "what a shame, there going to have to fly in a new one". But no! As soon as the helicopter was back together, the supervisor called out "all aboard". "But, but, but shouldn't they at least test the bloody thing!" my mind screamed. But being a new recruit and as no one else seemed bothered, I boarded the helicopter along with everyone else. It was my first helicopter ride and I didn't know that helicopters defy the laws of physics. They should actually screw themselves into the ground! It was a funny sensation as it hovered just above the ground and then we flew off. The view of the Gippsland coastline was spectacular **(Fig 17)**.

Figure 17
Eighty Mile Beach and Lake Victoria

Figure 18
Landing on the Ocean Digger

I soon discovered that helicopters were also very noisy – no ear phones in those days. So I had to shout at the bloke next me "what happened to the helicopter to cause the repairs". He shouted back "it caught fire on the way back from the rig". My heart sank and I reprimanded myself "bloody stupid question – keep your mouth shut next time". But we eventually arrived at the rig, the "Ocean Digger". It looked so small in the vast expanse of windswept ocean, just like a postage stamp floating in a lake. But as we came closer, the rig became bigger, and bigger and eventually the helideck was visible and it then looked like the postage stamp. Horrified, I thought "We're not going to land on that, are we!!!!" When we landed **(Fig 18)**, I stepped out onto the helideck and looked up at the rig. Compared to the land rig on Casterton-1, it towered up to the heavens. Then I stepped onto the walkway to get to the office. This was a metal grill and you could see down through to the ocean 30 metres below. I never was good with heights and I leapt back onto the helideck. Everyone laughed – what a wimp!

The Ocean Digger was the first offshore rig built in Australia. I think they ran out of money at the end because there were no doors on the toilets. No privacy for your morning constitutional. The generators were located on top of the accommodation so you had to be a heavy sleeper to get a good night's rest. Fortunately, I'm a heavy sleeper and I had the added advantage of being the well site geologist. I could lie in bed and listen to the generators. Yes, we're still running into the hole – back to sleep. Or we're drilling ahead at last – better get up.

Flounder-1 was drilled in September, 1968 and encountered a 35 metre gross oil column in sands in the Eocene intra-Latrobe Group. It was the first oil discovery not at the top Latrobe. I must have put a jinx on the discovery because Flounder-2 intersected a 19 metre gross oil column and Flounder-3, an 11 metre gross oil column. The reservoirs were deposited in an estuarine environment and were highly variable with individual oil sands being 1-5 metres thick. The evaluation of the geology proved difficult and it took many years, and another three appraisal wells, to determine reserves of 115 million barrels. The field was eventually developed and began production in 1984, so I wasn't a jinx after all.

On the return trip, we'd stay overnight in Melbourne. Exlog were the mudloggers for Esso at that time. They were located in Melbourne and always had great parties on a Saturday night. We would rock up from the rig, throw our gear in the corner, join the party and catch a flight back to Sydney on Sunday morning. I remember one Sunday morning waking up after an Exlog party with a terrible hangover. There was someone knocking on the front door but nobody else was capable (or cared enough) to answer it. So I crawled off the couch and staggered to the door. When I opened it, there were two Jehovah's Witnesses standing there. I suddenly realised that I was standing there in my underpants. Worse still, behind me there were bodies lying on the carpet, ladies knickers draped over the furniture and empty beer bottles lying everywhere. The Jehovah's Witnesses didn't even bother to try to convert me - they took one look inside and moved onto the next apartment. There was no hope of salvation here – we were all doomed to hell fire and eternal damnation.

In December, 1968, whilst drilling a production well on the Marlin Platform in the Gippsland Basin, the well took a kick from a shallow gas sand. Efforts to control the kick caused the gas to break out from the sea bottom surrounding the platform with foaming gas – a most dangerous situation. Fortunately the gas did not ignite and "Red Adair" was called in from Houston. His team assembled a huge amount of mud and equipment, and on New Year's Eve, the blowout was sealed and the Marlin Platform saved. Amazingly, no one was killed during this disaster.

My first offshore discovery was Mackeral-1 in April, 1969. The well encountered the top Latrobe with a good drilling break. We circulated bottoms up and over the shakers came this beautiful unconsolidated sand. Under fluorescent light, the sand lit up like a Xmas tree – it was oil bearing. So a core was cut but it was only partly oil bearing with the bottom part having no fluorescence and being water-wet. Log analysis indicated a seven metre oil column which was not very encouraging. However, we knew from Kingfish that the seismic time maps were not accurate in defining the structures in depth and so it was with the Mackeral discovery. Later appraisal drilling confirmed it to be a 300 million barrel oil field.

My next well site trip was to Turrum-1 in June, 1969. This well was located in the channel between the Marlin and Tuna fields and targeted the Late Cretaceous section below the Latrobe Group in a large anticline. We entered the Late Cretaceous section which was mostly shales and siltstones with thin, low porosity sandstones. The weather became progressively worse and became so atrocious that no further helicopter flights were possible. The rig was also running low on supplies, including the mud for the drilling operations. Some of the sandstones were full of over-pressured gas and since the mud was deteriorating, it came out of the hole all froth and bubbles. The gas detector was running at hundreds of units. Then I got a call from head office to cut a core. Now, I wasn't very experienced but I knew enough to know that this was not a good plan. Pulling a core out of the hole could swab in the high pressure gas sands and cause a blow out. The drilling engineer and toolpusher also thought it was a bad idea and after some furious discussions between Odeco, the owner of the Ocean Digger, and Esso, no core was cut. In any event, we had to stop drilling and circulate the mud in the hole until the weather cleared and additional mud could be resupplied to the rig.

At the end of this stint, I was transferred to the Glomar III, a floater. This rig was a ship with the drilling rig in the centre and the riser and drill pipe going through the bottom as compared to the Ocean Digger, a semi submersible rig with four pontoons supporting the drilling platform and drilling equipment. Naturally this vessel is not as stable as a semi and in bad weather it rolls like a drunken sailor and makes life on board pretty miserable. Now the Glomar III was drilling the Snapper-2 appraisal well on the Snapper Gas Field in July, 1969. In this field, the gas column in Snapper was 120 metres thick and I was to supervise the coring operations which were to core the entire 120 metres of gas bearing sand. The gas bearing Latrobe sands were at a depth of about 1500 metres so by the time the first core had been retrieved, measured, described and packed in boxes, the next core was on its way out. After two days with very little sleep, Esso took pity on me and sent some relief.

A frustrating episode in the Esso office occurred when BHP shares went up $1 (they were only $15 at the time) because a gas field had supposedly been discovered in the offshore Otway Basin. Esso owned 50% of these leases and was the operator with BHP owning the other 50%. It turned out that speculators had intercepted a morning report from the rig to the Esso office reporting the previous day's drilling results including the geological section penetrated. Included in this report was a 10 unit gas show which is not significant. However the speculators didn't know this and thought that a new gas discovery had been made by Esso and BHP. These rumours spread through the stock exchange and BHP shares rose. After this episode, BHP and Esso management decided that we had to code our morning reports to head office. The code was a simple transliteration of letters and numbers that a 10 year old child could crack. But it was a tedious process for us and an even greater nightmare for Burt Culp, the geologist at head office who had to collect this information from not just one but two irate wellsite geologists. After a few days, it was unanimously decided that "fuck this for a joke" and we all went back to the old system and to hell with the stock market.

In July 1969, the Americans landed the first men on the moon, a momentous event equivalent to the first flight by the Wright brothers in 1906. A new dawn was beginning and the space race with the Russians was to slowly evolve into a serious exploration of our solar system. It took a while but the spinoffs from this race cannot be calculated and included the accelation of computing power, satellites, GPS, mobile phones to name but a few.

My most exciting wellsite operation was on Mussel-1 in the offshore Otway Basin in August 1969. The well was located in the Great Australian Bight some 20 kms west of King Island. It started off routine – flight to Melbourne, flight to Burnie, drive to Smithtown and then helicopter to the rig – also the Ocean Digger. The helicopter took off from Smithtown and headed west to King Island. As we crossed the west coast of Tasmania, I looked out the window and all I could see was dense rain forest, then a very lonely beach extending for miles in both directions with no evidence of any human habitation, and then a dark, foreboding, wind-swept sea. I thought to myself "so this is what they mean by wilderness". On the way to the rig, we entered a thick fog and visibility was zippo. It was an unusual sensation sitting in a helicopter surrounded by snow white fog and I was very pleased to see the Ocean Digger appear out of the gloom.

There is something about drilling rigs that attracts real characters. On the Ocean Digger, the ballast control man was called Derek Desoto Phillips. Now Derek was your typical British Raj Englishman – he even had a "Jimmy Edwards" handle-bar moustache. We were having dinner one night and a couple of Schlumberger engineers had arrived on the rig. They were French and joined us for dinner. About half way through dinner, Derek stood up and asked everyone "Do you colonials know anything about the Battle of Waterloo?" He then used all the condiments on the table – Wellington was here, Napoleon there, Blutcher here and General Ney there – and over the next 15 minutes, proceeded to enlighten "us colonials" with a terrific rendition of the Battle of Waterloo. Everyone had a good laugh (and history lesson), except the Frenchmen who were seething and Anglo-French relations were set back another 10 years.

As it turned out, Mussel-1 was the driest hole I had ever monitored at that time. The gas detector was so dead you could put a ruler on it. It was so dead, I thought that it wasn't working and gave it a carbide test. When calcium carbide comes in contact with water, it generates acetylene gas. The carbide is put in the mud system and pumped to the bottom of the hole. When the acelylene comes to the surface it sets off the gas detector. It can also be used to check the lag times of cuttings coming to the surface. The gas detector actually was working but the well was devoid of any shows whatsoever – and we (me and the mudloggers) were bored out of our minds. We were blissfully unaware that Mother Nature was about to liven up our day.

The weather began to deteriorate and, being located in one of the stormiest seas in the world, we stopped drilling and pulled up to the shoe, that is, up inside the intermediate casing. This was prudent oilfield practice in those days because weather forecasting was not as accurate as today's satellite and computer driven forecasts – and we weren't quite sure what was about to hit us. The wind kept increasing and the waves got higher, and things started to get exciting. As we weren't drilling, there wasn't much for me to do so one of the mudloggers suggested I go down to the ballast control room for a view of the storm. So down I went and Derek explained to me the workings of the ballast control room where the buoyancy of the rig is controlled by balancing the weight of each of the four pontoons supporting the platform. The ballast control room was located beneath the main platform of the rig so you had a great view of the four pontoons and the riser guiding the drill pipe beneath the sea. By this stage, the waves were about twelve metres high and looked like apartment blocks rushing towards us. With the rig dipping into the trough before each wave, each wave looked like it was going to break over the top of us. The ballast control room also monitors weather conditions and the strain on each of the eight anchors holding the Ocean Digger in place. Derek informed me that the wind was now gusting at 100 miles (160 kms) per hour and one of the anchor chains had now exceeded its breaking strain. I stupidly asked "What do we do now?" and Derek, in true British tradition, remarked "I suggest a short prayer". After this, I left the ballast control room - firstly, so I wouldn't be there when the next wave tore it off the bottom of the rig, and secondly, to get my life jacket. I remember that the wind was so strong, I had to crawl on my hands and knees to get to the mudlogging unit and it took two of us to open the door. The next day, we discovered that the steel ladders had been torn off one of the pontoons. I had no idea that water could be that powerful.

This wasn't the only exciting thing to happen that year. In August, 1969, the Petrel-1 well in the offshore Bonaparte Basin blew out and caught fire. The Sedco 135G was badly damaged and it wasn't until February 1970 that a relief well was drilled and the blowout controlled. Although no-one was killed during the initial blowout, nine men were drowned when a supply boat sank during the relief operation.

By the end of 1969, Esso had discovered 11 fields in the Gippsland Basin, of which 5 were in production. The first homes in Victoria were connected to natural gas in April, 1969. The basin peaked at almost 500,000 bopd (barrels of oil per day) in 1985 and to date the basin has produced four billion barrels of oil and seven TCF (trillion cubic feet) of gas.

In the late 1960's, there was a minor oil boom going on and the stock market had a strong rise from 1966 to 1968. In 1960, the international oil price had been set at only $US1-80 per barrel but the Arab – Israeli War of 1967 created a lot of instability in the Middle East including the closing of the Suez Canal.

In early 1968, Arco and Humble made the giant oil discovery at Prudhoe Bay in Alaska. Humble Oil was a subsidiary of Esso and, as Prudhoe Bay turned out to be a 15 billion barrel oil field, shares in Esso rose on the US stock exchange. Most of the Americans in the Sydney office owned lots of Esso shares through their generous superannuation scheme and so they were all very "happy little vegemites".

Data Analysis

Experienced geologists were in demand and Esso, being a large company, did not pay high wages. So I was ultimately "seduced by the dark side", quit Esso and joined a consulting company called Data Analysis run by Mike Gehan, an ex Schlumberger engineer.

My next wellsite job was Hindhaugh Creek-1 in the onshore Otway Basin in western Victoria in September-October, 1969. The well-site was right on the coast. It was the middle of winter and I had the night shift. The wind came from Antarctica via the Great Australian Bight and howled across the rig lease all night long. It was bloody freezing!!! All I had was a miserable blow heater in the mudlogging shack but that was enough to attract every roughneck and roustabout on the lease – all huddled around that little heater. Of course, any heat that got past the assembled crowd wasn't sufficient to melt the icicle on the end of my nose. But I was by now getting smarter about "how to survive as a well site geologist". I arranged for the lady who cleaned my motel room to put on the electric blanket at about 6-30 each morning. So when I arrived at the motel all frozen and stiff after the night shift, I had a quick shower and then eased myself into this lovely warm bed – ahhhhh - and within 3 seconds, I was fast asleep. One Saturday morning, I woke up about 1-00 in the afternoon and flicked on the TV. There was an Aussie Rules game on. Now being a New South Welshman, rugby was our game and we called Aussie Rules "aerial ping pong". But having an open mind, I settled down to watch the game. It turned out to be a terrific game, fast, tough and skillful. It was Carlton versus Richmond, and Richmond won by three points. So I became a Tiger fan and remained so until the VFL became the AFL.

My next wellsite job was on Mount Emu-1, being drilled in the Darling Basin in western New South Wales from November 1969 to January 1970. This well was being drilled by the Planet Oil Company NL, a company listed in 1962 by John Fuller. So I hired a car and was on my way to the town of Ivanhoe. I told Avis that I was going bush and to give me an old car. But no, when I picked up the vehicle, it was a brand new Ford Falcon with a couple of hundred miles on the clock. I was cruising along at 70 miles per hour on my way to Ivanhoe when, out of nowhere, an emu ran across the road and I hit it point blank. Now slamming into an emu at 70 miles an hour is not recommended in the driving manual for a brand new Ford Falcon. There was blood and feathers everywhere and I reckoned that the engine had been dislocated from its mountings. The car was a write-off and they had to send another car. This next car was somewhat older than the one I'd just wrecked! Never a dull moment when driving in the country!

When I got to Ivanhoe, I stopped for lunch and a beer at the local pub. The locals were friendly and when I explained what I was doing there, they told me that I was crazy, looking for oil in such a god-forsaken place. So I told them "but that's exactly where you find oil!" When I got to the well, I dumped my gear in my room and drifted over to the mudlogging shack. I looked at the first sample and saw this horrible, nasty quartzite. "But this is basement" I exclaimed to the wellsite geologist I was relieving. He sympathised "I agree but try to tell that to the management". I tried but to no avail so on we went drilling through this quartzite at a very slow rate. As I discovered later, these quartzites were not basement but actually Devonian sandstones which had been buried to great depths, silicified and inverted by later compression. Nevertheless, Mount Emu-1 was a very dry hole. During the setting of intermediate casing in Mount Emu-1, I joined the logging engineer and tester to drive into Wilcannia for a few beers. When we got there, they wouldn't serve us in the pub. We were looking very scruffy in our rig boots and dirty shorts but the main problem was that we weren't wearing shirts. They did eventually serve us but we had to sit outside with the local aborigines who also weren't wearing shirts! It was summer, stinking hot and didn't improve my opinion of Wilcannia as a tourist destination.

It was with Data Analysis, that I attended my first APEA Conference in April, 1970. APEA was the Australian Petroleum Exploration Association and they had a conference once a year rotating from city to city. In 1970, it was in Surfers Paradise and it was magnificent. The opening cocktail party on Sunday night – well I had not seen so much sea food in all my life – lobsters, prawns, bugs, fish, scallops all piled high on beautifully decorated tables around the venue. Of course, the beer and wine flowed all night.

I was staying in a motel called Las Palmas right on the beach front. I'd become friends with a couple of blokes from Melbourne and we played billiards in the evening before dinner. One evening as I was on my way to the billiard table, a young lady jumped out from her room and implored me to remove a nasty insect from her wall. She was English and had been well and truly brainwashed by her Aussie workmates as to the vast number of poisonous nasties which infest our magnificent country. The insect in question was a harmless beetle which I removed without sending the poor thing into the afterlife. She thanked me profusely and I went on my way. She was attractive, petite and quite a knockout. As I wandered down the path, I thought "what's the matter with you – overcome your shyness and ask her out". So I went back and invited her to join us for a drink. To my pleasant surprise, she accepted so we went to join the others. As it turned out, she lived in Neutral Bay, literally a few kilometres from Cremorne where I lived. Her name was Maureen but she preferred Mo for short. So I took her phone number and called her when I returned to Sydney.

One of my best mates in the class of 1963 at university was Al Raam who lived in Cronulla. His parents were very rich and he used to invite me to go water skiing in Port Hacking on the odd weekend. I actually got good at it and, eventually, got up on one ski. After I contacted Mo on returning to Sydney, we spent many pleasant weekends as guests of the Raam's. Now although Mo loved the beach, she wasn't keen on getting totally immersed in seawater which is mandatory for water skiing. So she used to act at lookout and sounded the alarm every time I fell off. As time went by, Mo and I fell in love and became inseparable **(Fig 19)**.

Figure 19
With Mo at a party – I'm the one with the outrageous moustach

LONGREACH OIL

My next job was Chief Geologist with Longreach Oil. Chief Geologist! What a laugh - I was the only geologist! Longreach Oil was a small company which had acreage in the Eromanga and Galilee basins, onshore Queensland. Virtually nothing was known about these basins. There was very little seismic and what there was, was poor quality, and there were very few wells. Most of these were located based on the interpretation of aeromagnetic data. It was difficult to mature prospects and we needed something new.

Longreach Oil also owned an exploration permit in the offshore Sydney Basin. Having worked the onshore basin, I knew that reservoir was going to be a problem. However, a marine shale, the Kulnura Marine Tongue, was present in the Upper Permian Coal Measures in some of the eastern onshore wells. In theory, this marine transgression could have reworked the Permian sandstones to create better reservoirs. So the company decided to acquire some regional seismic over the permit. It was a great experience for me as offshore Sydney was very busy with shipping and there were many approvals to obtain – Army, Navy, Fisheries, Customs, Environment to mention but a few. We eventually got the approvals and the survey went ahead. Having never seen a marine seismic survey up close, I decided to go along to see it all first hand. That turned out to be a big mistake as we got hit with a bad storm and had to return to base. I was so sea-sick that I actually thought death was a serious option. I was cured for life and vowed that that I would never set foot on a seismic boat ever again and I never did!

The seismic was eventually acquired and processed. It was interpreted by Bill Nixon, a Canadian who had quit Esso and was consulting in Sydney. The seismic was not good but a structure which we called the Sealion Prospect, was mapped just off the northern beaches of Sydney. In fact, the westerly dipping Narrabeen Group at the Longreef golf course could be interpreted as the western flank of the structure. As Longreach Oil had no production or significant cash resources, we needed to farmout the well. That is, we had to find a big company to pay for the large cost of drilling the exploration well and try to negotiate a carried interest in the project. Given the current drilling results in the onshore part of the basin, this was not going to be easy.

I was contacted by John Strevens, one of the founders of the Australian Oil and Gas Company, the permit holder of the onshore part of the basin. He was retired but still had a passionate interest in the Sydney Basin. He lived at Avoca Beach just north of Sydney and invited me up to Terrigal to observe the oil and gas seeps. I was very sceptical, but early in the morning before the wind picked up, you could see faint oil slicks with gas bubbles off the rocks. This was all very interesting and indicated the probable presence of mature source rocks in the basin. But the presence of source rocks was not the problem in the Sydney Basin – it was the very poor quality of the reservoirs in the basin.

At the same time as all this was happening, an American geologist called Pete Temple had quit Esso and joined a small Australian company called Beaver Exploration. This company had farmed into a large permit (held by BP, Shell, Aquitaine and Todd) on the east coast of the north island of New Zealand, and they invited Longreach Oil to join them in the exploration of the area. This involved a field trip to the east coast basin of New Zealand.

So Pete Temple and I and another Australian geologist Warren Leslie, flew to Auckland, hired a car and drove to Napier on the east coast of the north island. I had never been to New Zealand so it was a great experience for me. It was an incredibly beautiful place. It was autumn and the silver birch, beech and poplar trees were all turning orange, yellow and red and the bush was a picture post-card. The other thing that really impressed me about New Zealand was their rivers. I was used to Australian rivers being red-brown and muddy but in New Zealand the rivers were all clean and crystal clear, just like a tooth-paste advertisement.

We drove south to Hamilton and had lunch there. I was amazed that the French wines on the wine list were cheaper than the Australian wines. But Pete was impressed and ordered a nice bottle of Chateau Neuf de Pape for what he considered “a song”. After lunch we drove to Rotorua and stayed there overnight. Rotorua was an incredible place - hot springs, boiling mud pools, steam everywhere. The geology was spectacular with major faults dropping Rotorua into a deep graben. But not a place to buy real estate long term – geologically the North Island is gradually breaking itself into two parts.

While on this field trip, Pete related an interesting story about field mapping in the Arctic when he was working his way through university. He and another geologist flew up to the area of interest in one of those planes with the skis instead of wheels for landing. They parked the plane as close to the outcrop as they could get, tied the plane down and hiked over to the outcrop.While mapping, the plane was approached by a large polar bear (is there any other kind?) who thought the plane was an animal invading his territory. As they looked on in horror, the polar bear proceeded to attack the plane and systematically tore it to bits. When the polar bear had finished “killing” the plane, it eventually wandered off confident in the knowledge that he was king of all he surveyed. After they were confident that the polar bear had gone, they crept down to the wreckage and fortunately, the radio was still in tact. They radioed the base, related their situation and were rudely told to stop wasting radio time with such a bullshit story. After much pleading, they eventually convinced base to send another plane. Now we all know that the Australian desert is unforgiving but at least our geologists when they are mapping in the field, don’t have to worry about being torn to pieces and eaten by large wandering carnivors!

The East Coast Basin is a thick Tertiary basin with a Pliocene coquina limestone then called the Te Aute Beds overlain by a thick sequence of marine shales. The underlying Miocene shales were heavily deformed and outcropped to the north. These outcropping shales had numerous oil seeps and were interpreted to be the source rocks for the basin. The coquina limestones outcropped in the southern side of the basin and had excellent porosities of 20-30% and permeabilities up to 2 darcies **(Fig 20)**. You could chip off a rock sample and blow through it.

Figure 20
Pete Temple inspecting the Pliocene limestones

From Rotorua, we drove south to Lake Taupo which was truly spectacular, and then across to Napier.

In 1970, Beaver Exploration acquired a 200 kilometre seismic survey in the northern part of the basin. The seismic was good quality and showed a series of reverse faulted anticlines with the objective limestones at reasonable drilling depths. The Beaver-Longreach joint venture drilled three wells, Ongaonga-1, Takapau-1 and Mason Ridge-1. The first two encountered the porous limestones but all three were dry holes – thus ending my New Zealand adventure. We figured that the Miocene source rocks generated their petroleum prior to the Pliocene being deposited and were never reheated sufficiently for a second oil expulsion. The geothermal gradients in the wells were very low caused by the cold Pacific oceanic crust being subducted under the east coast of New Zealand.

Longreach Oil had an associated company called Longreach Metals. Their geologist was Brian Cornish and he gave me a stock market tip called Poseidon. He had bought 5,000 shares on a tip that they had found a large nickel deposit in Western Australia. I looked at the company but the shares had gone from 20 cents to $7 each and I figured that I had missed the run. So I didn't buy any! When I returned from New Zealand, the Poseidon boom was in full flight. It is hard to describe compared to today's regulated market. In those days, you could buy shares and sell them in a week's time without ever paying for them. At lunch, we'd go to the Wentworth Hotel where one of the local bars had a ticker-tape machine and you could follow your stock prices while having a beer. I even went to the stock exchange once, the old one before it moved to Bond Street. The lift stopped at the third floor, the doors opened and three people fell into the lift – the gallery was so packed, we literally couldn't get out of the lift! It was a crazy time but my most notable stock tip was Tasminex. I was able to get into this stock at the prospectus stage and subscribed to 5,000 shares at 20 cents each. After listing, the stock slowly rose and after a few weeks eventually got to $1 a share. "I'm out of here" said I and sold, remembering my Oil Search experience. A short time after I sold, the company announced that it had a magnetic anomaly that "could be as big as Poseidon". The shares went to $90 for about one day and then quickly fell back to $1. That was one I missed but no complaints – I made some money. Brian sold his Poseidon shares at $200 each and left Longreach to start his own company. Poseidon shares actually got as high as $250 per share before they began their slide into history. Thus ended the Poseidon boom and with it the hopes and dreams of a lot of resource entrepreneurs. It also gave me another bitter lesson in share trading - I lost all the money I had just made!

Bill Nixon, between consulting jobs, was working on the onshore Gippsland Basin. He was convinced that the oil and gas fields didn't magically stop at the coastline. He had found an old well, Seaspray-1, which had very high resistivities and excellent oil shows at the top of the Latrobe sands but had never been tested. He put together a prospect report and gave presentations to many brokers, and managed to raise private funds to drill the well. The story was so good that even I put some of my own money (a very modest amount) into the well. The location of the well was at the intersection of two roads so access was no problem and the well was called Crossroads-1. It was drilled in June 1971 with lots of eager private investors anxiously awaiting the results. The well was tested and actually flowed some oil for a few minutes but then the water came in. Having a salinity of 200 parts per million (ppm), it was the freshest water Bill or I had ever seen and explained the high resistivities. Everyone was very disappointed but, hey – that's the oil business.

HONEYMOON

While all this was going on I had asked Mo to marry me and we were married in a registry office on 12th June, 1971. Mo worked for the Teachers Guild and we had our reception at their club. It was a great party and we all had a great time. I remember the "telegrams". Al Raam was my best man and he was reading them out. He said that there was a telegram from "Mor – a - cam – bee" and I thought to myself "where the hell is that" Mo was in fits of laughter and explained that it was from her family in Morecambe. One telegram said "Congratulations on the spudding in of Maureen No1" – that turned out to be prophetic as the Maureen Oil Field was discovered in the North Sea in 1973. There was another telegram from Mary in Wilcannia and it said "Why you marry white girl?" and I thought "who sent that? – I'll kill the bastard!"

After much fun and merriment **(Fig 21)**, we finally departed on our honeymoon. First stop was Gosford where we stayed over-night. In the morning, we discovered that we had left one of our suitcases behind so back to Sydney we went and started all over again. When we got to Surfer's Paradise, we stayed at the Las Palmas Motel for old time's sake – it was still there and hadn't changed much. The next day we headed north to Townsville.

When we got to Townsville, we got the ferry to Dunk Island where we were going to stay for a few days. The accommodation was basic but the island was beautiful. You could look back at the Queensland coast and see absolutely nothing! No high rise apartments, no sign of human habitation – it was so relaxing. There were oysters you could pick off the rocks, and I even caught a trevally which we had cooked that evening for dinner in the restaurant. We went back to Dunk Island in June, 1991 for our 20th wedding anniversary and it was so different. Large resorts, palatial swimming pools, taverns and all the trappings of a modern tourist resort.

Figure 21
Cutting the wedding cake

From there we drove to Gladstone for an overnight stay before going on to Cairns. Gladstone was boom city with all the coal mining and exports to Japan. We had to wait at a railway crossing for about half an hour while hundreds of trucks full of coal passed by. All the motels were full and we had to stay at one of the old pubs in the town. The room was very basic and the bed squeeked – not a good start for a honeymoon!

When we got to Cairns, we stayed a few days at Green Island. This was a coral island and was surrounded by pure white lime sand beaches. You could swim from the beach to the reef and view the coral. The accommodation was chalet style but the ants were everywhere and it was a battle to keep the little buggers out of place. From Cairns, we drove north to Mossman. At Mossman, there was a beach that stretched south as far as the eye could see. Mo was blown away by the complete lack of any development. It was like we were the only people on the planet. From there we drove to Daintree and the ecology of the area changed completely. We entered the Daintree rain forest and it was truly spectacular. In some spots the leaf canopy blocked out the sun, it became dark and really eerie - I almost expected *Tyrannosaurus rex* to come lumbering out of the mist.

From there, we returned to Cairns and went to Mareeba on the Atherton Tablelands. We stayed with Trevor and Marita as they had moved there from Broken Hill. My idea was, on our return to Sydney, to come down the inland road so Mo could see some real live kangaroos. This was a good plan except for the bit between Mareeba and Charters Towers. The map said that there was a road between Mareeba and Charters Towers – what it didn't say was that it was still being built! The road was mostly a dirt track and there were no petrol stations. We had to buy petrol off one of the many construction camps along the road. We got to Charters Towers very late that night, very tired and very hungry.

Our next stop was Roma where we were to stay at the caravan park. When we got there, the office was closed. One of the other residents said that the manager was at the local pub running the Friday night chook raffle. So down to the pub we went where he gave us the key and said let yourself in and fix me up in the morning. Ahh, life was so much simpler then!

On returning from our honeymoon, the Poseidon crash was in its death-throws and there was blood all over the stock market. I thought "this is not looking good" and on returning to work, I expected the worse. Sure enough, I was told that money had dried up, the directors were cutting costs and thanks but your services are no longer required. To make matters worse, I also had lost all the money I had made during the Poseidon boom. Fortunately Mo had her job and so I became "Mr Mum" for a couple of weeks.

DELHI PETROLEUM

The Poseidon crash had thrown a lot of geologists out of work so getting a job was not going to be easy. Fortunately Delhi Petroleum was looking for well-site geologists and, thanks to a tip from Mike Wiltshire I got in contact with them. Delhi was the operator for the Cooper Basin permits and following the Gidgealpa and Moomba gas discoveries, there had been a number of new discoveries (Toolachee, Merrimelia, Strezlecki, Della and Big Lake gas fields and the Tirrawarra, Moorari and Fly Lake oil fields) made in the basin between 1969 and 1971. Delhi needed wellsite geologists for the appraisal and development drilling programs required to bring these fields into production. The contract was tough – I was paid a daily rate for time in the field and office but I had to pay my airfares to and from Adelaide. But it was a job and the net pay was still good and, as it turned out, the experience was incalculable. My main area of work was the Tirrawarra and Moorari oil fields. I didn't know it at the time but the Tirrawarra Oil Field turned out to contain 80 million barrels of recoverable crude oil and turned Bridge Oil into a major Australian oil company.

In July, 1971, I flew from Adelaide to the Moomba Gas Plant and from there to the Tirrawarra-2 site in a single engine Cessna. The rig was located between two sand dunes. If you walked to the western dune, there was another dune to the west. If you walked to the eastern dune, there was another dune to the east. So not a lot of sight-seeing was possible. Now the Tirrawarra Oil Field was in the middle of the Simpson Desert **(Fig 22)** and it was very hot – just about every day in the 40's and one day in summer, the temperature reached 52°C. To make matters worse, the drilling camp air-conditioning didn't work – some bright spark forgot that the drilling rig was built in the US and ran on 120 volts, and bought local air-conditioners running on 240 volts.

Figure 22
Lonely drilling rig in a vast desert

The wild life was also interesting. There was nothing living out there above the level of insects (not counting ourselves of course although some would argue with that premise). The flies were – well "thick as flies" and worked during daylight hours. At sunset, the flies clocked off and the flying ants and moths clocked on. And I'm not talking a few moths – I'm talking millions coming from miles around attracted by the lights on the rig. We had no idea what the hell they lived on in that desert or how they survived without water but somehow they thrived. Each morning the ground around the rig was inches thick in dead moths. But the flying ants were the worst. They were so small that they could wiggle their way through the flywire screens on the mudlogging shack. So all my paperwork was covered in sweat and squashed flying ants. Unfortunately, my sweat and blood-stained reports were not enough to convince management that we needed air-conditioning on this god-forsaken well-site.

The oil reserves in the Tirrawarra Field were trapped in the Tirrawarra Sandstone near the base of the Permian at about three kilometres below the surface. The wells were drilled through the Mesozoic sediments, logged and cased, and then we were called out to log the Permian. Years later, oil was discovered in the Jurassic sandstones in the basin but in 1971, we never saw these sediments. Delhi was an American company and did not believe that oil could be generated in non-marine sediments. As the Jurassic was all non-marine sediments, they never bothered to properly log these sediments. If they had, these oilfields would have been discovered much earlier by the existing companies and the Australian oil industry would have been quite different.

In the onshore technology of the day, the mudlogging unit consisted of a simple hot wire gas detector **(Fig 23)**. There were no mudloggers - just the wellsite geologist. As this meant working 24 hours a day while drilling, the gas detector was fitted with an alarm. The background gas readings were normally less than 10 units, so the gas detector alarm was set at 10 units. The wellsite geologist slept in the mudlogging unit whilst the rig was drilling and if the well encountered any gas shows, the alarm would go off and wake him up.

Figure 23
Basic Gas Detector

It was primitive and not very effective. By the time the alarm went off, the well had probably drilled through the sand so there was no opportunity to cut a core. All you could do was organise a drill stem test. Also, if the well intersected an oil sand without a gas show, the alarm would not go off, and you wouldn't know until morning. Fortunately, this never happened to me.

The first well was an appraisal well on the Tirrawarra Oil Field. The casing shoe was drilled out and the samples to reach the surface were mostly cement. Eventually, the basal Triassic red beds appeared to be followed by the thick coal seams at the top of the Permian. Some of these were very thick, up to 30 metres of pure coal without any shale breaks. The Upper Permian was called the Toolachee Formation and was mostly coals, shales and siltstones. Below this was the Patchewarra Formation where the first fluvial sandstones appeared. When drilling into these sands, the bit drills faster than the hard, laminated shales and siltstones, and provides what is called a drill break. Once a drill break is detected, the drilling is stopped while the well circulates the mud and drill cuttings to the surface. This could be up to an hour, depending on the depth of the drill break. Once the drill cuttings reach the surface, they are examined by the geologist to determine their quality and detect the presence of any fluorescence. If the sands are gas bearing, this will register on the gas detector. If the sands are oil bearing, they will fluoresce under ultraviolet light from the microscope. If the sands and the shows are sufficiently encouraging, then a drill stem test is conducted. This involves pulling all the drill pipe out of the hole, removing the drill bit and connecting the test tools to the drill pipe and running the whole lot back to the bottom of the hole. If the test recovers gas or oil, then a core is taken. This involves pulling the test tools out of the hole, removing them from the drill pipe, connecting the core barrel and running back into the hole. The core bit cuts a circular ring in the rocks so that a core of rock rises into the barrel. After cutting 9 metres, the core barrel is pulled out of the hole, and the core is examined by the geologist at the surface. As you can imagine, all this pulling the drill pipe in and out of the hole involved a lot of hard work by the roughnecks so the wellsite geologist was not the most popular bloke on the rig during these operations.

On this well, after a lot of false starts, the well eventually intersected a sand with a 40 unit gas show. The sand contained a lot of clay and its porosity looked poor but the drill break was good and the gas show significant. As I was taught by Jim Cundell – "when drilling an oil well, if in doubt, always test" - so I decided to conduct a drill stem test, much to the groans of the roughnecks. As there was plenty of shale above this sand, designing the test tool configuration was relatively easy. The test tools were run in the hole and the packers set in the shale above the sand. The packers expand against the hole to hold the weight of the mud and allow the fluids in the sand to flow into the drill pipe. Normally a water cushion is placed in the drill pipe to allow these fluids to flow from the sand at a controlled rate without dropping the pressure too rapidly and causing phase changes in the reservoir. When the test tool is opened, the pressure difference causes the fluids in the reservoir to flow into the drill pipe and rise to the surface. There the flow rate is measured and you find out whether you have oil, gas or boring old water.

In this test, the water cushion was quickly expelled from the drill pipe and the well started flowing gas. This was my first successful drill stem test. While I had seen many oil and gas discoveries in the Gippsland Basin, these were offshore where open hole tests were not safe and were not carried out. The testing is always carried out after the well is completed and long after the wellsite geologist has departed. So as you can imagine, I was pretty excited about all this and watched the test with great interest. Once the gas started flowing, it is deliberately lit and burns into a prepared flare pit. The gas rate kept increasing and the noise was incredibly loud. Then the rate started to die and I cried out "Oh no!" as my immediate thought was that the sand was not extensive and the pressure was dropping. If this was the case, the reserves would be limited and non-commercial. But then the ground began to shake like an earthquake, and the rig was rattling and then "kaboom", a slug of orange flame and thick black smoke came out of the pipe, just like a flame-thrower. Then the gas was flowing again and the whole process repeated itself. What was happening was the gas contained condensate, a light petroleum which is gaseous at the high temperatures and pressures at the bottom of the hole. As the gas rose in the drill pipe, the temperatures and pressures dropped and the condensate liquified in the drill pipe and constricted the flow of the gas until the back pressure built up and blew it out of the hole. When the gas flow stabilised, it was flowing at 8 MMcfg/d (million cubic feet of gas per day), a very respectable rate given the poor visible porosity of the sand. Later core analysis indicated that the sand had an average permeability of only 3 millidarcies, so I had learned another valuable lesson. Poor permeable sands buried deeply with high reservoir pressures can still flow gas at commercial rates.

After the test, the well continued down to the Tirrawarra Sandstone. Although the Tirrawarra Sandstone was not oil bearing in this well, we cored it anyway to provide reservoir information for the field development. It was totally different to the other sands, having much better porosity and permeability due to much lower clay content. It was deposited in a high energy, braided stream environment similar to the current Coopers Creek, and provided an excellent and areally extensive reservoir in this part of the basin. Back in the Delhi office, I remember getting into trouble for not coring the top of the Tirrawarra Sandstone. In this well, there was a point bar (channel sand) overlying the Tirrawarra braided stream sands. So the drill rate increased slowly while drilling through the point bar and before the strong drill break at the top of the Tirrawarra Sandstone where I started the coring. Geologically speaking, I was correct but I'm sure the reservoir engineers were not impressed.

Collecting and describing a core is another complex operation. The idea is to collect the core into core boxes and label and measure the depth. The most critical thing is to make sure the core is the right way up. Each core box can hold about 9 metres of core in three rows. When the core barrel comes to the surface, it is suspended by the driller about 0.3 metres above the derrick floor. The core trays are laid out next to the barrel and the wellsite geologist collects the core and places it in the tray from the bottom up. A roughneck hits the core barrel with a sledgehammer and, as the core falls down and out of the barrel, the geologist pulls the bottom of the core towards himself using a geological hammer so that the core is the right way up. This is a tense operation as while each piece of core is being collected, your hand is situated under many tonnes of steel for a few seconds. It was therefore good policy to be nice to the drillers and not piss them off in any way whatsoever!

My nastiest experience collecting a core was during this time. I was getting pretty experienced by now and collecting cores was becoming routine. I was collecting this particular core and placing it in the core boxes. As I pulled a piece of core towards me, the core barrel hit the deck missing my hand by inches. I pulled my hand away and screamed "What the fuck!" and then pieces of core began falling on my head so I made a hasty retreat. What had happened was the core barrel had parted with the lower part falling onto the derrick floor (fortunately missing my hand) while the upper part was swinging around in the derrick unloading its contents all over the derrick floor and my poor bloody head. Being an experienced rig hand, I was wearing my hard hat so the only damage inflicted on me was severe bruising of my shoulders and arms, and a very loud ringing in my ears!

Normally, drilling is carried out using a simple mud system. Without going into a boring dissertation on the chemistry of drilling mud systems, normal muds are dirty, hot and alkaline but can easily be washed off at the end of the day. On the Tirrawarra Field, the Tirrawarra Sandstone reservoir contained some clay which was sensitive to water, swelled up in the pore space when coming in contact with the drill mud, thus reducing the permeability of the sandstone around the well bore. This is called formation damage in the industry and is a very common problem. When the first Tirrawarra well was drilled, it was initially drill stem tested at over 1,000 BOPD but after being completed and production tested, it would only flow at 300 BOPD. Obviously this had a serious effect of the economics of the development of the field so the operator decided to experiment with an oil-based mud system to try and reduce this formation damage. An oil based mud uses oil as the dominant liquid to produce the mud system. The idea is that only oil comes in contact with the clay when the reservoir is drilled, the clays do not swell and the formation damage is reduced.

This was all very fine unless you had to work with the stuff. It was bloody awful. The oil would contaminate the gas detector and the valves had to be continuously changed. Valves you ask – well this was 1971 – no silicon chips – no computers – everything was analog and most things had to be written down on pieces of paper. The roughnecks also hated it as they got covered in it while drilling and it was the devil to remove in the shower. To my horror, I discovered just what they were talking about. While collecting drill cuttings samples from the shale-shakers, the oil based mud was splattering everywhere and was all over the cuttings. While performing this task after every 10 metres of drilled hole, I eventually got covered, just like the roughnecks, with this ghastly oil based mud.

The method of removing oil based mud off human bodies was not very sophisticated. It consisted of tubs of a mixture of soap and kerosene which you used to remove the oil. This was bad enough, but the shower water came from a water bore and was very brackish. It was like rubbing sandpaper all over your body. After this particular episode, I never felt completely clean, so as soon as I got back home it was great to wash with normal soap and water. But I still didn't feel completely clean. So I tried a suggestion of one of the roughnecks – I went to the local gym and sat in the sauna. After getting up a good sweat, I stayed there as long as I could stand it. The roughneck was right and I felt terrific after this session in the sauna. And I felt clean at last!

The mud man who produced this ghastly concoction was called Clay, an American from Mississippi. I remember him to this day as he wore jeans and a white shirt, and spent his entire day organising the mixing of the oil based mud, measuring its properties and running around the mud tanks bossing everybody about and ensuring that everything was OK. So while the rest of us ended up with bits of his bloody mud all over us, he returned to the camp each night without a single spot of mud on his white shirt. For me it was comic relief as everyone on the rig hated him more than the wellsite geologist!

Being a wellsite geologist in those days was tough – no two weeks on and two weeks off. We were there for the duration of the well, generally four-five weeks. To make matters worse, it was Delhi's policy to cut a bottom hole core before the well was logged and completed. The bottom hole core was mostly in the Merrimelia Formation, a glacial sequence which was hard as concrete and cored very slowly. It was also very boring to describe being mostly dark grey shale.

At the end of the well after electric logs have been run and interpreted, the well is then completed for production. The various contractor personnel no longer required at the wellsite would be shipped back to Moomba base by single engine Cessna. This normally included me, the electric logging engineers, the coring engineer and sometimes a lot of extra gear. One time, we were returning to Moomba and were all packed into this tiny little plane. The pilot turned around and told us that we were over weight and we would have to do a rapid take-off. As we had no idea what this meant, we said OK. So he put the breaks on, revved the engine up until I thought the plane was going to shake itself to pieces, then he released the brakes and we took off down the runway like a rocket. At the end of the runway there was one of those boring sand dunes that I mentioned earlier, and we were hurtling down the runway at it at just below the speed of sound. I was in one of the back seats and was peering between the front seats at the windscreen. The sand dune got bigger, and bigger, and bigger, until it literally filled the windscreen and I thought "we're not going to make this – we're all going to die"! Now you may all think that I'm a terrible wimp and you'd be correct. But a knowledge of physics is not necessarily a good thing. All I could think of was the mass of the plane together with full tanks of fuel hitting a sand dune at just below the speed of sound – it was not a promising scenario. Fortunately the pilot knew what he was doing and at about one metre from the sand dune, he raised the flaps and we took off straight up like a Caribou (the plane that is, not the lumbering mammal!). All the blood drained from my head and I thought I was going to pass out, and where my stomach was – well, it wasn't in the plane. Amazingly, we survived and arrived at Moomba. But when we got out of the plane, there was much kissing of terra firma.

After a couple of weeks off, it was then back out to Moorari-2 in September-October. On my return from Moorari-2, I got to the front door of our unit and thought to myself "Is Mo going to be still inside or has she wisely decided that being married to an oil geologist was just plain crazy and taken off to greener pastures". But there she was with a big smile, a big kiss and later on a bit of you know what. She must have really loved me to put up with that first six months of our marriage.

Just before I finished my contract with Delhi, I witnessed one of the great natural disasters to hit the Australian oil industry. In 1971, it had not rained in the Cooper Basin since oil exploration began - about ten years earlier.

So we often located the drilling rigs on clay pans. These were large, flat, dried out lakes which required no site preparation. There was plenty of room for the drilling rig, mud tanks, generators, casing racks, contractors' equipment and camp – they were just perfect! At the end of 1971, a cyclone appeared in the Coral Sea and ran down the coast of Queensland. But instead of moving out to sea (as they'd been doing for the last 10 years), it came inland and dumped an ocean of water on the wrong side of the Great Dividing Range (or the right side if you were a pastoralist). The Cooper and Strezlecki creeks flowed across the basin flooding everything in sight, and even ended up filling Lake Eyre. The whole Cooper Basin was a sea of red brown mud. Landing strips were flooded, planes couldn't take off or land, trucks were stranded on washed out roads – it was a catastrophic mess of monumental proportions. Of course those rigs located on the clay pans remained there for 18 months before they could be retrieved, cleaned up and made fit for drilling again.

Christmas 1971 was one of our most memorable days. We were living in Mo's flat in Neutral Bay and it wasn't exactly large. We decided to have Christmas at home and invited a few friends – a few – it ended up being 14. Mo's mum, Mabel, was out from England, my mum was there, Audrey and her two children. It was a very hot day and after lunch, everyone went out into the backyard for some games. After much alcohol, this soon turned into a fight with the hose with most of us getting wet. After hosing Nigel, Mabel thought it would be safe to retreat into the house – but no – the hose came through the window and she was drenched. A truce was called and we all dried off and settled down to a very pleasant afternoon.

BURMAH OIL COMPANY OF AUSTRALIA

While well-site work put food on the table, it wasn't my preferred career path. So I saw an advertisement in the paper for a geologist in Perth for Burmah Oil Company of Australia (BOCAL). So I applied and got the job. Burmah Oil was a British company that had farmed into Woodside's permits on the North West Shelf. The Woodside (Lakes Entrance) Oil Company was formed in 1954 to explore the onshore Gippsland Basin. The company's exploration efforts were unsuccessful but in 1961, it hired Nicholas Boutakoff as Chief Geologist. Based on his ideas on the offshore areas off northwestern Australia (which he published in the 1963 APEA Journal) the company applied for and was awarded Permit to Explore 213H covering a vast area (360,000 square kilometres) of the North West Shelf of Australia in June, 1963. In 1967, a joint venture was formed between Burmah Oil (BOCAL) 16.67% and Operator; British Petroleum 16.67%; Shell 16.67%; Chevron 16.67%; Woodside Oil 25% and Mid East Oil 8.33% to explore this large area.

BOCAL drilled the first well, Ashmore Reef-1 in 1967 but it was a dry hole. In 1968, BOCAL drilled Legendre-1 which flowed oil at 1014 bopd and was suspended. Then followed eight dry holes but in 1971, the drilling of Scott Reef-1 resulted in a gas discovery which tested 8 MMcfg/d. North Rankin-1 followed in May 1971 and intersected a 550 metre gross gas column which tested at 12.8 MMcfg/d with 25bbl/Mcf condensate. The Angel-1 and Goodwyn-1 gas discoveries followed and collectively, cumulative reserves were then estimated at seven TCF gas.

In August 1971, Woodside Oil merged with Mid East Oil, and then merged with Burmah Oil Australia to become Woodside Burmah Oil NL. This company now had a 50% interest and operatorship of the North West Shelf permits and became a very formidable company in the Australian oil industry.

It was January 1972 when we all headed off to Perth. I cashed in Mo's airline ticket and bought two train tickets for her and her mum, Mabel who was staying with us on one of her numerous trips to Australia. In those days, the train trip was half the cost of the airfare. I flew over ahead of them and stayed in a hotel until Mo and Mabel arrived. One evening before they arrived, to fill in time, I went to the Summerville, an open air theatre at the University of Western Australia. The film was a Japanese film called the "Throne of Blood" – a Japanese equivalent of Macbeth. I really enjoyed the film, particularly the comparison of the feudal kings of Britain with the warlords of Japan. When Mo and Mabel arrived, we then rented a flat in Hensman Street, South Perth. While the accommodation was sparse, the area was good – lots of parks, a neat local pub called "The Windsor" and close to work – the BOCAL office was in Nedlands. It was here that Mo told me that she was pregnant and I felt like the cat who had just swallowed the canary – you couldn't wipe the smile off my face.

My first assignment with BOCAL was the Bonaparte Basin-Timor Sea area. The Aquitaine – Arco consortium owned most of the southeastern part of the area and had discovered the Petrel and Tern gas fields in the Bonaparte Gulf and the Puffin-1 and -2 oil discoveries in the Timor Sea. BOCAL had drilled Ashmore Reef-1 in 1967 and Sahul Shoals-1 in 1969, both very dry holes. Arco had drilled Heron-1 and Flamingo-1, so there wasn't a lot of well control over this vast area. Heron-1 was a very deep well with a total depth of 4,170 metres and a bottom hole temperature of over 200° C, stretching the electric logging technology of the day. But it did discover gas in poor reservoirs but was deemed to be non-commercial. Flamingo-1 was drilled in 1971 and has good oil and gas shows so there was some encouragement in the area. On the eastern side of the Sahul Platform, recent seismic acquisition had come up with some large, low relief structures. These were so gentle that you could not see them on one seismic line. We had to join the seismic lines together and then lay the paper sections out down the office corridor to see the rollover. Two of these were drilled in 1974 while I was working for Burmah Oil in the North Sea and resulted in the Sunrise-1 and Troubadour-1 gas discoveries.

Wellsite work with BOCAL was more civilised than Delhi and we worked a two week shift at a time. At the end of one of these wellsite assignments, a cyclone was approaching and the rig was evacuated. I remember the sky becoming an evil looking greenish yellow colour; there was no wind and the humidity was oppressive. When I returned to Perth, Mo said "we've been invited to Jurien Bay to stay with some friends who have a shack there". Great – I've just been evacuated from the rig to escape the cyclone and now I'm going to be driving back into it. The cyclone was forecast to pass close to Jurien Bay so when we arrived there, we "battened down the hatches" to make it secure. Then we had a great meal, lots to drink and went to bed. Fortunately, the cyclone took off in another direction and there was barely any wind during the night. The next day, we went fishing and caught enough fish for a good feed that evening. It turned out to be a great break after two weeks on the rig.

One of the most interesting things about BOCAL was that they had their own palynology laboratory and did their own age dating. They had palynologists, Barry Ingram, Malcolm Macleod, Gerry Orbell and Dick Lister on staff. The company had developed its own Mesozoic zonation system as many of the index species were not present in the northern hemisphere basins. Seismic events could then be tied to the wells giving greater accuracy to the predicted sections to be drilled in future prospects. During the preparation of the palynology samples, part of the sample was kept separate for a qualitative source rock analysis. The source rock kerogens extracted from the shales were classified as vitrinite and inertinite (poor or non-source); exinite, cutinite and sporinite (gas source), and alginite and sapropel (oil source). This was a terrific method as it was so cheap to obtain lots of very useful data. The better source rock intervals could then be confirmed by more expensive geochemical analyses on sidewall cores using the Rock Eval technique as it developed. But there's more! Based on recent overseas studies, BOCAL developed its own spore colour maturation profile for determining if the source rock intervals were immature (not hot enough to generate hydrocarbons); oil mature (generating oil) or gas mature (generating gas). This was superior to other methods such as vitrinite reflectance where samples could be contaminated by cavings from drill cuttings up the hole or reworking by older sediments being redeposited in younger sediments. The spores could be identified, the spore colour measured to determine maturation, and any older or younger spores could be rejected as reworked or caved. The amount of very useful information that came out of BOCAL's palynology lab made our job of evaluating the hydrocarbon potential of these very under-explored basins so much easier.

On the home front, Mo was starting to fill out and exhibit her "inner beauty". In addition, the nice thing about moving from Sydney to Perth was that land and house prices were very much cheaper than those in Sydney. We bought a block of land in Glenbank Crescent in Kallaroo for $4,500. We picked this block because there was a reserve on the north side. But when we looked at the map, we realised that this was the most northern block in Perth, excluding the town of Wanneroo. On the south side of our home was the Telethon home. This was auctioned by Stuart Wagstaff and we had a ring side view of the proceedings. The home was bought by a young couple Greg and Sue with two young daughters.

Kallaroo is about 20 kms north of the city and this was considered to be in the wilderness by the locals (who rudely said that we lived in South Geraldton); but to someone from Sydney, it was just up the road. Even more amazing, we bought a display home (with slight modifications) for $14,500. So our new home in Perth cost us the princely sum of $19,000.

We moved into our new house in September 1972 and it didn't rain until the following March so it was pretty tough getting a garden established.

Mo gave birth to Chris on 12th December, 1972. At this time, fathers were at last being encouraged to attend the births of their children. Prior to this, fathers were supposed to pace up and down in waiting rooms and then hand out cigars when told of the new arrival. So I attended a class for fathers as to what we should do to help, and then attended the birth. This turned out to be both a traumatic and wonderful experience. Mo was induced and had a difficult labour, and I was not the most popular person in the room. It was all MY fault that she was going through all this pain and she let me know it. All I could do was hold onto her hand and tell her to push when the doctor did. I have never seen anyone work so hard as a woman giving birth - it was a real eye opener. Eventually Chris was born and the nurse showed him to me. He was a purple mass of wringled flesh – without doubt, the ugliest thing I had ever seen. I wanted to shout "Hey Doc – put it back" but then he cried and began to breath – and he turned pink before my very eyes. I have never seen anything so amazing and beautiful in all my life. And Mo had completely changed – I was no longer the anti-Christ but her loving husband again as she held Chris in her arms. It was altogether a wonderful experience and I would not have missed it for the world.

Of course I had to celebrate so I went around to visit Tony Gates who was working in Perth in the mining industry. His wife Pauleen told me that he had gone to Colin Wilson's birthday party, and as Colin was also my accountant, I decided to join the party. It was a very boozy night with lots of food, beer, wine and rugby songs. I don't remember driving home but the next day, my clothes were all sprinkled down the hallway. I went round to the hospital to see Mo and Chris and she took one look at me and said "you look dreadful". She thought I should be in bed, not her!

At the end of 1972, Gough Whitlam was elected Prime Minister of Australia after many years of Liberal rule. Their election campaign of "It's time" was a winner and even I voted for him. However, the selection of Rex Conner as Federal Minister for Resources was to cause the Australian oil industry many problems, as will be discussed later.

As oil exploration moved offshore, all nations were busy fixing their offshore borders with their neighbours. Once agreed, these offshore areas could be leased to companies for petroleum exploration. In most cases this was easy – simply the edge of the continental shelf where extending into an ocean or the midway point between coastlines where there was no deep ocean. In other cases where there are islands, the boundary is between the island and the coastline. This allowed the UK to claim most of the English Channel as the Channel Islands belonged to the UK – much to the annoyance of the French. The Ashmore and Cartier islands also allowed Australia to claim most of the Timor Sea as the border with Indonesia was midway between these islands and West Timor. The difficulty was the border with East Timor which was then a Portuguese colony. Portugal was claiming the midway point between East Timor and Melville Island. All of this area had already been leased to oil exploration companies by the Australian government. Obviously, it was in BOCAL's interest for the status quo to remain. I was asked to undertake a structural geological study of East Timor and the Timor Sea to determine the position of Australia's continental boundary. If Australia could argue that Australia's continental crust extended to the Timor Trench, then that trench would be the border with East Timor. Fortunately, there were recent studies on the geology of East Timor which showed that the island consisted of deformed deep water sediments deposited off Australia's continental shelf and Asian shallow water carbonate overthrust from the north. The Australian plate was moving north and colliding with Timor and Papua New Guinea. To the east, the Indian Ocean crust was diving under the Indonesian archipelago (subduction) but Australia, being thick continental crust and less dense than the mantle, could not be subducted. It was colliding with Timor and eventually, over millions of years, Timor would eventually become a mountain range like the Owen Stanley Ranges in PNG. I was able to demonstrate that the Timor Trench was the boundary between Australia's continental shelf and the Eurasian Plate. Some of this work formed part of my first paper "Plate Tectonics and the Geology of the Timor Sea" which I presented at the 1973 APEA Conference in Canberra.

The dispute with Portugal continued but by 1974, Portugal began abandoning its colonies, Angola, Mozambique, and East Timor. An independence group called Fretelin declared independence and in the midst of all this chaos, the Indonesians invaded East Timor in December, 1975. The border dispute then continued with the Indonesian Government until 1989 when the 68,000 sq km Zone of Co-operation between Australia and Indonesia was created. In 1991, the Portuguese government filed an application in the International Court of Justice to overturn the treaty but this was unsuccessful.

Following the APEA Conference in Canberra, I was selected by Woodside Burmah to attend the Australian Administrative Staff College at Mount Eliza during May-June 1973. This was a great experience and introduced me to many management skills I was to use in later life. I also mixed with people from all different types of industries. Amusingly, they thought that BOCAL was a subsidiary of the British Oxygen Company. We were broken up into groups and given various tasks to complement the management theories they taught us. I soon discovered that the oil industry was very unique in the way it managed risk. Joint ventures, used by the oil industry to manage their enormous risks, were very rare in our industrial society. The reason is that joint ventures are very inefficient and slow in making decisions unless they consist of like-minded companies. As there are very few companies in the Australian oil industry, this can normally be achieved. But it only takes one difficult joint venturer to make the operator's life a nightmare and destroy the efficiency of the project. After dinner each evening, we would retire to the bar and swap experiences. There was one fellow who could play the piano so we often ended up singing around the piano – mostly "Sweet Georgia Brown". It was like being transported in time back to the 19th century! One Saturday, they let us back out into the world and a few of us took off to Melbourne to watch a VFL match at the MCG. Management courses can be very boring but Mount Eliza was just terrific.

BURMAH OIL NORTH SEA

In 1947, Shell and Esso formed a joint venture called Nederlandse Aardolie Maatschapij or NAM for short (thank goodness). In 1959, NAM discovered the giant 100 TCF Groningen Gas Field in the Netherlands. The gas is trapped in Permian Rotliegend sands of aeolian origin. This field assisted Western Europe in building natural gas as the main source for its power generation. The field has only produced 60% of its reserves and is expected to last for another 50 years. Shell and Esso continued the partnership by being awarded acreage in the UK and Dutch sectors of the

North Sea in 1969. In 1971, they discovered the Auk Field with oil in Permian Zechstein carbonates. The next well discovered the giant Brent Oil Field and set off an exploration boom which built the North Sea into a major oil province. I am very proud to say that I was a very small part of that boom.

In August 1973, I was transferred to work with Burmah Oil North Sea in London. We flew Scandinavian Airlines from Perth, Bangkok, Tashkent and Copenhagen and then onto London. We spent a couple of days in Bangkok and really loved it. The Thais were very friendly and so very polite, but it was Chris who was the big hit with his blond hair. We took a tour of the city including all the Buddhist temples. Then off to Copenhagen. We had a refueling stop in Tashkent and I was looking forward to this. I thought Tashkent was going to be like Samarkand but it turned out to be just another big industrial city. It was a great flight – past the Himalayas, over the Hindu Kush, across Afghanistan and into the Soviet Union. Flying over Afghanistan, the geology was terrific but there was little in the way of roads or towns. But as soon as you crossed the border into the Soviet Union, there were roads, towns, airports and factories everywhere.

When we landed at Tashkent, a couple of Russian soldiers with sub-machine guns entered the plane, took away our passports and escorted us to the terminal. All very intimidating! The temperature was over 40°C and there was no air-conditioning or shops in the terminal – we couldn't even buy a bottle of water. I said to Mo "if the Russians want to convert us to communism, they're doing a very poor job of it". So we were glad to get back on the plane, have something to drink and get cool again.

They say that the boundary of Europe and Asia is the Volga River and they are right! As soon as we got to the Volga River, the blue skies ended and the countryside was obscured by thick cloud. We just had an overnight stay in Copenhagen but compared to our luxurious hotel in Bangkok, the hotel we stayed at in Copenhagen was a falling down hovel.

Eventually we got to London. Mo's big sister, Rosemary, worked at the Eden Plaza Hotel so we stayed there. Of course Chris, being only nine months old, was the centre of attention. After London, we travelled north to Morecambe to stay with Mo's mum and dad, middle sister Jean and her three boys. They all lived in a large stone house called Bare Hall. Now

Bare Hall was built somewhere around the time of the War of the Roses so it wasn't the most comfortable place. But it was summer and we all survived. I wouldn't have liked to spend a winter there – and never again would Mo! Mo's dad, Arthur, invited me to the local Pub for a few pints of the local beer which I really enjoyed. But the Lancashire accent, that was a problem. I understood about every second word, and I suspect that they didn't understand much of what I said either. During my stay in the UK, I grew to love the English pubs and especially their beer. There is something unique about this British institution and in all my travels around the world I have never encountered anything that comes close to the ambience of an English pub. Alas, since they joined the EU, most of the pubs in London only sell European lagers. These days you have to get out of the big cities and into the countryside to find a really good pub and a decent English beer.

Morecambe was a seaside holiday town and was very popular. However, it consisted of a huge muddy tidal inlet, and no respectable Australian would be seen dead there for their summer holidays. But Jean's boys asked me if I'd like to go for a walk along the canals. This turned out to be a real treat. They were absolutely beautiful – quiet, lush green with every duck species in Britain – and the occasional boat drifting along without a care in the world. It beat the hell out of the seaside. Years later, they re-connected all these canal systems and created a whole new tourist industry.

Before we left Morecambe, I bought a dark green Rover. Mo called it "the tank". We drove back to London with it. Rather than being a "tank", it turned out to be a lemon!

The Burmah Oil office was in Salisbury House in Moorgate in the City of London. The first thing we had to do was find a place to live. Initially this was difficult as rental accommodation was very scarce. However, being an Australian on a two year assignment, I had no problem in being offered a wide range of rental accommodation. In 1973, the UK had a very onerous Landlord and Tenant Act which was very biased against the landlord and made it difficult to evict tenants. The irony of this situation was that it was easier for foreigners to rent accommodation in London that it was for the English themselves.

We selected a house in the town of Broxbourne in Hertfordshire on the A10 to Cambridge. It was perfect for me as it was a 30 minute train ride express to Liverpool Street Station and a five minute walk to Salisbury House. Broxbourne was located on the River Lee in beautiful countryside and was close to shops and restaurants. The local pub, the Crown, had a terrific beer garden sloping down to the river bank. It was at the Crown that we joined another English institution. On Sunday, Mo would put a roast dinner in the oven under a slow heat and we would go to the Crown, have a few drinks and let Chris run around in the beer garden. The Crown would close at 1-30pm and we would walk home just in time to sit down for lunch and watch "Match of the Day" on television. The Poms had it all timed to perfection.

As I was considered to be an expert in geochemistry, my next task with Burmah Oil was to attend the Sixth International Congress of Organic Geochemistry in Paris in September, 1973. In reality, I wasn't an expert, I just knew more about geochemistry than anyone else in Burmah Oil. I certainly wasn't going to miss out on a trip to Paris so I didn't tell anyone. I was all ready to go, got my airline tickets and realised that I had no accommodation. As it was August and the holiday season, this was not easy to find and I ended up with a very reasonably priced hotel on the Rue Pigalle. When I got to Paris, I discovered that the French are not like the English – they don't queue. The taxi rank outside the airport was like the Battle of the Somme. No place for children, the disabled or the elderly. After watching the carnage for about 10 minutes, I got the general idea and charged into the fray. I eventually got a taxi and ended up at this rather dingy hotel. Ah, no wonder the price was so reasonable. But it was clean and the concierge friendly considering my very poor French. After telling him "je suis Australien" he eventually took pity and spoke to me in English.

That evening, I went out for a meal and the concierge recommended a restaurant just up the road. So I wandered up the road and noticed all these beautiful ladies in short skirts everywhere. Strewth, I'm in the Kings Cross of Paris! After a leisurely stroll up the Rue Pigalle, I got to the restaurant. I sat down and perused the menu. Now my French is very basic and all I could recognise was escargot and boeuf bourguignon on the menu. It turned out to be one of the best meals I'd had at a restaurant. I had never tasted escargot but they were absolutely delicious. The boeuf bourguignon came in a saucepan which they plonked down on the table in front of me. But the flavour was magnificent. So my first meal in Paris, although the restaurant would not be found in any of the tourist brochures, lived up to my expectations.

After this meal, I wandered up the street and ended up at the famous Moulin Rouge. Being on my own and without a booking, I didn't like my chances of getting in but I thought I'd give it a go anyway. The entrance was full of bloody tourists including Japanese arriving by the bus load. The poor bloke who was trying to bring order to this chaos was very helpful. He said that no tables were available, but that, being on my own, I could sit at the bar and watch the show. A beer was the equivalent of $15 a glass (and this was 1973) but what the heck – I may never be in Paris again. The show was truly spectacular and I will never forget the finale with a tank rising from beneath the stage. The tank had a dolphin in it with bikini clad ladies swimming about with it.

The conference was held at the Institute Francais du Petrole in Ruell-Malmaison and was well attended. Papers were given by top men in their field: Welte, Correia, Tissot, Teichmuller, Hunt and Postuma. Research by Welte showed bimodal distribution of vitrinite reflectances with depth indicating possible different temperature thresholds for coal macerals. Tissot gave an excellent paper on source, maturity and migration of Silurian source rocks in Algeria. Research by Stahl on German natural gases showed that carbon isotope ratios could distinguish biogenic gas from thermogenic gas. Postuma gave a paper on the origin of sulphur in oils from the Lake Maracaibo Basin in Venezuela.

At the end of the conference, I was reassured to know that the work being carried out by BOCAL in the Perth office was of high quality by international standards. It was clear that we were well advanced in source rock kerogen analyses and maturation studies but were behind in crude oil geochemistry.

At the conference, I met Gerard Demaison who was working with WAPET on their geochemical studies on the North West Shelf. We had lots to discuss and both had the same problem. In 1973, the North West Shelf was classified as gas prone. No oil fields had been discovered except the enigmatic Barrow Island Oil Field which contained an unusual oil in poor Cretaceous reservoirs. Both BOCAL and WAPET had identified good oil prone source rocks in the Jurassic sections in the Dampier and Barrow sub-basins. These source rocks were thermally oil mature and we were both trying to answer the obvious question - Where the bloody hell was the oil? Gerard ended up back in Chevron's research centre in California and became a real expert in geochemistry co-editing AAPG Memoir 35 "Petroleum Geochemistry and Basin Evaluation".

After the conference, I spent a few days in Paris. I went up the Eifel Tower – a difficult and time consuming process as the place was again packed with those damned tourists. As a matter of fact, there seemed to be more tourists in Paris than Parisians. However, the highlight was the Louvre – I could have spent an extra week in this museum. The paintings were magnificent. Some were two metres high by eight metres long while the most famous painting of all, the Mona Lisa, was barely bigger than a postage stamp. But my favourite was Van Dyck – his portraits were so life like that it was like staring at a real person. And the eyes seemed to follow you.

No sooner that I had returned to London, than I was off again on a field trip to Ireland to study the Carboniferous outcrops on the west coast. I flew to Shannon airport and then by bus to the town of Kilkee. Half the towns in Ireland start with Kil which comes from the Gaelic Cil meaning church or monastery. When I got to the hotel, checked in and unpacked, I wandered down to bar and there were Peter Power and Frank Brophy who were also on the field trip – small world heh! On the first day, the weather was initially OK and we were on a rock platform studying an outcrop of very fossiliferous Early Carboniferous limestone. I've never seen that many crinoids packed together in one place and so well preserved. Suddenly a storm brewed up out of nowhere and hit us with fierce winds and torrential rain. We all got back to the hotel wet, cold, bedraggled and looking very miserable. Our lecturer was Irish and assured us that a pint of Guinness would fix us all up. So we straggled into the bar and he ordered pints of Guinness for all of us. To an Australian, black beer didn't look very enticing but when in Ireland, do what the Irish do! As it turned out, the Guinness was wonderful and the lecturer was right – after a few more pints, we were all fixed up. The rest of the field trip had better weather, and I was very impressed with Ireland. The coastline was spectacular with headlands and beaches like the coast of NSW. The countryside was so green it reminded me of New Zealand – without the sheep!

The Burmah Oil office, Salisbury House, was a very old building, even by London standards. It even had one of those lifts with the closing cage and was operated by a very old gentleman who asked you what floor you wanted. It was something straight out a Dickens novel. Even more amazing – the company gave us "luncheon vouchers" to use to buy lunch as a sort of employee benefit. The Burmah Oil geologists had already found a local pub that took luncheon vouchers to buy beer so lunch consisted of lots of beer with very little food.

In October, 1973, the Egyptians and Syrian launched a pre-empive strike against Israel with the Egyptians crossing the Suez Canal and entering the Sinai Peninsula, and the Syrians capturing the Golan Heights. The Yom Kippur War caused severe casualties on all sides and huge losses in tanks and planes. The Soviets and Americans initially tried to instigate a cease fire by withholding supplies from both sides but President Nixon was under domestic pressure from the Vietnam War, the Watergate scandal and pro-Israeli elements in the Senate. Eventually America began re-supplying Israel which allowed it to recapture the Golan Heights and Suez Canal before a ceasefire was finally imposed at the end of October. This infuriated the Arab members of OPEC who began to use oil as a weapon, reducing production and causing the posted oil price to rise to $US11-65 per barrel by the end of 1973 (it was $1-80 per barrel in 1970). With the embargo, spot prices rose up to $16 per barrel and a new era of the oil industry had begun. This was the First Oil Shock and caused a global recession combined with inflation, an unheard of scenario in economics. The Dow Jones stockmarket index fell 50% from 1150 at the end of 1973 to 550 at the end of 1974. The Watergate scandal eventually caught up with President Nixon who was forced to resign in August 1974.

On the 16th December 1973, I turned 30 and my thoughts drifted back to the Cuban Missile Crisis. "Hey – you've made it to 30 – so far so good!"

At this time, the Conservative Government was in power with Edward Heath as prime minister. He tried to curb the power of the unions and deregulate the economy but this only led to the disastrous coal miners strike in January 1974. The government was forced to implement power restrictions for industry – infamously referred to as the "Three Day Week". As a result, there was no heating in Salsbury House for 2-3 days each week. It was so cold in the office that I had to wear a fur coat and gloves to work – not the best for interpreting seismic sections! Eventually, the Conservative government called an election in February but they lost and Harold Wilson was returned to power (excuse the pun!). To make matters worse, the IRA had an ongoing campaign of bombings in London while we were there. Apart from the small but ever present danger, the most frustrating part of this was the many false alarms that a bomb had been left in a particular place. This was often on trains or train stations and had the effect of occasionally extending my 30 minute train ride from Broxbourne to over an hour.

In the office, most of us were working on the Viking Graben in the North Sea. Interpretation of the seismic had defined an enormous fault block called the Ninian Prospect. The structure was 16 kilometres long and 6 kilometres wide, so big it covered two UK blocks. BP operated Block 3/3, the north block and Burmah Oil operated Block 3/8 the southern block. BP drilled the first well (3/3-1) located on the very highest part of the fault block. They encountered a thin section of oil bearing Middle Jurassic sands but went into Liassic marine shales. They abandoned the well as being non-commercial. The fault block tilted to the west and was faulted down to the east. On the western flank, we could map a thickening section below the Base Cretaceous Unconformity. We interpreted this to be the Middle Jurassic sandy section and recommended that our well (3/8-1) be drilled further down the flank to intersect a thicker section of sand than encountered by BP in their well.

Burmah Oil had a very small staff in the London office and there were very few of us with much experience. Having the most well-site experience, I was designated to supervise the geological evaluation of Burmah Oil's first well on the Ninian structure. I also had to train a new recruit on the role of a well-site geologist.

The trip to the well-site was truly an adventure. In 1973, there was no oil field infra-structure in the Viking Graben, just the gas facilities in the Southern North Sea. So the first leg of this great expedition was by train from London to Aberdeen. Because it was a long journey (by European standards), they sent me first class which was very nice. I do remember, sitting in the same cabin with two businessmen, one of whom was complaining that in earlier times, British Rail would serve complimentary ports for its first class passengers. That was obviously the "good old days" because there was no complimentary port on my train.

Being winter, the journey was rather bleak with Edinbrough being the nicest city we passed through with its imposing but magnificent castle on the hill. Aberdeen was interesting with all the buildings made of granodiorite which gave the city a grey, cold look. I checked into the office and was told that I was to get the next day's fight to Sumburgh. Sumburgh – where's that you ask. It's the airport for the Shetland Islands, windswept islands to the north of Scotland (actually part of Scotland but you know what I mean). That evening I had a few drinks with the drilling manager. He was a Scot and knew the local dialect. However, I was to discover that the Scots spoke something that only vaguely resembled English. I approached the bar and asked for two pints of McKewans. I thought this was a perfectly reasonable request but no – this didn't work – they could no better understand me than I could them. So it was good old sign language – point at the pint glass – point at the McKewans tap and carefully raise two fingers the right way round. That worked and I rejoined the drilling manager who thought it was all very amusing.

The next day, I went to the airport for my flight to Sumburgh. When I went to board the plane, I discovered that it was a DC3. It had certainly seen better days and looked like it had been on the losing side of a fight with a Messerschmitt 109. But the DC3 was a great plane during WW2 and this one was still going strong. Only one problem – it was bloody freezing. At least in WW2, they gave you those warm, fleecy lined jackets you see in the movies. Not so for me - I had to do with what I thought was a nice warm jumper.

Despite all this, we made it to Sumburgh and my first visit to the Shetland Islands. The people who live here must be either really hardy or completely mad. This part of the islands was flat, tree-less and the wind whistled off the North Sea and cut through you like a knife through butter. I wanted to get back on the DC3 – it was warmer.

Before we got on the helicopter, we got some thermal underwear to put on and a very brief induction. Basically, the induction was that if the helicopter crashed into the North Sea and you survived the crash, you had approximately three minutes to live before perishing from hypothermia – thanks guys! Once we took off and headed across the North Sea, I realised that the induction was quite correct. The sky was grey, the wind was howling, the rain lashed the helicopter and the sea was a boiling mass of giant waves and white caps – nobody could survive down there. Fortunately, we arrived and safely landed on the Ocean Kokui. This was a massive rig, specifically built for North Sea conditions – it was bigger than any offshore rig I'd been on. It was also more comfortable than any rig I'd been on – I even had my own room and with a porthole no less. As it turned out, the porthole was a bad idea. The view was not pretty – it consisted of 15 metre high waves bearing down on the rig. After visiting the mudlogging shack, I returned to my room and papered over the porthole – ignorance is bliss OK!

Looking at the stormy conditions, I was curious as to how they could continue drilling. This was my first view of motion compensators on an offshore rig. They were an ingenious pulley system that compensated for the heaving of the rig and allowed for the driller to keep a constant weight on the bit while the drilling rig went up and down. In the North Sea, the motion compensators worked overtime.

Now, the Ninian structure was located at about latitude 62° which was about as far north as I could imagine. Being winter, it took getting used to the lack of daylight and a seemingly endless night. The sun rose at about 11-00am, a pale watery thing barely making it above the horizon, and then set at about 2-00pm. Being hidden behind the endless North Sea storms, it certainly didn't give out any warmth. As well, if you were busy in the mudlogging shack, you could easily miss an entire day!

The drilling went relatively smoothly and we eventually reached the Kimmeridgian shales, the main source rock in the Viking Graben and the seal for our main objective, the Middle Jurassic sands. These shales were thin so it wasn't long before we got our main drilling break. So it was circulate bottoms up and wait for the cuttings to arrive. When they arrived, there was a modest gas show and weak fluorescence in the sands. It didn't look promising but then we were drilling with very high mud weights (13 pounds per gallon) to prevent possible blowouts from unforseen and unknown geological conditions. So I made the decision to core.

Now coring on the Ocean Kokei was not as simple as coring on the Ocean Digger. The driller was from Louisiana, chewed tobacco and spoke with a southern drawl that defied interpretation by true English speaking peoples. The roughnecks were all Scots who, as described above, spoke their own curious version of the English language and certainly couldn't understand the driller. Tripping in and out of the hole was a major undertaking on the Ocean Kokei. However, coring was a much more serious matter as it was my arms and legs that were in danger. When the core barrel was suspended above the drill floor and I was trying to retrieve the core without losing my hand or having the core spill out everywhere, neither the driller nor the roughnecks could understand a word I was shouting at them. So it was back to sign language – up a bit – down a bit – gently, gently, don't hit the barrel that hard. Looking back, it was all very funny – we were all speaking English but without any communication whatsoever.

When the core came up, we laid it out and described it. The first core was interbedded sands and shales, and while the sands were oil bearing, the porosities didn't look that great. But there was no obvious water contact and as the drilling program called for coring to the oil-water contact, we continued to core. The second core came up with much better quality fluvial sands and these were also oil bearing – no sign of water. I was trying to keep calm but this was looking like a major oil discovery. So we kept coring and eventually cored a total of 360 feet (110 metres) of gross oil bearing Middle Jurassic section. The last core was in the top of the Liassic Shales and we still hadn't encountered the oil-water contact. I was ecstatic – this was beyond my wildest expectations. The trainee geologist didn't understand my excitement – he thought this was a normal result from drilling an oil exploration well. I took him aside and told him to savour this moment in his life, and that if he lived to be 90 years old, he would never see such a thing ever again.

Ninian was also the highlight of my professional life – I never saw such a thing ever again - either. It eventually came on stream at 500,000 barrels per day and produced over two billion barrels of oil. I feel very privileged to have been the wellsite geologist on the discovery well for this truly magnificent oil field.

After Christmas 1973, I went to the operations office in Aberdeen for our ongoing drilling campaign. Being the middle of winter, I went to work in the dark, it drizzled all day and then I came home in the dark. I spent New Years Eve in Aberdeen and had my first haggis. Although I knew what haggis is made of, I'll try anything once. It was very bland and not something I would eat again – unless it came with lots of chilli! During this time in Aberdeen, the Transocean 3, a jackup rig in the southern North Sea broke up during a storm and collapsed in a heap on the sea bed. However, one of its legs broke loose and, being full of air, floated away. It was carried by the current northwards and being in the early evening, disappeared from sight. Now there were about a dozen drilling rigs in the UK and Norwegian sectors of the northern North Sea, and all were put on alert to the jackup leg floating towards them in the dark. It was a tense night but fortunately, the leg didn't hit anything and was never seen again. It presumably sank somewhere in the North Atlantic.

Back in the office, we had to evaluate the discovery well. With the next well, we needed to intersect the oil-water contact. In addition, the wedge of sediments beneath the Base Cretaceous Unconformity seemed to have two sequences. The lower one was definitely the Middle Jurassic sands intersected in 3/8-1 but the upper sequence was downdip from 3/8-1 and could have been sand or shale and this would have a big effect on potential oil reserves. The geological team recommended a five kilometre step out to achieve these objectives. The engineers all fell about laughing – they wouldn't have a bar of such an insane proposal. They thought all geologists were mad anyway. Management agreed with them (that geologists were mad that is, certainly not our recommendation) and so we drilled 3/8-2 as a 2.5 km stepout.

Figure 24
The North Sea lake in summer. Note the rig on the horizon

So off to the well site once again. This time it was summer and a completely different environment. Instead of the sun rising at 11-00am, it rose at 2-00am and set at 11-00pm. It never got really dark – the sun disappeared below the horizon and there was a bit of twilight, and then it rose again. But at least it was bloody warm! The view from the helideck was magnificent. The North Sea was very calm and by doing a 360° turn, I could count four other rigs from the Ninian location – drilling on the Brent, Hutton, Cormorant and Alwyn oil discoveries **(Fig 24)**.

Without going into great detail, the 3/8-2 well was the same as the first well – we cored another 360 feet of gross oil column and went into the Liassic Shales. We had learned nothing new except that we had one bloody big oil field. The next well, 3/8-3, was our original location and this well finally found the oil-water contact and determined that the upper sequence of the seismic wedge was Late Jurassic shale and not sand. One up for the geologists – it's so good to know that we're not completely mad all of the time.

But it wasn't just all work. Burmah Oil was a very social company and the Poms love their rugby. So it wasn't long before I was invited to Twickenham for a match between the Barbarians and the Wallabies. So off we went on Saturday morning to Twickenham. Obviously this included a few beers before the match. Then we bought more beer to take to the match. Just before we got to the gate, they gave me the beer to hold. As I approached the turnstile, there was a big sign announcing "No alcohol to be taken into this ground". Thanks guys – you've really dropped me in it. But with my mind racing as to how I was going to get the beer into the ground, the bloke in front of me got to the window. He was at least 90 years old and had a large can of Watney's beer under his arm. Great, I'll watch what he does and follow suit. He walked up to the counter, plonked the huge can of beer down on the counter and asked for "one adult". He was given a ticket, picked up his beer and went on his merry way. I do not know why that sign was there – perhaps it was to increase local employment. It certainly wasn't obeyed as there was more alcohol at Twickenham that day, than in the local brewery. The match was great but the Wallabies lost - but what the heck, we all had a great time.

My next social event was an invitation to Cardiff Arms Park. In the Burmah office, there were two Welshmen, Jeff Clarke and Barry Houson. They invited me to a rugby match between Wales and England. I agreed and asked who else from the office was coming. They said "just you". I put this down to ancient Celtic-Saxon rivalry and thought nothing more about it. However, when they said we were to meet at Paddington Station at 8-00am, I became suspicious. It's a two hour train trip to Cardiff and the match didn't start until 1-00pm. When I arrived at Paddington, I realised why the rest of the office had declined the invitation. On the platform were 10,000 Welshmen and one Australian. The train was packed to the rafters so we crashed the first class carriage. The ticket inspector was not amused but somehow Barry convinced him that if British Rail couldn't accommodate their customers, then their customers had to make their own arrangements. But we kept a low profile for the rest of the journey.

On arrival at Cardiff, all became clear to me. The train arrived one minute after the pubs opened, so 10,000 Welshman (and me) got off the train and straight into the pub. There we drank for two hours and then went to the match (with more beer of course). Cardiff Arms Park is truly one of the world's great rugby grounds. I thought we Aussies were big beer drinkers but the Welsh leave us for dead. There were people at the ground with kegs of beer – yes, kegs of beer. This was a level above Twickenham (and Australia!). The ground was buzzing and the beer was flowing. Then a band marched out onto the ground, set up in the middle and the wizened old conductor raised his baton – and 50,000 Welshmen fell silent. The band started to play and everyone at the ground began singing. Now we all know that the Welsh are good singers but this was different. All 50,000 people at the ground were singing and it was magnificent. They sang in Welsh Gaelic so all I could do was to enjoy the atmosphere. But then they began to sing "Lloyd George knows my father" and as I knew the words to this song, I joined in. Eventually the singing stopped, the band left the ground and the match started. It was a great match which Wales eventually won – and this was an excuse to go back to the pub and drink more beer. Eventually Jeff and I decided "enough is enough", so we got on the train back to London and had a much needed Indian curry before staggering home to our wives.

Our house in Broxbourne had a very narrow but long backyard which backed onto a canal. In the middle of this yard was a luxurious green willow tree. In autumn it turned into a beautiful array of orange, red and brown colours and dominated the backyard – like a landscape painting. In winter, it lost all its leaves which blew all over the place and were impossible to clean up. To every Ying there's a Yang! The house and backyard were great for entertaining. One summer afternoon we had a BBQ and invited our neighbours and some friends from the office. Jeff Clarke asked if he could bring his dog and I said OK. Now Jeff's dog was a Rhodesian Ridgeback which he had brought back from Libya where he had worked in the days of King Idris before the revolution and Qaddafi took power and nationalised their oil industry. When he arrived, I told him that he was only allowed to bring his dog, but certainly not his horse – the dog was enormous! The poor thing was cooped up in his flat in London and when he saw our backyard, he thought he was back on the Plains of the Serengeti. He took off like a rocket down our garden until he got to the brick wall at the end where he came to a screaming halt, digging up half the lawn in the process. It was like a Disney cartoon and was the talking point of the afternoon.

Jeff Clarke went on to bigger and better things and ended up President and CEO of a Canadian company called Coho Energy Inc in the 1990's.

On Fridays, we used to retire to the local pub for a few beers after work. One time, it developed into a big night and just before midnight I staggered down to Liverpool station to get the last train home. On the way home, I fell asleep and was woken up by the train guard at the end of the journey. "Where am I" I asked the guard. "Bishops Storford" he replied. "Where the hell is that" I asked but he was not amused. He told me that the next train back to Broxbourne was at 5-30 in the morning. So I had to phone for a mini-cab to get me back to Broxbourne – all in all a very expensive exercise and I was thankful that the train wasn't going to Cambridge. I didn't get home until 3-30 in the morning and Mo was very suspicious. She didn't believe my unlikely story about being to Bishops Storford and back (I don't think she knew where it was either!). I had a lot of sweet talking to do before I dug myself out of that very large hole.

Back in the office, we were busily evaluating the Norwegian sector of the Viking Graben. The Norwegian Government had gazetted this area for applications by private companies and Burmah Oil wanted to make some bids. The geology was similar to the UK side and in those days we were only looking for the big fault blocks which were so successful in the UK sector. However, Dell Taylor, a BOCAL seismic interpreter who had also been transferred to London, came up with an area of bright seismic amplitudes higher in the section and clearly visible even on the paper sections of that era. Bright amplitudes on a seismic section can be an indicator of gas bearing sands since they have low density and higher reflectivity. Now oil was our main objective as gas prices were very low so we weren't that interested. In addition, the seismic amplitudes covered such a vast area, 750 square kilometres, of the Norwegian sector that it simply couldn't have been hydrocarbons. Wrong! This prospect was eventually drilled in 1979 and is known today as the Troll Gas Field with reserves of over two TCF gas and two billion barrels oil. It was still producing 125,000 boepd at the end of 2017.

Burmah Oil bid on two blocks which had a number of large fault block prospects. However, several weeks later, the Norwegian Government cancelled the bidding round. They took up all the industry's bidding documents, formed their own national oil company Statoil and drilled all the prospects themselves – cunning bastards! After that, they earned the permanent nickname of the "Blue Eyed Arabs".

After this experience, we concentrated our efforts on the UK side of the North Sea. Drilling was very active and there were lots of wells to review. However, most of them were confidential as the information was not released for five years under the confidentiality provisions of UK legislation. So the normal procedure was to exchange wells. If you had a confidential well, you could exchange the information for another well held by another joint venture via a Well Trade Agreement. Our problem was we had Chevron as a joint venture partner. Chevron was a very large company (one of the Seven Sisters) and had acreage all over the North Sea. As they generally already had the information on the wells we wanted to trade, they would veto the exchange agreement. Our solution was to attend the scouting lunches every month to see what we could find out. As it turned out, the other UK companies operating in the North Sea had the same problem. If they had Shell, Esso, BP, Mobil, Texaco, Gulf or Chevron as partners, they were also unable to exchange well information. So we got to know all the other UK operators and would exchange geological tops and oil & gas shows for each other's wells. Of course, this activity was in breach of the confidentiality terms of our operating agreements. We all had to very careful to keep all this activity secret from our joint venture partners – or we would have been instantly dismissed. We would use the formation tops and thicknesses to produce our maps, and then remove the well information from the maps afterwards. But it greatly assisted in our understanding of the geology of the North Sea and managed to partially overcome the information monopoly by the oil majors in the North Sea during the 1970's.

On Chrismas day, 1974, Cyclone Tracy hit Darwin. Most of the houses were made of wood and were built on stilts. This allowed the cool afternoon breezes to waft under the houses and keep them cool. Whilst this actually worked, the design was certainly not cyclone proof and the town was virtually wiped out.

I digress to give you a bit of the history of Burmah Oil as it has an Aussie flavour. This is mostly taken from "The Rush That Never Ended" by Geoffrey Blainey. An English gentleman called William Knox D'Arcy was a mad punter and in 1882 was talked into buying shares in the Mount Morgan gold mine in Queensland. The gold was very fine grained and set in ironstone as opposed to the normal quartz veins. Nobody was sure how to get it out but this didn't stop D'Arcy – by 1886, he owned 36% of the company. Eventually, the gold was mined commercially making huge profits for D'Arcy. After the gold ran out, they then discovered a rich copper deposit below the gold.

Now D'Arcy had to do something with all his money so he decided to explore for oil in Persia. He was granted a 480,000 square mile concession in 1901 and drilled several wells without success and by 1904, was about to give up. At this time, Lord Fisher, First Sea Lord of the Admiralty, decided to switch the Royal Navy from coal to oil – a brave decision given that, at that time, Britain had lots of coal but no oil (a situation that was to reverse 70 years later). Britain needed to find a source of oil for its navy. The Burmah Oil Company was formed in 1886 and developed the oil fields in Burma (now Myanmar) which had just become a province of the British Empire. In 1905, the Admiralty had negotiated a long term contract with Burmah Oil to supply oil from Myanmar to the navy. It also convinced Burmah Oil to acquire D'Arcy's concession in Persia. Exploration continued without success for another three years before they discovered the giant Masjid-i-Sulaiman Oil Field in 1908. To develop the discovery, the Anglo-Persian Oil Company was formed in 1909 with the shareholders being Burmah Oil and the public. The Anglo Persian Oil Company became British Petroleum and Burmah Oil ended up with 22% of their shares. As a matter of interest, the D'Arcy Exploration Company, then a subsidiary of BP, was renamed the BP Exploration Company in 1957.

Burmah Oil used this investment in British Petroleum as collateral to borrow $US1 billion to purchase an American company called Signal Oil & Gas. This gave Burmah Oil an interest in and operatorship of the Thistle Oil Field in the Viking Graben (Block 211/16) to compliment its Ninian Oil Field. Things were looking good for the company. Then, in early 1974 the stock market crashed and blue chip shares fell between 40-50%. The collateral value of Burmah Oil's BP shares fell below their debt liability and the American banks called in the loan. At the same time, Burmah Oil was incurring substantial losses from its tanker operations and the company was in financial trouble. On 31st December, 1974, Burmah Oil sold its BP shares to the Bank of England for 179 million pounds. Whilst the company survived and built up its downstream Castrol operations, it was eventually bought by British Petroleum in 2000.

Because of its 22% shareholding in BP, Burmah employees had rights to the BP Social Club. The BP office was close to the Burmah office and we used to go there every now and then. Outside the entrance to the BP office there was a well head as a sort of memorial statue. It was labelled with a plaque that stated that this single onshore well from Persia had produced over 100 million barrels of oil – just staggering! BP was also having its own success in the North Sea at this time. In 1970, they discovered the giant Forties Field with oil in Paleocene sands at 2,100 metres. The field was estimated at that time to have potential reserves of over one billion barrels of oil - and they owned it 100%. However, the development costs were about $US1.0 billion, a lot of money in the early 1970's. They financed it themselves but their engineers had some tense, career defining moments as the very expensive platform (first of five) was lowered into the North Sea and anchored successfully in place. In November 1975, the Queen inaugurated the first production from the field. Because in those days women were not allowed on ships (bad luck I think), the Queen had to get special permission to go onto the platform. As the oil actually belonged to the crown, she must have thought that very amusing. Production from Forties peaked at 500,000 bopd in 1979 and when it was sold to Apache in 2003, original oil-in-place was estimated at 4.2 billion barrels. It is still producing today.

During our stay in the UK, the IRA continued their sporadic bombing campaign. But in 1975, they started bombing pubs in the evenings after work when they were crowded. As I would go for drink after work on

Friday, it was always in the back of your mind. We did have one close call. One evening in September, we had drinks with Rosemary in Trader Vic's bar at the Hilton Hotel in London. A week later, the Hilton Hotel was bombed by the IRA killing two people and injuring another 63.

In January, 1975, I took some holidays and we all went skiing in Austria. We flew to Munich, then a bus to the resort of Kitzbuhel. Mo, being a good skier, hit the slopes and I took Chris to the kiddies slope with a toboggan. This was also about my level of skiing ability. We had a lot of fun and Chris loved the toboggan and wanted to go faster – just three years old mind you!

On one Saturday evening in March 1975, we threw a fancy dress party. This was a real hit. Everyone dressed up. The winner came as a postbox but the highlight was Barry Houson who came as a cleaning lady. Half way through the party, we decided to go to the local pub for a quick pint. After settling down with our beers, Barry got up and started to collect all the empty glasses in the pub and wiping down the tables. He was so ugly, the locals weren't really sure what was going on but we were all in fits of laughter. Bob, my brother in law, came as a convict including ball and chain – he looked great. Now Bob was a Major in the British Army and towards the end of the evening, he disappeared without saying goodbye. We thought it was very strange for him not to say goodbye until the following morning when we heard the news that King Faisal of Saudi Arabia had been assassinated. Then we knew exactly where Bob was - on his way to the Middle East, if not already there.

In April 1975, the Vietnam War came to a rapid end with the fall of Saigon. I will never forget the people hanging off the helicopters as the last Americans left their embassy. The western world was in recession and so it was rather a depressing time for the Americans. But the oil price was still high and the oil business was bubbling along.

In May, 1975, we also took a trip to the island of Jersey. When I say "we" that was me, Mo, Chris and Nana. We drove "the Tank" to Weymouth and caught the ferry to Jersey. We had a great holiday but Nana played a trick on me. The first day on the beach, she went into the ocean for a swim and said how good it was. So I did the Aussie "He-man" thing by running and diving into the ocean. Now the air temperature was about 28°C but the island of Jersey is in the English Channel and the water temperature was hovering just above zero. Mo told me afterwards that I bounced off the ocean – I was certainly back on dry land within a few seconds.

WOODSIDE PETROLEUM

My term with Burmah Oil North Sea was over and we returned to Australia in September 1975. Whilst on our return to Australia, we stayed at a resort on the island of Mauritius, in the middle of the Indian Ocean. Mauritius was a beautiful island with volcanic hills covered in dense rain forest and surrounded by pristine coral reefs. The locals were a mixed bag of natives, Chinese, Indians and Africans providing a colourful and vibrant society. In the capital, Port Louis, they had one of the best markets we had ever seen. We drooled over the spices, nuts and other exotics but all to no avail – we couldn't buy a thing that could be taken back into Australia. Our resort was located on the northern side of the island and we were accommodated in individual chalets. One evening, we were having dinner and at the table next to us, there were two South African couples on holiday. One of the blokes was very drunk and had to be carried back to his room. Being a good Samaritan, I helped carry him back to their chalet. They invited me back for a drink which naturally I accepted. Mo was tired and wanted to go to bed so I went on my own. Their wives had also gone to bed and we were joined by two of his mates. We were having a few beers when one of them produced a couple of bottles of Tequila. These South African were serious drinkers of Tequila and did it the traditional way, straight with a taste of salt on the back of your hand and a suck of lemon. I'd never tasted Tequila and I didn't this time either because it has no taste – just a warm feeling on your insides as it slips down. We were having a great discussion on who had the best cricket and rugby teams, Australia or South Africa? Because of apartheid, Australia and South Africa hadn't played each other for a number of years so the discussion, although vigorous, was purely academic. Suddenly one of the South African's eyes rolled over into the back of his head and he fell off his chair. This was obviously normal behaviour for Tequila drinkers since his two companions didn't turn a hair – they just picked him up and put him to bed. The discussion carried on until the next South African's eyes rolled over into the back of his head and he too fell off his chair. By this time, I was getting into the swing of this Tequila drinking and I pitched in, picked him up and put him to bed. After this, the two of us carried on but I must have been the next one to fall off my chair for the next thing I remember was waking up in my bed, partially clothed in a torn, blood-stained shirt and covered in scratches. Mo was glaring at me and wanted to know who had given me the scratches! So we back tracked to the South African's chalet and discovered a garden with a giant cactus in it. On this cactus were shreds of my shirt. After talking to the last man standing, he related that I hadn't fallen off my chair but had stood up, declared that I was going home, staggered to the exit and fallen down the stairs. Being very drunk, this didn't slow me down at all. Apparently I got up and headed off back to my chalet, walked into the cactus and somehow found my way home. By this time, I had the worst hangover ever but Mo had her revenge. She subjected me to the worst thing that can ever happen to a bloke with a hangover – she took me shopping! By the end of the day, I swore that a drop of Tequila would never ever again touch my lips, a vow that I have not broken to this day, not even a margarita.

We arrived back in Australia in August 1975. The country was not in good shape. The economy was shot, the currency was falling, and there was very little oil exploration. While we were away, the Whitlam Government had made Mr Rex Connor Minister for Mines and he did not get on well with the Australian oil industry. He was a staunch union man and hated big business. He was determined that the Australian government should have more involvement and control over its oil industry. The following is taken from "Where God Never Trod" by Rick Wilkinson. Rex Conner formed the Petroleum and Minerals Authority (PMA) in August 1974. The first task the PMA undertook was to purchase 50% of Delhi Petroleum's interest in the production from the Cooper-Eromanga basins and 25% of their interest in the exploration permits. The Queensland and Western Australian governments challenged the formation of the PMA on the grounds that it was unconstitutional and a violation of state's rights. This challenge was upheld by the High Court and the PMA was disbanded. The PMA's interests in the Cooper Eromanga basins were bought by the South Australia Gas Company, a subsidiary of the South Australian Oil and Gas Corporation. The South Australian government achieved what the Federal government could not – it had its own oil and gas company. The South Australia Gas Company was eventually privatised in 1988 to become Sagasco Limited with the SA Government retaining a 32% interest. Sagasco Limited was taken over by Boral in 1993 for $816 million. The upstream part of Boral was split off as Origin Energy Limited in 2000.

Rex Conner also formed the Pipeline Authority and took over from AGL the building of the natural gas pipeline from Moomba to Sydney during 1974-76. His dream was to build a national grid of pipelines connecting the giant gas fields offshore Western Australia, via the goldfields to Moomba and thence to the major Eastern States markets. Now in hindsight, this was not such a bad idea and most of his dream has now come true. But incredibly, a Pakistani agent called Tirath Khemlani was hired to borrow $4 billion petrodollars to fund the Pipeline Authority. When this leaked to the press, all hell broke loose and the government was nailed to the wall. Then came the unthinkable – "the Dismissal". The Governor General sacked the Whitlam government. This was a momentous event in Australia's short history and was all very new to us. But Richard Nyein, a BOCAL engineer who came from Myanmar, was very agitated. Where he came from, such events were always followed by civil war – and these were in turn always followed by a military dictatorship. Fortunately this didn't happen to Australia. I think we Aussies are far too cynical about our politicians to ever go to war over them. Malcolm Fraser was invited by the Governor General to form the next government, a decision which was overwhelmingly ratified by a later election. As it turned out, the Fraser Government was no better than the Whitlam Government and failed miserably to improve the economy, despite having an overwhelming mandate to do so by the electorate.

On my return, the Burmah Woodside office had been moved from Nedlands to Allendale Square in the city. Allendale Square was a new high rise office block in the middle of the city and I wondered what they had done with the palynology lab. But low and behold, it had also been transferred to Allendale Square. So all the gases used in the preparation of the samples got vented straight out into the atmosphere. Can you imagine what the HS&E Department would say about that today?

I was given the task of evaluating the offshore Canning Basin. This was a very boring assignment as, based on existing seismic, the Mesozoic-Tertiary section dipped monotonously to the northwest. While I suspected that most of this dip was due to multiples, it didn't help to unravel the structuring. There was also very little well control; Bedout-1, Leveque-1 and Lacepede-1 had been drilled by BOCAL in 1970-71 and East Mermaid-1 drilled by Shell in 1973. My study formed the basis of my next paper "The Offshore Canning Basin" in the AIMM publication "Economic Geology of Australia and Papua New Guinea Part 3: Petroleum" in 1976.

It was a very broad brush paper and not very illuminating but I did the best of what little information I had.

Then came my most interesting and challenging assignment, the Inner Dampier Sub-basin. I was working with Lauchie McLean on the Rosemary-Legendre Trend and the then undeveloped Angel Gas Field. The structuring of the Rosemary-Legendre Trend was very complex and to make matters worse, the seismic was poor and full of multiples. In addition, there were serious problems with the different vintages of seismic. In the era before GPS, seismic surveys were located by Shoran, a system of shore based navigation. This was passable if you had some offshore islands (as in the Browse Basin) to provide triangulation to verify your position. But in the Dampier Sub-basin, there were no offshore islands and the seismic had to be positioned using just two stations. This meant that you were never sure exactly where your survey was and different vintages mostly did not tie. The Legendre-1 well, drilled in 1968, had flowed 1,000 bopd from Early Cretaceous sands and we were desperately trying to find an up-dip location. But the sands were poorly imaged by the seismic and, together with the multiples we could not confidently map an up-dip prospect. We did, however, define a prospect called Lewis which was drilled in late 1975. I went up to the well-site during the drilling of the objective section but the well was unsuccessful. It was my first experience with over-pressured shales and it took many attempts and wiper trips to obtain all the electric logs.

The next door neighbours to Greg & Sue were John & Sylvia. They were struggling financially as John was not working full time. One weekend, John asked me to come over to give him some help. He had a live sheep in his back yard and he wanted me to help him kill it and cut it up for meat. I told him that this was probably illegal in the suburbs of Perth but being ex British army, he didn't think that this was relevant. So we tied the sheep to the Hills clothes hoist, I held its back legs and he got out his favourite machete and with one fell swoop, sliced its head clean off. Strewth – I'm glad I wasn't a poor enemy soldier in the Malaysian jungle while John was creeping about. So we then skinned the sheep, kept the fleece and cut up the meat which he popped in his freezer. John also knew how to cure the fleece by treating it with alum crystals and then we hung it on the Hills hoist to dry out. That was definitely one of my more memorable weekends.

Woodside was trying to evaluate the recoverable gas reserves of the Angel Field. Three wells had been drilled on the field in each of the years 1971-73 but the seismic was poor and mapping uncertain. The original wells had been cored so my first task was to describe these cores in more detail, determine the environment of deposition and input this into the seismic mapping. When I went to the core store, I discovered that the cores were still wrapped in the original sleeves – I was the first person to see them after the wellsite geologist wrapped them up six years ago. I opened the cores and was immediately hit by the smell of oil. On closer inspection, the sands had good oil fluorescence despite being full of natural gas. It appeared that the field had once been full of oil but this had been displaced when the structure was filled with a later migration of gas. The oil would have migrated up-dip to the south - right where the Legendre-1 oil discovery was located. Again, I cursed for not being able to map that structure. With better seismic processing, GPS navigation and the advent of 3D seismic surveying, Woodside was eventually able to map the Legendre structure and the commercial field was developed in 1997 with production starting in 2001 and peaking at 46,000 bopd in 2003. Cumulative production was 48 million barrels. Ah well, that was another one that got away (although fortunately not from Woodside)!

One of my more unpleasant experiences occurred in early 1977. One Sunday morning, the police knocked on our door and asked if our next door neighbour Greg was home. They told us that his wife Sue had been killed in a car crash on her way home from work early that morning. It turned out that Greg was not home because he was out looking for Sue when she didn't arrive home. The police left and sometime later, Greg returned home. I wandered over to tell him the terrible news but he took one look at my face and completely broke down. I took him back to our house and we comforted him as best we could. He was left with two young daughters – life can be so very cruel.

In a macabre coincidence, our friend Roy who was a fireman, was the first on the crash scene and had the unpleasant task of recovering the body. He recognised the car so it really shook him up as it was for him no longer a routine third party accident.

Because of the over-pressured shales in the Dampier Sub-basin and Browse

Basin, Woodside developed a method of over-pressure prediction called the d exponent. This used detailed velocity analysis of the seismic data to predict zones of low velocity interpreted to be over-pressured shales so the drillers could raise the mud weight to optimise the drilling rates. It also helped to develop better casing programs which made huge savings on the drilling costs. Richard Nyein, Lauchie and I co-authored a paper "Occurrence, Prediction and Control of Geopressures on the Northwest Shelf of Australia" which Richard presented at the APEA Conference in Sydney in 1977.

Mo accompanied me to this conference but being 8 months pregnant, was not able to fly. So she went by train. On return from Sydney, Alex was borne on 22nd May, 1977. But he decided to come three weeks early. When Mo told me it was time, I bundled her into the car and was about to set off when I noticed that I didn't have enough petrol to get to the hospital. To make matters worse, it was a Sunday and I had to find a rostered petrol station. After much panicking and a very rapid filling of the car with petrol, we eventually made it to the hospital in time. Again, I was present at the birth but this time I was more prepared. For some reason, they made me wait outside until the contractions increased and I almost missed the birth. Alex was born with black hair and looked almost human. I was getting pretty good at fatherhood but it was not to be. When I first met Mo, she wanted six kids. So when I proposed to her, the only condition was that we only have four kids. After Chris was born, she quietly told me that perhaps three kids would be fine and after Alex was born, she cried "that's it – no more kids".

In the northern part of the Bedout Sub-basin, east of the North Turtle Ridge, Bob Duke had mapped a prospect he called Coxon Shoals. It was an anticlinal structure with Late Triassic sands as the main objective. The section deepened to the north so any Triassic source rocks, if present, should have been in the generation window. We prepared a technical report together with a risk analysis but the North West Shelf joint venture would not approve the prospect for drilling. Bob Duke was doing a part time MBA degree and applied the Monte Carlo technique to the risk analysis. It was the first time I'd seen this technique used in risk analysis. Later, the prospect was farmed out to British Petroleum who drilled the prospect in 1980. Phoenix-1 discovered gas in the Triassic sands but the reservoirs were poor and nobody needed another gas discovery on the North West

Shelf in 1980. So the well was plugged and abandoned. After a very long wait of 34 years, a joint venture consisting of Apache Energy, Carnarvon Petroleum, JX Nippon and Finder Exploration drilled a second well on the Phoenix structure. Phoenix South-1 was drilled in May-August 2014 and encountered four oil bearing sands in Triassic sandstones of the Keraudron Formation. A follow-up well, Phoenix South-2, was drilled in November-December 2016 and as well as confirming the oil discovery at Phoenix South-1, also discovered gas and condensate in deeper sands. This field has not yet been developed due to low oil and gas prices that occurred after 2014 but more will be said of this later.

Eventually, Burmah Oil's financial difficulties progressively worsened and, in August, 1976, it was forced to sell its controlling interest Woodside Burmah to Shell and BHP. The company was renamed Woodside Petroleum Limited in 1977. BHP took over the administration while Shell took over the technical operations of the company. So we had Shell as our new bosses. Initially, things went on pretty much as normal. I was working with Henry Askin, a Shell secondee to Woodside. We were trying to map structures along the Madeleine Trend in the Dampier Sub-basin but again, the seismic in this area was even worse quality than the crap we had in the rest of the basin. Henry was a geophysicist and was convinced that the problem with the seismic quality was in the processing. Woodside had just drilled Withnell-1 in 1976 at the southern end of the trend. This well drilled a pile of Jurassic shales, most of which were over-pressured. We were evaluating the results of this well and noticed that the sonic velocities in the Jurassic shales were very low, even lower than expected from the over-pressuring. We met with GSI, the seismic processing company, to look at their velocity analyses. We discovered that they were ignoring the low velocity points and were picking a velocity trend which they thought was more in line with reality. Without any geological input, this was the right thing for them to do. However, when we showed them the velocity profile for Withnell-1, they were stunned. So GSI began to reprocess the seismic data using the low velocity profile and the seismic quality began to improve. It took many more years and a lot of research, before the multiple problems were reduced and seismic quality improved to where new and more subtle oil discoveries could be made.

I also worked on the Browse Basin where the team had developed two interesting prospects. The Caswell Prospect was a gentle anticline in the centre of the basin. The Early Cretaceous sands were very deep and it was a very difficult well to drill. Due to a combination of over-pressured shales and loss circulation in interbedded fractured limestones, we ended up drilling in a 6 inch hole at total depth. There were oil shows but no effective reservoir and the well was plugged and abandoned after spending an awful lot on money. The Bassett Prospect was in the northern Browse Basin and was a poorly defined anticlinal structure – the seismic was appalling. I remember Elwood Horsman, a geologist with Chevron remarking at the joint venture meeting that "Bassett is a dog!" Despite this vote of no confidence, the well was drilled in 1978 but it only tested the Late Cretaceous Puffin Formation sands and did not reach, as we know today, the main objective Early Cretaceous Brewster Sands. This deeper prospect was eventually evaluated in 2013 but more of this later.

After this, I worked in the Beagle Sub-basin where a prospect called Jarman had been developed. In the Beagle Sub-basin, the Jurassic section was very sandy as opposed to the successful Dampier Sub-basin where the thick Jurassic shales allowed the controlling faults to seal. In my assessment, prospects such as Sable, Ronsard and Picard had been unsuccessful because of this lack of lateral fault seal. The Jarman Prospect, however, was a closed anticline based on the seismic of the day, and this did not require any faults to seal for it to be successful. Jarman-1 was drilled in early 1978 and the formation tops came in amazingly close to prediction but it was still a very dry hole.

There are many stories in the oil industry about how large oil fields were missed and discovered later by different companies. I have already mentioned how Esso almost missed Kingfish, the largest oilfield in Australia. BOCAL has a similar story. Madeleine-1 was drilled by BOCAL in 1969 and encountered good oil shows in the Late Jurassic Angel sands. The electric logs indicated a possible transition zone but the well was plugged and abandoned. After further mapping of the seismic, a second well, Walcott-1, was drilled in 1979 but it was also a dry hole. As seismic processing and quality improved, further seismic mapping revealed an up-dip location to Madeleine-1. The Wanaea-1 well was drilled in 1988-89 and encountered a 103 metre gross oil column in the Angel sands. The Wanaea and Cossack fields turned out to contain over 300 million barrels of oil reserves and are still producing today. BOCAL missed it by about one kilometre.

In the Woodside office there were three people, including Bob Duke as mentioned earlier, who were doing part time MBA degrees to improve their management skills. Mike Chaney was a geologist with Woodside and went on to become one of Western Australia's most successful businessmen. The third person was Ian Whiteley who convinced me that I should do the same. So I enrolled in February, 1978 and began the course. It was terrific. A lot of management is just plain common sense but the MBA gave you all the skills to apply it – management accounting, economics, taxation and statistics.

In April, 1978, Cyclone Alby developed off the North West Shelf. It followed the normal track down the coastline to Exmouth and then off into the Indian Ocean. But then Cyclone Alby decided to "do its own thing", turned 90 degrees and headed straight for Cape Leeuwin. I was in Sydney for a joint venture meeting and when I woke up in the morning, the first thing on the news was "Cyclone Alby hits Perth". I rang Mo to see if everything was OK and she told me that the cyclone destroyed the jetty at Busselton but in Perth, there was just a lot of wind. She also said that our swimming pool was now full of red dust!

Eventually, it became very clear to the old BOCAL employees that there was no career path in the new Woodside. Morale was so low that Shell was called the "Yellow Peril". This actually got written up in the press and further alienated our situation. Our response to all this was to have long lunches but this only led to a drinking problem. The most momentous occasion occurred after a long lunch in Shafto Lane. We had carried on drinking into the early evening and eventually got thrown out of the bar. One of my compatriots said he'd give me a lift home and I was too drunk to say "No thanks – I'll catch a cab and you should come with me". We returned to Allendale Square to get his car but it was locked. The security guard was certainly not about to open the door for two very drunken yobbos. But my companion was not only big but as it turned out very strong as well. He grabbed the door and ripped it open with his bare hands, shredding the rails. The security guard fled in panic to a safer position and we went down the lift, picked up his car and drove home. The next day I went to the office with a very large hangover and in the full expectation of being sacked.

Somehow we survived but for me this was the last straw. I had to get out of Woodside or I'd become an alcoholic. So I chatted to Mo and said I was going to leave Woodside and try my luck consulting but that this would involve some financial risk. I'll never forget what she said to me – "As long as we're all together, we can live in a caravan". Wow! This was another one of the many reasons why I love this gorgeous woman. So with that unconditional vote of support, I strode into the Exploration Manager's office and handed in my resignation. I have never seen such a reaction – his smile went from ear to ear, he jumped up from behind his desk, rushed over to shake my hand and wished me all the luck in the future. No company has even been so glad to get rid of me. As I walked out of the Exploration Manager's office (you will note that I can't remember his name), I congratulated myself on what a great decision I had just made.

Consulting

Magnet Metals Group

My first client after leaving Woodside was with a company called Magnet Metals, a small Australian company controlled by Alan Burns and Derek Gascoyne. They had just acquired the offshore permit WA-58-P, a very large permit which covered in inner parts of the Dampier and Beagle sub-basins. Don Walker was the seismic interpreter and had mapped a number of prospects in the area. The one that interested me was the Bruce Prospect, a Triassic structure on the southern flank of the Cossigny Trough. There was a thick zone of bland reflectivity on the seismic which was interpreted as shale thereby creating seal for the objective Triassic reservoirs. This was the first time I had seen seismic data processed specifically to determine lithologies. Today, it is done routinely. Without any nearby well control, it was reasonable for the processing company to interpret the low velocity section as shale.

Being a small company, Magnet needed to farm the drilling out to a larger company. I prepared a farmout report but, as the stock market was buoyant, Alan decided to float a new company to drill the well. This was my first introduction into the corporate side of the petroleum industry. Stirling Petroleum NL was listed on the ASX after raising the funds to drill the well.

Stirling was the first Western Australian company to drill an offshore well and I was the wellsite geologist. Bruce-1 was drilled July-August, 1979 using the "Southern Cross" drilling rig **(Fig 25)**. The well encountered oil shows in the Mungaroo sands below about 50 metres of claystones in the Cossigny Member. The highly porous Mungaroo sandstones were very thick and had so few shale breaks that they gave the bland, low velocity zone on the seismic section. The oil shows were not significant and the well was plugged and abandoned. However, the well did indicate the possible presence of pre-Jurassic oil prone source rocks in the Cossigny Trough.

Figure 25
Southern Cross drilling Bruce-1

As a matter of interest, WA-58-P covered the area that now contains the Wandoo Oil Field. In 1979, we couldn't see the structure. The 1970's seismic was very poor and the shallow sedimentary section was completely dominated by multiples. It was not until the late 1980's when seismic acquisition techniques improved and increased computer power allowed superior seismic processing to remove the multiples, did the Wandoo structure become even vaguely visible.

In 1979, four exploration permits were awarded to Esso (2), Hudbay and a consortium of Mobil, Phillips, BP and MIM on the deep water frontier Exmouth Plateau area to the west of the North Rankin Gas Field. Technological advances in the oil industry now allowed deep water discoveries to be commercially developed, particularly with the current oil prices. This was the beginning of deep water exploration in Australia.

By this stage, I was so busy consulting that I had to give up my MBA studies. Although I never got the degree, the two years I spent were invaluable to my later management positions. In addition, my consulting company was beginning to earn good money and my tax bills were funding the social welfare costs for entire suburbs in western Sydney. My accountant, Colin Wilson, suggested I take advantage of a new tax avoidance scheme. It was very complex and would have involved a lot of tax auditing, so I decided against it. I preferred to pay my tax bills and sleep at night. This turned out to be one of my better decisions. It was the infamous "Bottom of Harbour" scheme. Simply put, a company would be stripped of assets and profits before its tax liabilities fell due, leaving it unable to pay. After the assets were stripped, the company and its records would be put in the hands of fictitious directors and the records sent to the "bottom of the harbour". Over 6,000 companies and up to an estimated one billion dollars were involved in the scheme. In 1982, the government introduced some very nasty retrospective legislation to recover these tax liabilities and this resulted in a lot of unhappy and financially diminished people, and probably contributed to the downfall of the Fraser Federal Government in 1983. The findings of the Costigan Royal Commission into this tax avoidance scheme read like a fictional gangster novel and included the use of the Federated Ship Painters and Dockers Union to provide the dummy directors, violence against ATO agents and even suspected murders. Soon after, Colin Wilson committed suicide and it was a very sad occasion. But to this day, I'm still not sure - "did he jump or was he pushed?"

Indonesia

My next consulting job was with Robertson Research to prepare a report on the Tertiary carbonates of Indonesia for Pertamina, the national Indonesian oil company. I was to work three weeks in Jakarta and one week back in Perth. I flew to Jakarta and stayed at the Ariaduta Hotel which was in walking distance to the Pertamina office. This was my first trip to Jakarta and I was horrified by the abject poverty of most of the people. But they mostly seemed to be happy, a fact I didn't initially understand. After working there for a number of months, I realised that the reason they were happy was that the Indonesian economy was improving and their standard of living was improving. The fact that their standard of living was improving from a very low base was obscured by my very fortunate situation of being born in Australia.

I did find the traffic rules in Jakarta to be very confusing. As a pedestrian, the green light was something that you didn't take any notice of whatsoever. You had to count to at least ten for the accepted number of cars to go through the red light before you even thought about crossing the road. If you were in doubt, you just watched the locals and walked across when they did. Once, I got a lift back to my hotel with one of the Pertamina geologists. As he was approaching the intersection to my hotel, I said to him "you need to turn left at the next intersection". Perhaps he didn't understand my accent because we ended up stopping in the middle lane. When the lights turned green, he turned left across a lane of moving traffic. No one tooted – this was perfectly acceptable driving behaviour in Jakarta.

My work on the Tertiary carbonates was very interesting as nearly all the fields in Indonesia were sandstone reservoirs. Miocene reefs were present throughout Sumatra and Java except that their reservoir quality was poor with the exception of the Arun Gas Field which seemed to be a one-off. The Arun Gas Field was discovered by Mobil in 1971. The initial well encountered a 300 metre gas column at a depth of ~ 3,000 metres in a Miocene Batu Raja reef. The reservoir pressure was an unexpected 7,100 psi and the well blew out, destroying the rig and killing a number of workers. The field has produced over 15 TCF of gas since 1975 via an LNG project.

The best reefs were of Pliocene age and located in the Vogelkopf of Irian Jaya where there were a number of nice oil fields. The best of these was the large Walio Oil Field which was initially defined by air photo interpretation. Drape over the reef caused a gentle topographical high and this could be seen on the air photos from the drainage patterns of the rivers. Petromer Trend obtained a contract over the area and acquired seismic over the topographic high. Interpretation of the seismic defined a reef prospect which was drilled, and low and behold – a 300 million barrel oil field. I wish finding oil was always that easy – but then it wouldn't be any fun!

As part of my study, I went to Balikpapan on the east coast of Kalimantan.

This was a major oil town servicing the many oil and gas fields in the Kutei Basin. This trip was another adventure. The airline was Pelita Airlines, an internal airline servicing the many islands of Indonesia. At the airport we piled into this Focker Friendship and after we were seated, they loaded the Schlumberger tools down the back of the plane. As we were about to take off, the bloke next to me started to pray. This was very disconcerting – did he know something I didn't? But we took off safely and headed across the Java Sea to the infamous island of Borneo. We crossed the southeast corner of Borneo and flew over the vast rain forests of that island. They seemed to go on forever and I thought that it's no wonder they're still finding planes that crashed here during WW2. Eventually we landed at Balikpapan and as the wheels touched the tarmac, everyone on the plane cheered and clapped. They definitely knew more than I did!

I also went on a field trip with Pertamina to various carbonate outcrops on the island of Java. We travelled by bus and it was a real experience. I thought the traffic in Jakarta was bad but on the country roads, there didn't appear to be any rules at all. Buses, trucks, cars and bicycles all travelled without care for life and limb, and nobody seemed to know how many lanes there were or which lanes were going in which direction. Actually there was one rule - the largest vehicle had right of way! The other fascinating thing was that there was a village every couple of kilometres confirming that Java is the most densely populated island in Indonesia. However, the traffic didn't slow down but rumbled through these villages at maximum speed (which fortunately for the local villagers, wasn't much). After the first day, we stayed at a small town on the way to Surabaya. We arrived in the early evening and after dinner I went straight to bed after a long day. The next morning I woke up and threw open the curtains and there, dominating the landscape outside my window, was a massive volcanic cone – truly spectacular.

I finally completed the Robertson Research report for Pertamina. The main conclusions were that carbonate potential existed in the Middle Miocene Baturaja Formation in the North and South Sumatra basins, in the Kujung Formation in the Northeast Java Basin, and in the Kais Formation in the Salawati and Bintuni basins in Iryan Jaya. However, prospects were poorly defined and Pertamina needed to invest in acquiring a considerable amount of new seismic if this potential was to be realised. I also noted that there were untested reverse faulted anticlines in eastern Java but the main objectives were probably clastics rather than carbonates. Again, these structures were poorly defined and required additional seismic. These will be discussed later on.

After completing this assignment, I returned to the Magnet Group to undertake a study of the Bonaparte Basin. One of their subsidiaries, Lennard Oil, has acquired an extensive acreage position in the basin. Both Arco and Esso had spent a decade exploring this basin without commercial success and had finally relinquished all their permits in 1978. In the following year, they approved a paper on the basin by Kraus and Parker in the AAPG. This paper was a detailed evaluation of the petroleum potential of the basin which they described in glowing terms. Obviously not everybody at Arco agreed with the relinquishment decision.

In November, 1979, the agreement to develop the North Rankin and Goodwin gas fields was signed between Sir Charles Court for the Western Australian government, Ross Harrison for Woodside Development and Ric Charlton for Shell on behalf of the North West Shelf joint venture. It was to consist of a domestic gas development followed by an LNG export phase. Due to the sheer size of the project, Woodside had to sell down its interest to Mitsubishi and Mitsui who bought a 16.67% interest and leaving Woodside with 16.67%. First gas sales to domestic consumers occurred in July 1984 under contracts totalling 414 TJ per day for twenty years. However, it wasn't until July 1989 before the first LNG shipments to Japan took place.

Cities Service

One of the best well-site jobs I ever had was with Cities Service. They were drilling Swan-2 in the Vulcan Sub-basin and needed a second well-site geologist. My contract was two weeks on and two weeks off but the contract stated that I still got paid during my two weeks off – I just love this oil industry! It was early 1980 and the trip to the well location was a major undertaking. I flew from Perth to Meekatharra and from there to Wyndham. There we changed over to a helicopter and flew to Kalumburu Mission in the Kimberley before refuelling and flying to the rig floating in the middle of the endless Timor Sea. After 21 hours, we landed on the rig – hell, I could have flown to London quicker than this! On my first stint, the well was drilling through the Early Tertiary and it was very uneventful. On my second tour, things had changed. When I got to the mudlogging shack, the well was drilling through the Late Jurassic. Describing these cuttings was interesting – they were black shales but the rock chips were bleeding gas from micro-fractures and oil was running across the tray. These were some of the best source rocks I had ever seen – and they were generating petroleum right now! There was however one nasty problem. After drilling through 2,000 metres of these black shales, they became progressively more over-pressured. Eventually, the mud weight was up to 15 pounds per gallon but the mud was coming out of the hole frothing gas, and the gas detector was reading over 2000 units. My memories of Turrum-1 came flooding back and again, I would go to bed with my life jacket on. Now the Timor Sea was not like the Gippsland Basin or the North Sea – it was dead calm and the weather was beautiful. But it still had its dangers being full of very large sharks and poisonous sea-snakes and I didn't fancy swimming about in it. So every morning, I would plead with head office to plug and abandon the well. There had been some gas shows in the Cretaceous but the sands were thin and not considered to be commercial. The main objectives in Swan-2 were the deeper Jurassic Elang and Plover formations which were supposed to be up-dip from Swan-1. However, from the amount of Jurassic shales Swan-2 had just drilled through, it was clear that this was not the case and the crest of the prospect was somewhere else. We also had no idea how thick these Late Jurassic shales were going to be. The cost of the well was way over budget and if the well did intersect any over-pressured gas bearing sands, we were in danger of blowing out and sinking. So eventually, the decision was made to plug and abandon the well so we ran electric logs and headed home.

On the return journey, the helicopter had mechanical problems and we had to stay at Kalumburu Mission for several hours until a new one could be flown in. The Kalumburu Mission had to be one of the remotest Aboriginal communities in Australia, but had one notable asset - huge orchards of mango trees. I asked if these were sold during the season as there was no way that the local population of the mission could have eaten them all. But unfortunately the mission was too remote and the mangoes rotted on the ground – what a waste. One of the local Aboriginals asked if I was the "rock doctor" and took me out to a place where he claimed there were green rocks. I always imagined that the Kimberley was a desolate area but flying over it in a helicopter, I discovered that it was covered with large trees, rivers and water falls, a beautiful landscape but so remote and inaccessible. It was even better on the ground. He stopped the landrover by a creek of crystal clear water. There were large trees overhanging the creek and everything was so peaceful and relaxing. But sure enough, my companion was right - the outcropping rocks were green. My metamorphic geology was becoming a little rusty but even I could recognise a classic epidote facies.

When this contract expired, I had some time on my hands. The Federal Government had just gazetted some areas for industry applications in the offshore Carnarvon Basin. From my days at Woodside, I thought that this was a terrific basin and was still intrigued by the presence of thick Jurassic oil-prone source rocks in the basin. There were lots of gas fields discovered but where was the oil? One of these blocks contained the Tryal Rocks-1 well drilled by WAPET in 1970. From my previous studies of this basin, I recalled that this well was drilled by a jack-up with water depth limitations and the crest of the structure was in water depths greater that the jack-up's capability. The well was therefore drilled on the flanks of the structure but still had good oil and gas shows at the top of the objective Barrow Group sands. So I approached two companies, Mesa Australia Limited and Minora Resources, to fund me to undertake a technical report on this block and, if the recommendations were accepted, bid on the acreage. The companies accepted my recommendation and bid on the acreage. Their bid was successful and the permit was awarded as WA-149-P. I will have more to say about this permit later.

During 1979, the international political scene had been steadily deteriorating. Vietnamese troops invaded Cambodia to remove the infamous Pol Pot regime; Lord Mountbatten was assassinated by the IRA; and the Shah of Iran fled the country during a period of civil unrest and the country fell to a religious fanatic called Khomeiny. Soon after, Iranian militants seized the US Embassy taking 53 hostages which they held captive for a staggering 444 days. And to top it all off, at the end of the year, Soviet troops invaded Afghanistan. Definitely a year to remember! And all the time, the oil price kept on rising and rising….

PANCONTINENTAL PETROLEUM

Since my consulting business was looking up, we decided to upgrade to a better house. We bought a very nice house in Karrinyup and while the settlement was going through, we went to the 1980 APEA Conference in Surfers Paradise.

When we arrived at Coolangatta Airport, the airport terminal consisted of something best described as a "tin shed". It was here that we met Tom Stephens, President of the South West Texas Bank and, due to a shortage of taxis we shared one to the Chevron Hotel. Fortunately, the Chevron Hotel was superior in luxury to the airport terminal and considerably improved Tom's first impressions of Australia.

At this conference, I was approached by Jim Walasko, a director of a Canadian company, Pancontinental Oil Limited. This company was a subsidiary of the Australian company Pancontental Mining Limited who had recently discovered the giant Jabiluka uranium deposit in the Northern Territory. The mining company wanted to expand its oil operations to include Australia and they wanted to hire me to run the new company. I was stunned – I had no experience in management so why me? But such opportunities don't come along every day and after some financial negotiations, I accepted the position. We had to return to Sydney and had to sell the house we had just bought in Karrinyup. The Western Australian government collected a lot of stamp duty and we never got to live in it.

Pancontinental Petroleum Limited ("PPL") was incorporated in May 1980 as a wholly owned subsidiary of Pancontinental Mining Limited. The PPL board consisted of Tony Grey as Chairman, John Rowntree, John Landragan and Richard Miles. I was initially Chief Executive Officer but eventually became Managing Director after proving that I could handle the job. Tony Grey was an excellent chairman both at board meetings and at annual general meetings of shareholders. The board meetings were very professional and financially strict, and I obtained valuable experience at running public companies during my time at PPL.

My immediate tasks at PPL were firstly, to put together a portfolio of exploration acreage; secondly, to hire some professionals to operate that exploration acreage; and thirdly, to complete the prospectus and get the company listed on the ASX. As the oil price was rising and the oil industry booming, the first two of these tasks were not going to be easy. Pancontinental Mining had already negotiated a farm-in deal with Magellan Petroleum to earn equity in their Amadeus Basin permits in the Northern Territory by undertaking a 3,000 km seismic survey and a ten well exploration program. This was a tough deal but the Amadeus Basin was a proven petroleum province with two existing fields, Mereenie and Palm Valley, and was very under-explored. PPL was to operate this enormous program. PPL also had farm-in deals with XLX Pty Ltd, owned and run by one of the most eccentric of Australia's oil industry entrepreneurs, Ian Sykes. One permit EP-111 was in the northern Perth Basin in Western Australia and the other, ATP-243-P was in the Surat Basin in Queensland. Again, PPL was to operate these permits.

The first thing I did was to approach Mesa Australia Limited, the operator of WA-149-P, the permit discussed earlier. I negotiated for PPL to earn a 10% interest in this permit by contributing to the exploration program. The joint venture then consisted of Mesa, Offshore Oil, PPL, PetroEnergy, Pan Pacific Petroleum, Weeks Australia, Golden West Hydrocarbons and Nomeco.

Through Mesa Australia Limited, I also tried to negotiate an interest in the Whicher Range gas discovery in the southern Perth Basin. The original well was drilled by Union Oil in 1959 and tested gas from the Permian Sue Coal Measures. In those days there was no market for the gas and the well was plugged and abandoned. Mesa Petroleum had taken up the area and was preparing to drill an appraisal well on the structure. The drilling consortium (Mesa Australia Limited, Offshore Oil NL, Western Mining Corporation and Monarch Petroleum NL) was full and there was no room for PPL. Whicher Range-2 was drilled in April-July 1980 and as well as being a very expensive well, it also failed to encounter commercial reserves of gas. "The force was with me" that day.

As discussed earlier, oil had been discovered in 1979 in the Jurassic sediments of the Eromanga basin in the Dullingari North-1 well. This set off a massive exploration effort to discover more of the same. A permit numbered ATP-267-P was located next to the Delhi-Santos operations in southwest Queensland and was owned by an American company called

Banner Petroleum. It was a very large permit but the western side was close to the mature Permian source rocks thought to be required for this petroleum system to work. Banner Petroleum was farming out and wanted to be carried on all future exploration plus a royalty. Acreage was at such a premium, I agreed to this deal and PPL became the operator of the permit. The agreement was signed in March 1981 between Banner 30%, PPL 17.5%, Offshore Oil 17.5%, Sydney Oil 17.5% and Bligh Oil & Minerals 17.5%. This deal became infamous in the Australian oil industry as "The Banner Carry" and was to cause many oil industry accountants to curse my name for all eternity. I will say more about this later.

The permit to the north of ATP-267-P was ATP-299-P. This was a permit with a lot of potential and was operated by Hartogen Energy, at that time a successful and well run company. I negotiated a 10% interest in this permit for PPL. PPL also picked up a 10% interest in ATP-298-P, a permit to the east of ATP-299-P, and where there were a lot of wells with oil shows including Kyabra-1 and Kyabra–2, drilled by Lennard Oil in 1980-81.

The final block in the portfolio was the Alton production licence PL2 in the Surat Basin by a farmout through Moonie Oil.

Being the operator of so many permits, PPL needed experience geoscientists fast. We recruited Andrew Svalbe as Exploration Manager, Tony Wright as Chief Geologist and Eugene Urschell from Canada as Chief Geophysicist. We got Richard Schroder back from Norway and also recruited John Gorter. It was a formidable team. We also needed a drilling engineer but most of these were working international on very large day rates. Then I discovered that Charlie Mann was working for Getty Oil. I had worked with Charlie in the North Sea and had a high regard for him. He was on staff with Getty Oil but as they were not drilling in the immediate future, I worked out a deal with Getty Oil to sub-contract him for our drilling programs.

The most frustrating task was the preparation of the PPL prospectus and my experiences at Mount Eliza and my MBA studies were to prove invaluable. The preparation of a prospectus for a public offering is a tedious and time consuming process. It is a legal document and everything in it must not only be true and correct but must also be able to be proved to be true and correct. So every statement in the prospectus had to be backed up with more paperwork confirming its correctness. There were other minor frustrations such as having to get permission from each company the technical team used to work for to be included in their resumes in the prospectus. This was going to be far too time consuming so we used the term "worked for a major international oil company" a lot in the prospectus.

Before the team came on board, PPL had the requirement to conduct a seismic survey in EP-111. This was not only part of the deal with XLX but also an obligation under the terms of the permit. United Geophysical had a crew in the Perth Basin and we got them to conduct this survey at very short notice. Being flat out in Sydney getting PPL listed, this seismic survey was not well supervised. This came back to haunt me many years later.

The main task was the large seismic survey to be acquired in the Amadeus Basin. Eugene and Richard were busy organising this but prior to being able to acquire this survey, we had to get approval under the Northern Territory Aboriginal Land Rights Act ("NTALRA"). This was not going to be easy. A few months before, the drilling of the Noonkanbah well in Western Australia had only been effected by much confrontation and acrimony. The Noonkanbah story is very complex and has had a lot written about it. This is my understanding of what happened. Exploration permit EP-97 was awarded to a consortium operated by Amax Petroleum in 1976. The aboriginal group at Noonkanbah was granted leasehold title over the area from the Western Australian Aboriginal Land Trust in 1977. Therefore, in fact Amax's petroleum title pre-dated the Noonkanbah aboriginal title. Amax went to enormous trouble to determine who the real traditional owners for the area were, and they were not part of the aboriginal group who held the Noonkanbah lease. Amax had dealt with the traditional owners and had obtained approval to acquire seismic surveys over the area without any problems at all. Sacred areas were avoided, damage to the environment was minimised and the seismic surveys went smoothly. Amax had also obtained approval for the drilling of Fitzroy River-1 from both the traditional owners and the Noonkanbah leaseholders but at the last minute the latter changed their minds and threatened to stop the drilling operations. The drilling rig had been contracted and was already on its way to the site. Further delays would have caused Amax considerable unnecessary costs and Western Australia's sovereign risk status was under threat. At this point, the Western Australian state government stepped in and forced the issue by providing a police escort for the drilling rig and ancillary equipment. Fitzroy River-1 spudded in August 1980 and was plugged and abandoned as a dry hole but the entire episode was a public relations disaster. Oil industry relations with Aboriginal groups were at an all time low.

Under the NTALRA, we were not allowed to contact the aboriginal groups directly but had to deal with the Central Land Council. Magellan employed an Aboriginal Liason Officer called Bob Liddle and they suggested we use him to assist in the approval process. Magellan had already negotiated production agreements in 1979 for the development of the Mereenie and Palm Valley oil and gas fields whereby the Central Land Council (CLC) received a lump sum payment of $500,000 plus royalties of 1.5% from production revenue. Having never negotiated with aboriginal groups or land councils, I was very grateful for his assistance. Nevertheless, I was rather nervous when we arrived for our first meeting with the CLC. When we entered the room, it turned out that the only aboriginal in the room was Bob Liddle. The CLC was entirely represented by white administrators and lawyers. I sighed with relief – at least now I could ease my way into these aboriginal land rights negotiations and hopefully not make too many mistakes during my rapid learning curve.

The negotiations did eventually involve discussions with the aboriginal traditional owners but as Bob Liddle knew them all, these negotiations were conducted in a positive atmosphere. One of the points for discussion was "Why does Pancontinental Mining have to prepare a huge environment management plan to mine the Jabiluka Uranium deposit covering a small area while PPL can conduct seismic operations over thousands of square kilometres with no such plan?" Not an unreasonable question! After much discussion regarding the difference between exploration and development, we managed to finally get an agreement which laid out the conditions for sacred site protection, environmental protection and employment of the local people. PPL paid the CLC to hire an anthropologist and varied the seismic program many times to avoid any sacred sites.

After all these negotiations were completed and the regulatory approvals obtained, PPL acquired 2,795 kms seismic data between November 1980 and April 1982. **(Fig 26)**

One interesting aspect of acquiring this seismic was that under our agreement with the CLC, there was to be no alcohol brought onto Aboriginal land.

Figure 26
Seismic trucks in the Amadeus Basin

Now the personnel on seismic crews in the middle of the desert do enjoy a nice cold beer at the end of the day, and so they were not very happy about this. Morale on the crew was low and production was only about 8-10 kilometres/day. I went back to the CLC and managed, with the most stringent conditions attached, to have alcohol allowed on the seismic crew. Morale increased and production went up to 12-14 kilometres/day.

In September, 1980, Iraq launched a full scale invasion of Iran with the most likely objective being the prevention of the spread of the Islamic revolution. The war lasted for eight years and included attacks on the Iranian Kharg Island oil terminal, on oil tankers and mines were also laid in the Persian Gulf. Amazingly, the oil price gradually fell while all this was going on. In July 1988 the US cruiser Vincennes mistakenly shot down an Iranian airliner killing 290 civilian passengers. These incidents may have contributed to Iran and Iraq eventually making peace and the war ended in August 1988 after the death of at least half a million soldiers and civilians on both sides.

On a more pleasant note, the years 1979 and 1980 turned out to be a great time for gas discoveries in the Australian oil industry. In the offshore Carnarvon Basin, the giant Gorgon Gas Field was discovered by WAPET and this field has now been finally developed as an LNG project. In the Browse Basin, the giant Brecknock Gas Field was discovered by Woodside and in the frontier Exmouth Plateau, the Scarborough gas discovery was made by BHP Petroleum. Both these fields still remain undeveloped. These three gas discoveries turned out to have massive reserves totalling in excess of 25 TCF gas. Onshore, the Woodada Gas Field was discovered in the northern part of the onshore Perth Basin. This field commenced production in 1983.

PPL listed on the Australian Stock Exchange in October 1980 after raising $11 million. The oil price kept rising until it reached $US40 per barrel in early 1981. This set off an oil industry boom that was so frenzied that oil geologists with experience were in such high demand that we became the subject of some rather amusing cartoons **(Fig 27).**

Figure 27
A very important geologist being escorted to the bar

An investor magazine in New York had forecast that the Amadeus Basin would become the next Middle East and PPL shares quickly rose to 95 cents per share. I had only been with the company for nine months and was now Chief Executive Officer of a $100 million company!

In January, 1981, I went to Calgary to visit the Pancontinental Oil office and meet the directors and technical team. I flew from Sydney to San Francisco and then Air Canada to Calgary. I was amused that Air Canada had to make their announcements in both French and English. I always thought that the English won the war for Canada! But I was horrified by the announcement as we were coming in to land that "the temperature in Calgary is -45°C". But that can't be right, I thought - the air liquefies at -45°C! I went through immigration, collected my bags and then went to catch a taxi. As I exited the terminal and took my first breath of Canadian air, I thought my lungs were going to freeze. It actually hurt to breathe! It's funny but my first thought was "how the bloody hell did the native Americans manage to live here in these temperatures without centrally heated homes"? I managed to thaw out my lungs in the very warm taxi on the way to the hotel and got a good night's sleep. The next morning I had slept in, so I had a quick shower and walked to the Pancontinental office. The streets were frozen and it was difficult to walk without slipping over and when I got to the office, the receptionist took one look at me and burst out laughing. My moustache, still wet from the quick shower, had frozen and I must have looked like Jack Frost. It was then that I was told that no-one in Calgary (except dumb Aussies) walks outside in these temperatures – there are walkways connecting all the buildings so you can get anywhere in the downtown area without going outside.

Returning to Sydney from Calgary was via San Francisco. I stayed overnight and visited a business acquaintance Ray Needham the following day. He introduced me to a restaurant called Tony's that served the most fabulous spare ribs I have ever eaten. Since the QF4 flight to Sydney wasn't until 10-00pm, I had some time to kill in the evening. I was staying at the Hyatt Regency and decided to spend some time in the bar before collecting my bag and heading for the airport. The bar had a dance floor and after 5-00pm, a band started up and the locals began trickling in all dressed up in their finest clothes and began ballroom dancing. It was all very elegant and provided me with a wonderful way to pass the time waiting to catch my flight.

Getting back to Australia, PPL's 1981 drilling campaign got off to a bad start. We drilled Ballymena-1 and Maxima-1 in ATP-243-P; and Blacktop-1, Mireeka-1 and Roccabright-1 in PL2 in the Surat Basin, all dry holes. We also drilled two wells, Wallaby-1 and Dingo-1, in the Amadeus Basin. Whilst Dingo-1 was a gas discovery, this was of little use in the remote Amadeus Basin – we were looking for oil.

The Dingo-1 gas discovery is an interesting story. The Dingo structure was a large simple anticline and the seismic quality, as opposed to other parts of the basin, was excellent. The main objectives were the Early Cambrian Arumbera Sandstone sealed by the Middle Cambrian Candler Formation evaporates. We drilled through the salt, intersected the top of the Arumbera Sandstone but alas, no oil or gas shows and I wanted to stop the well right then and there. However John Gorter convinced me that we should drill deeper – we needed information on the entire Arumbera section and also to determine the source potential of the underlying Julie Formation. This was a potentially important petroleum system which was very poorly understood. The well was drilled deeper, went through an intra-formational shale seal and intersected the lower Arumbera sands with good gas shows. The well drill stem tested at rates up to 5 MMcfg/d. First blood to John Gorter!

The timing of the Dingo-1 gas discovery was also very unfortunate. It was immediately after the Azaria Chamberlain case where the baby Azaria was allegedly taken by a dingo during a camping trip. Nobody believed such an outrageous story and her mother Lindy was tried and convicted for murder. The case caused a huge media frenzy. PPL was given the 1981 bad taste award by one of the local newspapers and this greatly upset me as we didn't name the prospect – it had been on Magellan's prospect inventory for years. Again, John Gorter came to the rescue – quick as a flash, he said "But Dingo stands for Deep Investigation of Natural Gas and Oil". That just made my day. I wanted to respond but the board advised me that media reporters have a very short attention span and therefore, we should not try to explain ourselves as this would only continue the debate. We did not respond and, sure enough, the board was right and there was nothing more in the papers the next day and the matter was never heard of again. Years later, further evidence was found confirming Lindy Chamberlain's story, her sentence was overturned and she was released in 1988. She also rightfully received $1.3 million in compensation in 1992.

The 1981 APEA Conference was held in Adelaide. These conferences can be very busy with lots of meetings and opportunities to discuss. I did get to see two papers: Keith Norlin gave a paper on the Woodada-1 gas discovery in the onshore Perth Basin and Delhi Petroleum gave a paper on the Dullingarri North-1 oil discovery in the Jurassic of the Eromanga Basin. Mo also attended the conference but I had to return to the office at the end so we didn't get any opportunity to tour the Barossa Valley.

During 1981, the Blina-1 oil discovery was made in the large onshore Canning Basin in Western Australia. One of the interest holders was a subsidiary of Occidental Petroleum, a large independent American oil company. The CEO of Occidental was Armand Hammer, a charismatic American oil entrepreneur. He made a press release that Blina-1 was the most significant discovery since the North Sea and this started another boom in oil stocks in Australia.

Doing business in the oil industry in the early 1980's involved a lot of lunches so we had our favourite restaurants in Sydney. One of these was the Pitts, a very elegant restaurant where the waitresses wore fishnet stockings. I was taking Dick Burgess of Nomeco (our joint venture partner in ATP-267-P) and Monte Currie, one of his American associates for lunch and chose the Pitts. As we entered I told them that I hadn't been to the Pitts for a while (which was true). The young lady greeted us and said "Good afternoon Dr Warris, good to see you again. Your usual table?" Great! They must have thought I lived in the place.

Being the operator of most of its permits, everyone at PPL worked very hard. So at the end of the week, the company put on Friday afternoon drinks in the board room. On one of these occasions, I was challenged on my fitness so I did 40 push-ups on the boardroom table. It then became traditional for me to do this every Friday before drinks. Sometimes we would kick on afterwards but generally I would head off home and leave them to it. Never in all my time at PPL was this occasion ever abused.

Being back in Sydney, after living in tiny Perth, took a bit of adjustment. Housing was twice as expensive and during the early 1980's, interest rates were ~ 18%. So a significant part of my net pay went in paying the mortgage. In addition, the top tax rate was 68% so there was little point in paying high salaries. So we were "paid" with very large expense accounts. These were the days before the Fringe Benefits Tax which was introduced in 1986. Mo and I dined at the best restaurants, drank the best wine and the entire family travelled around the world once a year first class. Life was great but we couldn't save anything.

PPL had two very rich Jewish investors as shareholders. They lived in New York and on one of our round the world trips, I contacted them and arranged to meet them for dinner. They took us to the Cygnet Restaurant which I discovered later was one of the finest restaurants in New York. We got a taxi which dropped off outside this very dingy, blackened building. But when we went inside, it opened up into the most palatial restaurant. We had a very pleasant evening discussing world events, the international oil business and eventually PPL. They even paid the bill – for which the PPL shareholders were eternally grateful!

From a family point of view, our best holidays were spent camping at Kangaroo Valley. This was a beautiful spot located southwest of Sydney on a tributary of the Shoalhaven River. At Easter, Mo would take off with Chris and Alex on Thursday afternoon, get there early, find the best spot and set up camp. The rest of us would arrive after work and join in the fun. There was nothing in Kangaroo Valley except a toilet block but there was a large tree with a strong branch extending out over the river. Naturally, there was a strong rope tied to this branch. Chris and Alex (and all the other kids) spent hours swinging from that rope and splashing into the river. I believe even I had a go every now and then. Kangaroo Valley was such a great camping spot. We would cook lots of steaks and sausages, drink lots of wine, sing songs round the campfire and the kids could run riot and be perfectly safe. And when everyone left on Monday afternoon, there wasn't one single piece of rubbish left behind.

One Easter, we decided to visit another venue, Hill End with our friends the Fitzgerald's. This was an old gold mining area located in a deep valley, an ideal camping spot. Gold was discovered there in 1850's and by the 1870's, the area had a population of 30,000. But in the 1980's, it was a tiny tourist town. When we arrived, it had just started to rain. We set up camp but it was wet and cold and we went to bed early. During the night, we woke up with wet sleeping bags and went out to investigate. The river had risen after the rain and we had to move to higher ground. In this area, we all had to stay close to the river as the area was littered with abandoned mining shafts, some of which were not clearly visible. But a number of mining shafts went horizontally into the cliffs. We all went exploring one of these shafts but before we entered, I gave everyone a geology lesson. The lesson began "Nobody touch anything!" After a certain distance into the shaft, I noticed the wooden supports were looking a little the worst for wear and figured that it was too dangerous to go any further. Amid much grizzling from the kids, I eventually hussled them all out. But you had to admire the old miners who blasted their way into these mountains with pick, shovel and timbers in search of gold and a better life.

On my return from one of my visits to Pancontinental Oil in Calgary, I had arranged to visit Dick Burgess in Nomeco's office in Jackson, Michigan. This involved flying from Calgary to Toronto and connecting to Detroit on a DC 9. I think it must have been the first DC 9 built because it looked very old and tired. As it sped down the runway to take off, I thought it was going to shake itself to pieces before we got airborne. Somehow it managed to take off but the same process repeated itself when we landed. Dick picked up a very relieved me at the airport and drove me to Jackson. It was a spectacular drive through thick forests concealing the huge population which lived in this part of the USA. The next day I went to their office, met their technical staff and discussed the various plays that they were exploring in North America After work, Dick took me home for meal at his place. He had a lovely house surrounded by trees and as we sat in the garden before dinner, all the squirrels came around to inspect the new visitor - so cute and very tame.

Our 1982 drilling program also did not bring the results we hoped for. At the end 1981, the Delhi-Santos joint venture had discovered the Jackson Oil Field in ATP-259-P (Naccowlah Block) which adjoined ATP-267-P to the west. This was a major oil discovery with recoverable reserves of 30 million barrels. In ATP-267-P, a large structure, the Kihee Anticline, had been mapped from existing and new seismic. It was a Jackson look-a-like and everyone in the office was very excited. Kihee-1 was spudded with great anticipation and while it had good oil shows, these were not initially commercial and the well was plugged and abandoned as a dry hole. We were all very depressed. Someone cheekily commented that "the rocks are smiling at us" but it didn't help. In fact, matters actually got worse. PPL also drilled the Kihee North-1 and Noccundra-1 wells in ATP-267-P, both of which were dry holes.

In the Amadeus Basin, drilling results were mixed. We drilled Mount Winter-1 and West Walker-1 and while the latter well tested gas at 3.5 MMcfg/d, gas was of little commercial use in the Amadeus Basin at that time. Even the drilling of these wells was painful. At the end of the Mount

Winter-1 well, PPL was in the process of moving the drilling equipment to the West Walker location. Half way through this operation, it began to rain and rain heavily. As it hadn't rained in this area for years, this was unexpected and financially disastrous. At one stage, we had one third of the drilling equipment at the Mount Winter location, one third at the West Walker location and the remaining third bogged in between. Months afterward, the local pastoralists would ring PPL up and ask when we were going to drill again so they could increase their stock levels!

PPL had spent a fortune on seismic and had drilled 12 wells with only two non-commercial gas discoveries to show for it. At the beginning of the drilling campaign, I had told the board that with modern seismic and permits in proven petroleum basins, PPL should achieve a 25% success rate. I remember walking into the next board meeting and thinking to myself – this is it – they're surely going to ask for my resignation after these uninspiring exploration results. But the PPL board was patient and they continued to support the original exploration strategy.

March 1982 was the 50th anniversary of the opening of the Sydney Harbour Bridge. The government opened the bridge for one day for pedestrian to walk across to celebrate the occasion. We joined with the Svalbe's and walked across as well.

The 1982 APEA Conference was held in May in Sydney at the Hilton Hotel. Being held in our home town, it was a time to catch up with old friends and see what deals were out there. Wapet gave a paper on the giant Gorgon gas discovery and Dr Playford gave a talk on the Blina oil discovery and reef potential in the Canning Basin.

In 1982, war broke out between the UK and Argentina over a bunch of rocks in the south Atlantic called the Falkland Islands. The UK won but the only positive result was the demise of the military dictatorship in Argentina. At the time, I thought that this war was really about the oil rights as the size of the continental shelf controlled by these islands was enormous. As it turned out, in 1995, the Falkland Islands government launched a licensing round and permits were taken up by Shell, Hess and Lasmo. Due to low oil prices during the late 1990's, exploration was patchy and intermittent but as oil prices increased, a UK company called Rockhopper Exploration made a significant oil discovery in 2010. In October, 2011, Premier Oil took over as operator in a $US1 billion deal to develop the Sea Lion Field.

The commerciality of this field, with estimated reserves of 350 million barrels, is still being evaluated.

At the beginning of our ATP-298-P joint venture, there were only five companies, PPL, Bligh Oil, Moonie Oil, Sydney Oil and Nomeco. Over the next few years, the number of companies involved increased to twenty following numerous sales and farmouts. There was one more farmout and while preparing the legal agreements, we were informed that a joint venture could not have more that 20 companies otherwise it was deemed to be a partnership by the Australian Taxation Office. The only exceptions were accounting and legal firms. We looked at trying to change this rule to also exclude oil exploration joint ventures but it needed the law to be changed by an act of parliament. They couldn't even hold the interest in trust because unless the company was on the title, it couldn't claim the exploration expenditure as a tax deduction.

In February, 1982, I gave a talk to PESA on "The small Australian explorer, its role and problems in the petroleum exploration industry" using PPL as an example. It was during 1982 that two of the most expensive wells of the time were drilled. Offshore Oil's West Barrow-1/1A and Citco's Vulcan-1/1A/1B cost about $30 million each – a lot of money in those days!

From its peak of $US40 per barrels in 1980, the oil price gradually declined to $US28 per barrel in 1984. In July 1982, the Penn Square Bank in Oklahoma City with numerous bad energy loans went bankrupt taking a number of other banks with it. This was a portent to a future, much greater disaster but we did not recognise it. Because of the falling Australian dollar, the price of oil had remained steady at around A$40 per barrel until 1985 and oil exploration in Australia remained strong.

During one of our visits to the USA, I arranged a meeting in Houston with Tom Stevens who I had met at the 1980 APEA Conferences. Tom was President of SW Texas Bank which was a major lender to the US oil industry. Following our meeting, we had dinner at the exclusive Houston Petroleum Club. The dinner was in aid of the book launch of the story of Charles Henry Keplinger and Tom arranged for Mo and I to attend. The Petroleum Club was located in a very posh suburb of Houston and as we were driven down the main street, we passed some very expensive homes lining the streets. The security at the club was very interesting with one of

Tom's aides having to talk into a rubbish bin to pass security and enter the lift.

At the end of 1982, PPL started to investigate possible gas markets for the Amadeus Basin gas. We had discovered Dingo and West Walker and Magellan had the large Palm Valley gas field. But these gas accumulations were worthless without a market. At this time, the Northern Territory government was well advanced in negotiations with the Queensland government to buy coal, transport it to Darwin and build a coal fired power station. Wouldn't it be better for the Northern Territory to build a pipeline from the Amadeus Basin to Darwin and build a gas fired power station? We contacted Williams Brothers to investigate the feasibility of such a project. We then met with Ian Tuxworth, the Minister for Mines and put to him all the benefits of our proposal. Firstly, it would be an all Northern Territory project; secondly, the Northern Territory would get the royalties from the gas rather than Queensland from the coal; thirdly, a coal fired power station would be very expensive whereas a gas fired power station could be built in units which could be added to as the market increased; and fourthly, the Northern Territory would not be at the mercy of unions disrupting the mines or ports – we tried everything we could think of. At first we were not optimistic as the coal fired gas project was so far advanced. PPL had calculated that burning coal, at the then price per tonne, was the equivalent of a $2 per Mcf gas price. Given that the price of Palm Valley gas was then $1-70 per Mcf, price was not a significant factor. But I give the then Northern Territory government the credit for recognising the overall benefits of the gas option and the enormous courage to switch projects at such a critical stage. During 1983, Williams Brothers completed their feasibility study and submitted their submission to the Northern Territory Electricity Commission. A 1,537 kilometre long 12" pipeline was built from the Amadeus Basin to Darwin and the gas fired power option became a reality in 1987. Although, PPL was not part of the final development, I like to think that the company had a major influence on the implementation of this project.

The most significant discoveries on the North West Shelf during 1983 were Harriet-1, Jabiru-1 and East Spar-1. Harriet-1 was drilled by Occidental Petroleum, an American company, and encountered 23 metre gross oil column in Early Cretaceous Flag sands. Following difficulties with their takeover of Cities Services and generally falling international oil prices, they sold their interest in the permits to Bond Corporation in 1984. Harriet was the first of many oil and gas discoveries which became the Varanus Island joint venture eventually operated by Apache Energy.

In 1983, the Jabiru-1 oil discovery was made by BHP Petroleum in the Timor Sea. The first well encountered a 57 metre gross oil column and potential reserves were estimated to be 620 million barrels. However, the next three appraisal wells were dry holes much to the shock and horror of the joint venture. Estimated proved reserves were then down to only 13 million barrels and it was only due to the introduction of the Resource Rent Tax that was designed to allow the development of marginal fields that the joint venture continued to appraise the discovery. The next three appraisal wells were successful, estimated reserves increased to 110 million barrels and the field began production in 1986.

East Spar-1 was drilled by WMC. This prospect was interesting in that there was no time closure, just an amplitude anomaly. Detailed velocity analysis of the seismic delineated a depth structural closure and it was on this basis that the prospect was drilled. However, it wasn't until 1995 that the field was eventually developed at a cost of $250 million. Sales gas was transported via the $400 million Goldfields pipeline to mines in the Pilbara. WMC owned 30% of East Spar and 62.5% of the Goldfields pipeline. The field eventually produced a total of 240 BCF gas.

During 1983, PPL's drilling campaign turned around. In the offshore Carnarvon Basin, the South Pepper, North Herald and Chervil oil discoveries were made by Mesa in WA-149-P. In the Eromanga Basin, PPL drilled Regleigh-1 in ATP-298-P, and Nockatunga-1 and Pigeon Lake-1 in ATP-267-P, of which Nockatunga-1 was a significant oil discovery. PPL was also involved in the Tintaburra oil discovery in ATP-299-P operated by Hartogen. In the Amadeus Basin, PPL drilled Undandita-1, Tempe Vale-1, Finke-1 and Tent Hill-1, all dry holes. In the Perth Basin, PPL drilled Diamond Soak-1, also a dry hole. By the end of 1983, PPL had drilled 24 wells and had made five commercial oil discoveries for a success rate of 21% - things were starting to look up.

Like ATP-298-P, ATP-299-P also had lots of joint venturers, fifteen in all – Hartogen (operator), PPL, Home Energy, Texas Gas, Lennard Oil, Peko Oil, Eromanga Energy, Laurel Bay, Nomeco, Plymouth, PetroEnergy,

Silverton, Versatile, Bralorne and Monarch. The Tintaburra-1 oil discovery was made in January 1984 and encountered a 10 metre oil sand in the Hutton Sandstone. Initially, we weren't sure how extensive the field was going to be given the thin oil column. However further appraisal drilling proved up a 1.7 million barrel oil field.

Production from the Nockatunga Oil Field began in April, 1985 and was soon followed by the Winna (October 1985), Thungo (1986) and Maxwell oil fields (1987). Initially, getting the oil to the Brisbane refinery was going to be a problem. The Jackson to Moonie pipeline was built by Santos and completed in 1984 at a cost of $125 million. It was designed to take up to 16,000 bopd but the tariff to use the pipeline was prohibitive and most other oil producers could not afford to use it and had to truck their oil to Brisbane. This sparked Arthur Nommensen and John Bonner to build a refinery in the town of Eromanga. They formed Inland Oil Refinery in 1984 and built and commissioned the Eromanga Oil Refinery by 1986. It has been processing ~ 1,250 bopd from local oil producers ever since. However, PPL was spared these problems by the nature of the Jackson oil which had a high wax content. Soon after the Jackson to Moonie pipeline began operations, difficulties were experienced due to the wax depositing on the sides of the pipeline and restricting the flow. To reduce the wax content of the oil travelling down the pipeline, Santos bought PPL's low wax Nockatunga oil from the Murta reservoirs at a price that equated to a much lower tariff.

Once development of the Nockatunga Oil Field began, the Banner carry began to get complicated. Some of the companies who signed the original agreement with Banner Petroleum had farmed out to other companies. These incoming companies had to take on their share of the Banner carry but all the past payments had to be adjusted for the joint venture accounts. So every time a new company came into the joint venture, the accounts had to be adjusted and these calculations were very complicated. After considerable work, PPL's financial manager, Paul Fuller, prepared a table and a very complicated formula for calculating the Banner carry in the joint venture accounts. He presented this at an Operating Committee Meeting to obtain approval from the joint venture that the proposed accounting formula for the Banner carry was fair to all companies. At the end of his presentation, he asked for comments. There was stunned silence! This was not good – PPL needed the joint venture to agree to this accounting method.

Surry Bogg, managing director of Bligh Oil & Gas, came to the rescue. He thanked Paul for all his efforts in devising this accounting method and added "I don't have the slightest idea of what you just presented but Bligh Oil & Gas approves the accounts". It took all my self control to stop myself from bursting out laughing and so, with a straight face, I asked for approval of the accounting method. It was approved unanimously.

The 1983 APEA Conference was held in Melbourne in March. The most interesting paper for me was Thomas & Brown's paper on hydrocarbon generation in the northern Perth Basin. This was to become my bible on the Perth Basin in a few years time. Mo also attended the conference and we had some excellent meals with our friends in the industry.

In November 1983, the Hawke Labour government came to power and their policy was to ban uranium mining. Pancontinental Mining had a serious problem. In addition, the stock market was falling and the world, including Australia, was in recession. Now PPL had a serious problem. Our share price had fallen below its par value of 20 cents a share and being a limited company, could not make share issues below this price to raise money. The company now relied on Pancontinental Mining for any future short term funding. But with the ban on uranium mining, this was not going to be a sure thing.

Things only got worse. One of our joint venture partners, a Canadian company called United Canso, was struggling under the deteriorating international financial conditions and could no longer continue to fund its Australian obligations. Under the joint operating agreement, we could have sued them, but there was no point in doing that if they had no money. So the joint venture agreed to let them withdraw and the remaining parties would take up their interest. To our amazement, they wouldn't agree to this – they wanted to keep their interest but not pay their way. Now we had to sue them! The legal dispute went on for a year and was extremely acrimonious. Our joint operating agreement was old and poorly worded. Their lawyer had a field day and gave me hell. No agreement could be reached and we then had to go to court. As we walked up the steps to enter the courtroom, they finally settled. It had cost our joint venture a million dollars and aged me about 10 years. But what really annoyed me was that they had achieved nothing. They had also spent a fortune on legal fees and had ended up with exactly what we had offered them in the first place. A year later, I met their lawyer at a Xmas party. He greeted me like a long lost friend. All I wanted to do was strangle the bastard. It was then that I realised that the law has nothing to do with justice. The law is only about the law. It's a game in which the side with the smartest, trickiest or most expensive lawyer wins. And lawyers are just actors – there really is no malice in what they do to the innocent.

After this annoying, lengthy and expensive distraction, I finally got back to what the shareholders paid me for – running the company. I realised that it not just the cost of the lawyers but also the lost management time that had been detrimental to PPL. I vowed that in future management roles I might have, I would do my very best to avoid legal disputes like the plague.

A Canadian couple Hank and Doreen Laska came to Sydney and rented a house at Vaucluse. Hank was CEO of a company called Diamond Shamrock and they were also friends of the Urschel's. Hank and Doreen threw the most fabulous Boxing Day parties at their house in Vaucluse. From their balcony you could glimpse the Sydney to Hobart yacht race. They had a great swimming pool with an overhanging balcony. We would throw money into the pool and the kids would jump off the balcony into the pool to fetch money. After a few beers, some of the blokes (including me) had a go as well. Later on, I was attending a very fancy cocktail party by a very large banking group. I was talking to one of the senior managers who finally recognised me. He said "I know you – you're the fellow who jumped off the balcony into the pool at the Laska's!" Oh well, I sighed – scratch that bank off the list of financers for the South Pepper-North Herald development.

In 1983, Mesa was sold to WMC who renamed it Westminco and this company took over operatorship of WA-149-P. They drilled six wells in WA-149-P during 1983-85 of which only South Chervil-1 was an oil and gas discovery.

PPL had the opportunity to invest in oil exploration in Indonesia through a joint venture with a company in Thailand. I flew to Bangkok and stayed at the Meridian Hotel. In the taxi from the airport to the hotel, the taxi driver asked me if I'd like a girl for the evening. I politely replied "no thanks". After a few seconds, he then asked me if I'd like a little boy for the evening. Good grief I sighed but I was in Bangkok. The next day we met with the Thai company managers who explained a very complicated deal to me. There was also a Hong Kong based company who had a carried

10% interest in the deal. I figured that this company probably belonged to one of the Indonesian generals who decided which groups would be awarded the production sharing contracts. On returning to Sydney, my two day visit to Bangkok must have set off all the "drug smuggling" alarm bells at the airport. Security took me aside and went through my bag with a fine tooth comb. They were very polite and joked with me about me looking like a really shifty character. At the next PPL board meeting, we decided the whole deal reeked of corruption and voted against participating. And no – I didn't take the taxi driver up on the little boy!

In ATP-267-P, PPL was experimenting with a new technique called soil gas geochemistry. We had lots of small structures in the permit and needed some method of upgrading these prior to commiting to the expensive cost of drilling an exploration well. John Conolly of Sydney Oil had a simple system of collecting the samples and analysing them. The theory is based on the assumption that shale seals are not perfect and minute amounts of natural gas leak to the surface above the deeper hydrocarbon accumulations. We used to run these geochemical surveys along the existing seismic lines but during one survey, samples were collected to the south of any existing seismic lines. This survey picked up a significant anomaly and based on this, additional seismic was acquired over and around the anomaly. Because of the difficult financial situation, the Maxwell-1 exploration well was not drilled by PPL until 1987, long after my departure. Maxwell-1 was an oil discovery in the Early Cretaceous Murta Formation and is still producing today.

PPL had a dispute with Ian Sykes of XLX Petroleum over our Surat Basin farmin and exploration program. Now Ian was one of the great characters in the Australian oil industry. If you wrote a letter to his company, he would reply on the back of the letter, put it back in the envelope and write "return to sender" on the envelope. This way he saved on stationary and postage! Another of the great stories about Ian Sykes was the East Winorah-1 well XLX Petroleum drilled in the Eromanga Basin in Queensland. He was his own wellsite geologist to save money. The well was located on a property where the pastoralist ran some pigs as well as cattle. Ian did a deal with the pastoralist whereby Ian would supply food scraps from the drilling camp to feed his pigs in return for the free use of his phone to call in morning reports to the Mines Department. The man was a legend! So it was with some trepidation that I hopped on a plane to Melbourne to visit with him and sort out our dispute. I got a taxi to his office in Bell Street, Coburg. It was difficult to find until the taxi driver figured out that it was a petrol station. So I asked the attendant if Ian Sykes was in and he said "in the back". Ian had a tiny office at the back of the garage and it was cluttered with boxes of files. We negotiated a revised deal which Ian proceeded to hand write on a sheet of paper. We both countersigned the letter, he gave me a copy and that was that. No expensive lawyers. I was beginning to really like Ian and appreciate his management style. Even more amazing – he bought me dinner! It was only a meal at the local Chinese Restaurant, but I was still impressed.

The PPL Amadeus team had mapped up a prospect in the Waterhouse Range some 50 kms southwest of Alice Springs. Waterhouse-1 was drilled by Exoil in 1965 but had failed to reach the objective Arumbera Sandstone. PPL wanted to drill a second well on the surface anticline. I asked Bob Liddle to accompany me to negotiate access with the Central Land Council. Bob told me that the Aboriginal name for Waterhouse Range translated as "the Range of Doom" and was a taboo area. This did not sound promising. Initially, the Central Land Council would not allow access but eventually in 1985 after I had left PPL, Waterhouse-2 was drilled.

The 1984 APEA Conference was held in Hobart. It was at this conference that John Gorter and Richard Schroder presented their paper on the hydrocarbon potential of the Amadeus Basin including a review of PPL's exploration there. John Gorter also presented a paper on the source potential of the Ordovician Horn Valley Siltstone. After the conference, we flew back to Melbourne to connect to our flight to Sydney. Somehow, Ansett got our seat allocation messed up and we had to fly back economy. This didn't bother me until we boarded the plane to discover that the business class section of the plane was half full of pilots and hostesses. Obviously, Ansett was run for the benefit of its employees and not its customers. But it got worse! Mo was sitting in her seat when someone who was already late for the plane, claimed her seat. So Ansett was going to throw her off the plane. Now I am normally a quiet sort of a bloke but I was furious at the way Ansett treated their passengers and caused a terrible scene and told them to get airport security to physically throw me off the plane. Mo eventually calmed me down and she agreed to get off the plane mainly because passengers we had met in the lounge were going to miss their connections. At least Ansett looked after her because she arrived home that evening well and truly pissed. I never flew Ansett again. Their management style may eventually have caught up with them as the company went into administration in 2001.

The ban on uranium mining had caused Pancontinental Mining to reassess its future strategy. They were not in a position to fund PPL's ongoing financial commitments. With its share price below its par value of 20 cents, PPL, being a limited company, could not raise additional funds by issuing shares below this price. I had to find another way to fund the company. I went to the USA to find funding there but their industry was in worse shape than ours. International oil prices had begun to fall in late 1981 and by 1983 the oil price was down to $US28 per barrel. The Australian oil industry was protected by our falling exchange rate (the Australian dollar was floated at the end of 1983) and this kept the oil price in Australian dollars fairly constant. Our oil industry was to remain buoyant until early 1986.

Through Ian Whiteley, who was into financial consulting at this time, we negotiated a deal with Pelsoil to put funds into PPL in return for issuing it shares in PPL. That would have resulted in control of the company passing from Pancontinental Mining to Pelsoil. The PPL board rejected this proposal and told me that Pancontinental Mining would continue to support PPL financially. I knew that, in the longer term, this simply wasn't possible. I decided to resign as managing director and return to Perth. This was a devastating decision for me and was the low point of my professional career. PPL had one of the greatest technical teams I have ever had the privilege of working with and, after a slow start, had been technically a very successful company. If the company had been originally set up as a No Liability company, I believe we could have raised the money to develop the Nockatunga oil fields and participated in the development of the South Pepper-North Herald oil fields. PPL could then have gone on to become one of the great Australian oil companies (or been taken over by ExxonMobil!).

PPL was eventually sold to Charterhall Australia Limited in July 1987 but after the stock market crash of 1987, the assets were sold off: Carnarvon Basin assets to Ampolex; Amadeus Basin assets to Magellan Petroleum and the Eromanga Basin assets to Santos. It was a great deal for Ampolex. The South Pepper-North Herald fields were developed in 1987 using the Vicksburg jack-up as a production platform. Three horizontal wells were drilled on South Pepper and one on North Herald with all four wells tied back to Airlie Island via a 25 kilometre pipeline. These were the first horizontal wells drilled in Australia and proved the enormous value of this technology. Development costs were $40 million and production began in 1988, peaking at 12,000 bopd in 1989. Cumulative production turned out to be 20 million barrels. They also inherited my original prospect – updip Tryal Rocks. This was drilled in 1998, 19 years later as John Brookes-1 which encountered an 85 metre gross gas column in the Barrow Group sands. Better late than never! The field began production in September, 2005 at 240 TJ per day. By this time, Ampolex (including PPL) was owned by ExxonMobil.

RETURN TO PERTH

After PPL, we returned to Perth in August 1984. After selling our house in Sydney, we had enough money to buy a house in Carine and a beach house in Falcon, just south of Mandurah. Our house in Carine was in Hartington Way, the "Best Kept Street", and the primary and high schools were just at the bottom of the street. Both Chris and Alex went through Carine High School and received excellent educations. They went onto the University of Western Australia with Chris receiving BSc in Chemistry and BE in Mechanical Engineering, and Alex receiving a BSc with Honours in Geology. We also saved a sandy brown dog from "Death Row". He was a mongrel part corgie with one ear up and one ear down. We called him Sandy and he quickly became part of the Warris family **(Fig 28)**.

On one side of the house was a laneway and on the other side were Evie and Theo and their two daughters Lydia and Tarnya. They were great neighbours and organised the best barbeques with lots of friends and relatives and enough food to feed the 62nd Siberian Army. Theo, originally from Macedonia would also cook these superb Macedonian sausages, monstrous coiled things, each big enough to feed an African lion. Evie and Theo's barbeques were simply "not to be missed".

The beach house was a two story fibro shack at 4 Panamuna Drive just a short walk to the bay or over the dunes to the surfing beach. It was owned

Figure 28
No I won't give that ball back!

by Bill Gardiner and Ian Whiteley and we initially bought Bill's share. Then after Ian decided to sell his half, we ended up owning the whole property in August 1985. We would go down there generally every second week. Mo had everything organised with military precision – the car was already packed before the boys got home from school on Friday. As soon as they arrived, she would give them a snack, then bundle them into the car and be off before the peak hour traffic got too bad. I would leave the office at about 4pm (with a clean change of clothes for Monday) and drive down separately. By the time I arrived at the beach house, the car would be unpacked and the boys would be in the surf. The whole process would be reversed on Monday morning – it was like a long weekend!

In the 1980's, Falcon was a tiny village made up of holiday shacks and beach houses. It did have a great pub, the Cobbler's Arms. On Friday evenings we would wander over for a few drinks in the beer garden where Chris and Alex could play around. We would participate in the "chook raffle" and play "chase the ace". After all that, we would buy a pile of fish and chips from the local shop and head home for dinner. Chase the ace was a crazy game where every time you bought a drink, you got a ticket. At about 7pm, the pub would draw out a number and whoever had that number would pick a card out of a pack of cards. If you got the ace of spades, you won the prize money. At the start, this would rarely happen but each time the selected card would be removed from the pack thereby increasing the odds and the prize money would jackpot. One time the prize money got to ~ $5,000 and with most of the cards removed, people came from everywhere, even as far as Bunbury and the pub was packed.

On Saturday morning a few minutes after the sun rose, we would hear the shuffling of bare feet and the clinking of spoons on bowls as the boys had their breakfast. Then they would be off surfing and the house would descend into silence again. By about midday, Mo would ask me to check on the boys to make sure that they hadn't drowned or been eaten by a shark. So I would climb up the stairs over the dunes and survey Cosier's Beach. And there they would all be – about a dozen of them plus Sambo all waiting for the next wave. Sambo was a red setter who thought he was in charge of security and he would swim around with them hour after hour. Nobody had the heart to tell Sambo that sharks just LOVE dog meat! Sandy also loved the beach house. He would come with us down to the beach, try and catch the shore break, dig large holes in the beach and fetch a ball – although he was not very good at giving it back.

Even I got into the surfing spirit and I bought a second hand wave ski. Of course wave skis were not allowed on Cosier's beach with the real surfers so I would carry it down to Falcon Bay and Surf "the Point". The Point was a reef on the western side of Falcon Bay and would only break when there was a decent swell. It's hard to describe the thrill of surfing – it's not like snow skiing where you can control your speed. Once you catch that wave, an annoying thing called gravity takes over and you slide down the face of the wave at incredible speed before riding the wave to the shore. The downside to all of this fun is that the Indian Ocean has a nasty habit of sending a king wave every now and then. When this happens on the Point, the wave breaks somewhere east of Mauritius and rolls over you like a tsunami. You quickly learn to hold your breath for a long time.

The beach house had a small area of lawn which was reticulated by a water bore. Of course only going down there every two weeks meant that in summer, the lawn turned very brown. I could have installed a timer for the reticulation but the bore had a problem. When not running, it would develop an air lock which required going down into the cellar, unscrewing the cap and filling it with water to remove the air lock. Only then would the pump start. One Saturday morning I walked down to the bore in a T-shirt, shorts and bare feet and climbed down the ladder into the cellar to start the pump. As I got to the bottom, something moved and I shot up the ladder like a Roadrunner cartoon. There was a four foot dugite at the bottom of the bore. The poor thing must have fallen in and there was no way it could get out. It would have starved anyway so I killed it, apologised to its spirit and buried in the back garden. But it did give me an awful fright.

At this time I bought our first computer. It was a Texas Instrument computer and very basic but was sufficient to play games. It had Space Invaders, Pacman and an hilarious game called "Hunt the Wampus". The boys would play Space Invaders and Pacman and were amazingly good but my reflexes were not in the same league so when the computer was free, I would play Hunt the Wampus. This was a game of stealth (not speed) where you entered a series of rooms and had to kill a monster (the Wampus) hidden in one of the rooms by firing an arrow at it. You could only move forward and sideways and you won the game by killing the Wampus with the arrow. If you missed, the Wampus would eat you and you of course lost the game.

It was also about this time that I met Keith Thompson who ran a company called Recon Exploration. This company ran geochemical surveys like the one that was run over the Maxwell Field. He was an American who lived in Dallas but he was to become a partner in many of my petroleum exploration ventures in the future.

Strata Oil

I had negotiated a consulting contract with Jacia Pty Ltd, the management company for Strata Oil NL. This company operated the Woodada Gas Field and Exploration Permit EP-100 in the onshore Perth Basin.

The Woodada Gas Field was discovered by American independent wildcatters Dan and Dudley Hughes who drilled the discovery well, Woodada-1, in 1980. It was a complex field with the gas trapped in Permian limestones, now called the Beekeeper Limestone, on a structural nose plunging to the north. The southern part of the field was stratigraphically trapped by updip Permian shales. Woodada-1 tested 6.9 MMcfg/d from the Permian limestones but after acidisation, the well flowed 32.3 MMcfg/d. Hughes and Hughes then drilled Woodada-2, -3 and -4 during 1980-81 and

Production Licences L4 and L5 were granted in 1983 with production of gas beginning in the same year.

Exploration Permit EP-100 was awarded in 1977 and the permit holders were Hughes & Hughes 65%; Strata Oil 26.95% and the remainder held by small and private interests. This permit had a very strange joint operating agreement which did not require parties assigning to a new company to have that party sign a deed of assumption. So the operator did not know from one day to the next, who his joint venture partners were. My first task was to determine the trail of farm-outs and sales from the original joint venture so that transfers could be executed and the correct joint venture companies be registered on the title. This turned out to be a very difficult and time consuming process and in the end, a complete paper trail could not be constructed. But we had enough information to demonstrate the legality of the current permit holders who were duly registered on the permit title by the Mines Department.

During the development of Woodada, Woodada-3 had recovered oil during a drill stem test of the Permian limestones. This was interpreted as evidence for a possible oil leg beneath the gas column in the field. A Canadian company, Hudbay Oil & Gas, farmed into the permit and drilled Woodada-5 and -6 to discovery this oil leg. The 1983 recession was over and there was a share rally in progress which continued to October 1987. Strata Oil shares soared during this drilling campaign but in the end, no oil leg was found and after drilling Peron-1 and East Logue-1, Hudbay withdrew from the joint venture in 1983. Strata Oil took over the operatorship and drilled Woodada 8 to 10 during 1983-84. They also took over the management of gas production from the field.

Prior to all this, in 1980, the State Electricity Commission of Western Australia (SECWA) had signed a twenty year contract with the North West Shelf joint venture to supply domestic gas to Perth. During the following year, after years of negotiation, contracts between Japanese trading houses and the North West Shelf joint venture were executed for the supply of LNG to Japan. Now the North West Shelf development could get underway. This included the building of LNG tankers, the construction of the LNG plant at the Burrup Peninsular, the building and installation of the North Rankin Platform, the drilling of the production wells and the construction of the North Rankin to Burrup sub-sea pipeline and the Dampier to Bunbury land natural gas pipeline. Even at that time, all this cost billions of dollars and required extensive financing arrangements and government guarantees.

The Dampier to Bunbury natural gas pipeline is a 660 mm diameter pipeline which had an initial capacity of 360 TJ/d. It was built by SECWA and when commissioned in 1984, was the longest natural gas pipeline in the world, extending some 1,600 kms from Dampier to Bunbury. The Woodada gas field had been producing since 1983; however, under the financing arrangements for the North West Shelf development, all Western Australia's domestic gas supplies had to come from the North West Shelf. I was present at a meeting between executives from Strata Oil and the Minister for Mines David Parker who very politely told us that the Woodada Gas Field could no longer continue producing and would have to shut down. Strata Oil was not impressed particularly as David Parker was also the Commissioner for SECWA and may have had a conflict of interest. Strata Oil seriously considered suing the WA Government but under the Petroleum Act, the Minister has sweeping and absolute powers and their legal advice was not to proceed down that path. The Mines Department terminated Strata Oil's pipeline licence connecting the Woodada gas field to the WANG natural gas pipeline and that was the end of the field until sales re-commenced to SECWA in late 1989. In 1991, David Parker was implicated in the WA Inc Royal Commission and he left Australia.

One of the most amusing experiences I had with Jacia was attending the Strata Oil Annual General Meetings. Jacia Pty Ltd was a private company owned by the Stowe family. It was the management company for not only Strata Oil but also Haoma Gold, both companies listed on the ASX and Ric Stowe was a director of both companies. Kevin Parry was a significant shareholder of both companies and sent his lawyer to attend the AGM's and ask very direct and embarrassing questions of the directors regarding these management contracts with Jacia and their inferred conflicts of interest. In my experience, most AGM's were quiet, boring affairs but the Strata Oil AGM's were more fun than a barrel of monkeys!

As Strata Oil and Hughes & Hughes were joint venture partners in the Woodada Gas Field, Strata Oil had developed a business relation with Dan and Dudley and had invested in some of their petroleum opportunities in the USA. These were not performing to Strata Oil's expectations so they sent me to the USA to evaluate the various fields. My first meeting was with Dudley Hughes and his technical team in Beeville, Texas. They were chasing deep gas in the Eocene Wilcox sands, and while they had made discoveries, the gas flows had rapidly declined. The reservoirs were tight and had to be fracced. Put simply, this involves pressuring the sands until they fracture and then pumping fine sand into the fractures to support them and stop them from closing. Sounds easy, doesn't it? But it's not. In those days, the fracture direction was not easy to control and the fractures could go anywhere and not necessarily into the sands. Even worse, they could go vertically down into the water bearing sands below the gas column and that would be the end of your well. Most of the fraccing operations had negative results and this was, in my opinion, the main reason for the poor performance of the Strata-Hughes discoveries.

My next meeting was to be the following afternoon in New Orleans. At the last minute, the joint venture operator contacted my hotel and cancelled the meeting. As I had a firm itinerary and my flights and hotel had already been booked and paid for, I went anyway. That evening I had dinner at Pascal Manale's, a restaurant recommended by the concierge. I was not disappointed. The waiter recommended the jumbo prawns which were enormous and drowned in a delicious spicy sauce which was accompanied by hunks of fresh bread. Fortunately the restaurant supplied a very large bib which saved my clothes and prevented any embarrassment.

My next meeting was with Dan Hughes and his technical team in Jackson, Mississippi. Dan Hughes and his team were mostly chasing the Smackover, a Jurassic limestone sealed by salt and sourced by the worldwide Late Jurassic source rocks. The problem here was mostly lack of exploration success. The traps were gentle anticlines difficult to define by seismic, and the limestones had highly variable reservoir quality. The wells were also deep (4,000 metres) and expensive. Because of the very high hydrogen sulphide content in the petroleum (some of the gas fields have 25% hydrogen sulphide), the wells have to be completed with very expensive chrome plated casing – ouch!

Before continuing, I decided to visit Natchez and catch up with Monte Currie who I had met with Dick Burgess on one of his trips to PPL in Sydney. I hired a car and drove down to Natchez. It was a pleasant drive through miles and miles of open farmland and endless forests. The town of Natchez is on the Mississippi River and it was the first time I had seen this river. We were all taught in schools that this was a mighty river but nothing prepares you for the sight. It is so wide, you can barely see across to the other side. It is also very busy with lots of barges going up and down, although the ones going up were barely moving as they struggled against the current. Natchez is a very pretty town and has an interesting history. It was one of the few towns not razed to the ground by Sherman's infamous and destructive march through the South during their Civil War. So most of the original stately plantation homes were not destroyed and have been meticulously preserved for history.

Now Monte was a bit of a practical joker and decided to play a joke on Mo. During dinner that evening, he rang her and asked, in his best southern drawl, if she was Mrs Warris. She said yes and he went on "I'm Sheriff Adams of Natchez County and I have your husband under arrest" He was so good that Mo was initially taken in until she heard the laughter in the background. Then she told Monte what for after which he handed me the phone. She gave me an earfull too and then we chatted about the home front and how the boys were doing before gaving her a rundown on the trip.

My next meeting was in Wooster, Ohio so I flew to Columbus, hired a car and drove to Wooster. It was November, 1985, freezing cold and the countryside was covered in snow. The play here was the Ordovician Clinton sands but the reservoirs were poor and the gas production had declined to a plateau of around 30,000 cubic feet per day per well. I was shocked – I thought that they had the decimal point in the wrong place! This was not going to make Strata Oil into a major oil company. On my way back to Columbus, there was a thick fog and visibility was barely 100 metres. I was being the good citizen and sat on 55 miles per hour, the then speed limit due to energy restrictions in the USA. But I was the only one doing this speed and most concerning of all were the semi-trailers passing me at 70 miles per hour. In this dense fog, these trucks would have had absolutely no chance of stopping in the event of an emergency. On my return to Perth, I prepared a negative report recommending that they sell their US assets and concentrate on their core area, the Perth Basin. However, they did not take this advice.

ABORIGINAL LAND INQUIRY

In 1983, the Western Australian government decided to carry out an Inquiry into Aboriginal Land Rights in Western Australia. This was in response to the Northern Territory Land Rights Act enacted in 1976. They appointed Justice Paul Seaman QC to conduct the Inquiry. His Report was submitted in September, 1984. Based on this report, the then Burke Labour government decided to prepare a Western Australia Land Rights Act. Peter Purcell and I were APEA's representatives on the drafting committee for this legislation. The first stumbling block in the drafting of this act was the definition of an Aboriginal. In the NT legislation, an Aboriginal is defined as a "member of the Aboriginal race of Australia"; however, the definition proposed for the WA legislation was, and I cannot remember the exact wording, to the effect of a person who had affinity with the Aboriginals of Australia. This of course could be anyone and the definition was vigorously debated. The committee continued for several months but eventually the proposal to have a WA Lands Rights Act was abandoned by the state government.

The main event in the oil industry during 1985 was the discovery by Wapet of the Saladin Oil Field in the Barrow Sub-basin. Saladin-1 encountered a 17 metre gross oil column in the Early Cretaceous Barrow Group sands and tested at 5510 bopd. After drilling another seven appraisal wells, the gross oil column turned out to be 30 metres, recoverable reserves were estimated at 30 million barrels and the field was developed. The field was eventually abandoned in 2014 after producing more than 100 million barrels over three times the original reserve estimate.

The 1985 APEA Conference was held in Perth. There were many good papers at this conference. Chas Lane and Bob Watson gave a paper on the Woodada Gas Field; Alan Tate discussed the deposition of the Dupuy Member of the Barrow Group; Adrian Williams and Don Poynton outlined the geology of the South Pepper Oil Field; Ed Kopsen and Gerry McGann reviewed the potential of the Barrow and Dampier sub-basins; plus there were papers on the Eromanga Basin, Merlinleigh Sub-Basin and western Officer Basin. The most notable social event at this conference was that Recon Exploration put on a breakfast on Rottnest Island. They picked us up in a helicopter at the end of Langley Park and we flew to Rottnest Island, we had a great breakfast and returned to the conference. The only woman at the breakfast was Mo as testimony to her enduring presence at

APEA conferences, and her sparkling personality of course!

In August 1985, it was Mo's 40th birthday. I organised all our friends to meet at the Chinese restaurant on Roe Street in Northbridge on Saturday night. It was BOYG so I organised all the drinks and I also organised a male stripper. I cleared this with the restaurant and they set up a curtain to separate our party from the decent people in the restaurant. We all had a great time with good food and lots of booze, and then the stripper arrived. He was a tall, athletic, good looking bloke and obviously a professional dancer. But he hadn't danced before anyone like Mo before. She played up something shocking and even removed his boots with her teeth. We made so much noise that the other people in the restaurant were peeping round the curtain to see what all the fun was about. It was a very rowdy occasion and for Mo (and the rest of us) it was certainly an evening to remember.

During 1986, Australia produced more oil (660,000 bopd) than it consumed (600,000 bopd) but unfortunately, this did not last long and by 1987, we were in deficit again. Although our production peaked at 800,000 bopd in 2000, it never again exceeded consumption and, with production down to 375,000 bopd in 2013, the gap is steadily getting wider **(Fig 29)**.

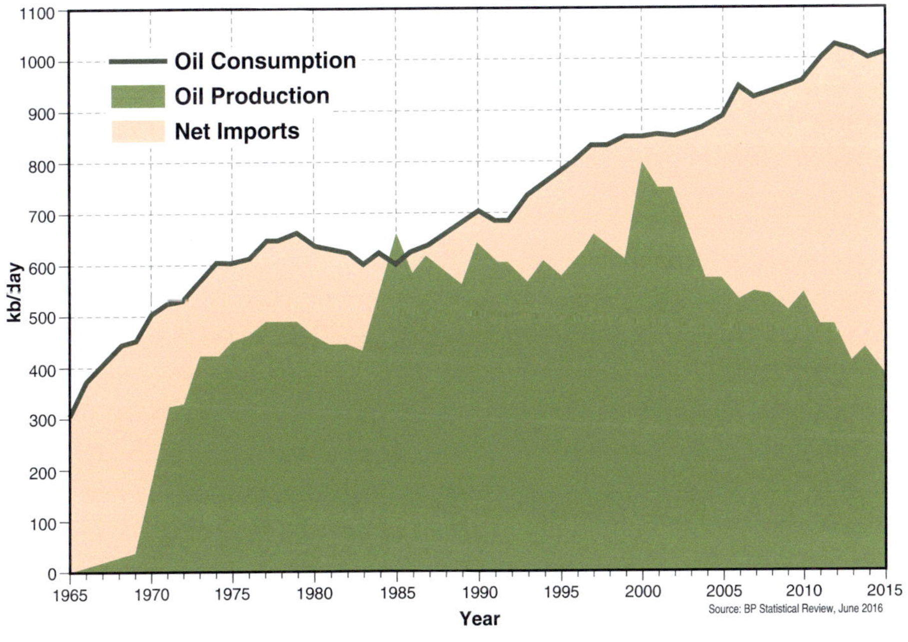

Figure 29
Chart of Australia's oil production

BARRACK ENERGY

I decided to move on from Strata Oil and took on a management position to run a company called Barrack Energy Limited. At the time I came on board, this company was called Western Resources Limited and was a subsidiary of the Barrack House Group. Western Resources Limited had an interest in L7 and the adjoining EP-96 in the onshore Perth Basin. Production Licence L7 contained the Mount Horner Oil Field, and in May 1986, the Barrack House Group took over Western Resources and changed its name to Barrack Energy Limited.

Western Resources also had an interest in PPL 51 in Papua New Guinea and the directors wanted to sell off this acreage. I had to evaluate this permit to ensure that we weren't giving up something that might be very valuable. PPL 51 contained the Puri-1 oil discovery so the permit certainly had excellent exploration potential. Puri-1 tested 1610 bopd but watered out after 23 days. The oil was trapped in the Miocene Puri Limestone sealed beneath the Kuku thrust fault. Potentially there was another compressional anticline to the south of the Puri anticline beneath the Era Thrust. This prospect was complex and very poorly imaged by the poor quality seismic. A company could spend a fortune just to find out if a drillable structure even existed, let alone the cost of drilling. Barrack Energy sold the permit to a new float, Kundu Petroleum, who raised $27.5 million in 1987. Kundu Petroleum did not achieve any success with the permit and the company was delisted in May, 1990.

Mt Horner-1 was drilled by WAPET in 1965 and discovered oil in fine grained sandstones in the lower part of the Kockatea Shale (later called the Arranoo Sands). Mt Horner-2 was drilled by WAPET in 1965 on a separate structure and was a dry hole. Due to poor reservoir quality and low reserves, WAPET relinquished the permit in 1973. In 1976, the area was granted as EP-96 to XLX NL, a company owned by Ian Sykes. XLX acquired a 60 kilometre seismic survey in 1977 and farmed the permit out to Pacific Basin Exploration Pty Ltd who acquired a further 137 kilometres of seismic. Between 1980 and 1981, the Pacific Basin joint venture drilled Mt Horner-3 to -5 and an exploration well Georgina-1. Mt Horner-3 was essentially a re-drill of Mt Horner-1 and was completed in the same sands. Mt Horner-4 and -5 found oil in the Basal Triassic Sandstone. These wells were also put on production but quickly went to 98% water cut. The permit was renewed for a further five years in November, 1981 and the joint venture drilled Mt Horner-6 which was on the wrong side of the fault and a dry hole. The joint venture companies, now including Western Resources Limited, were granted Production Licence L7 in May, 1984. No further worked occurred until Barrack Energy took over the operatorship in 1986.

During the mid-1980's, the international oil price had been steadily declining from a high of $US38 per barrel down to $US26 per barrel and the international oil industry was in depression. In 1985, Chevron announced a takeover offer for Gulf Oil. The oil industry was stunned – not only was this the largest takeover of all time, but Gulf Oil was one of the Seven Sisters! How could one of the Seven Sisters, the feared Titans of the oil industry, be vulnerable to a takeover? It was totally unthinkable! But the takeover went through, thousands of people were laid off - and then there were only "Six Sisters".

Resource Rent Royalty was introduced for Barrow Island during 1986. It was an agreement between the WA state government, the Federal government and Wapet. The Barrow Island Royalty Variation Act 1982 replaced the state well head royalty and the Federal excise system. The old system made Wapet's proposed water flood program for the field uneconomic. The water flood program successfully maintained oil production from Barrow Island and ensured maximum oil recovery from the field. The field is still producing almost 5,000 bopd.

During the early 1980's, the falling exchange rate had kept the oil price in Australian dollars fairly constant at around A$40 per barrel and Australian oil exploration remained strong. But disaster for the oil industry struck in January, 1986. The Saudi Arabian government had been reducing their oil production to balance OPEC production in the face of falling demand. Their revenues were falling and they warned the other OPEC members that this could not continue. These warnings went unheeded and so in December, 1985, they began to increase their oil production to regain their market share. The result of this was that the oil price crashed to $US8 per barrel. The international oil industry was thrown into turmoil. Over the next few years, exploration budgets were slashed, petroleum developments were shelved and hundreds of thousands of jobs were lost, 175,000 in Texas alone. Financial officers were put in charge of the international oil companies which became totally risk averse. But the saddest result was that, in order to save costs, virtually everybody over the age of 45 was made redundant (I was 42 at the time) and all that experience was lost to the industry. The international oil industry had effectively lost its memory. A survey of AAPG members after the crash found that one third of them were unemployed or working outside the industry (ie driving cabs). The oil industry was to totally change and the dividing line between the "old oil industry" and the "new oil industry" is January, 1986.

It is amazing that comets throughout history have been the harbingers of doom for the human race. Guess what? Halley's Comet appeared in the skies during February 1986. Harbingers of doom! It certainly was for the oil industry.

If this wasn't bad enough for the international energy industry, in April, 1986, the number four reactor at Chernobyl in the Ukraine exploded. This resulted in the release of a radioactive plume which spread over the western part of the Soviet Union and Eastern Europe and caused the evacuation of over 300,000 people in the Ukraine, Belarus and western Russia. It was the worst nuclear power plant disaster and set the international nuclear power industry back 20 years. And after the oil price crash in January, popular belief was that the world was awash with cheap oil so who needed nuclear power anyway?

In April, 1986, the APEA Conference was in Adelaide. Following on from the oil price crash, this conference turned out to be "the Last Supper". The oil industry had collapsed, there was no future and everyone was determined to have one last good time as the "Titanic" oil industry slowly sank beneath the waves. There were two salesmen from a contractor (who shall remain nameless to protect the guilty) who (as it was to be their last expense account), bought a case of champagne, filled the bath with champagne, hired two hookers and decided to have a good time. When the hotel waiter came in the next morning to fill the mini-bar, he asked what they had used. They replied "everything" – the mini-bar was completely empty – even the angostura bitters. Apparently, they had a REALLY good time!

But we were not to be outdone. John Schulstad from Western Geophysical decided to invite the old Pancon guys to a lunch for old time's sake. This was at Checkmates, one of those tits and bums restaurants for which Adelaide was so famous. On the way there, we picked up a stray – a young stockbroker who didn't know anybody at the conference, and so he joined us for lunch. It was a great lunch! The food and service was excellent, the girls sexy and the booze flowed. Towards the end of the afternoon, I had a scantily dressed young lady sitting on my lap and I discovered that she was part owner of the restaurant. So I negotiated a deal that if we could fill the restaurant on Tuesday, we could have a free lunch on Wednesday. She took one look at how much booze we had drunk and replied "yes, but food only". I agreed. In the meantime, the young stockbroker was so drunk that he actually offered to pay for half the lunch. John readily agreed. When he got his share of the bill, his jaw dropped and you could see what was going through his mind - "how the bloody hell am I going to explain this to my boss?" When I got back to the hotel, Mo took one look at me and told me to smarten up as we were going out to dinner. Then she noticed the glitter all over my suit. She told me to take off my clothes and threatened "if there is any glitter in your underpants, you're in deep shit!" It was, without doubt, the best business lunch I ever had during my career. During dinner that night and all the next morning at the conference, we told everyone what a great restaurant Checkmates was and gave it a glowing recommendation. Sure enough, Checkmates was packed to the rafters for lunch on Tuesday. The young lady left a message that we could have a free lunch on Wednesday. We didn't take up the offer on the basis that we had such a good time on Monday, that we didn't want to risk spoiling such a wonderful memory.

In addition to these excesses, Recon Exploration put on a flight one evening to see Halley's Comet. The idea was to get above the polluted atmosphere to get a clearer view of the comet. In reality, we were still millions of miles from the comet and the view wasn't much better than on the land.

Following the oil price crash, Barrack Energy decided to concentrate on the Mount Horner Oil Field and not lay me off! If more oil could be found, this field, with its shallow reservoirs and proximity to the Kwinana refinery, could be commercially developed even with the low current oil price of $US13 per barrel. But before we started, Barrack Energy decided to take an acreage position in the northern Perth Basin. During my contract with Strata Oil, I quickly realised that the northern Perth Basin had excellent petroleum potential and was very much under-explored. That was, after all, my recommendation to Strata Oil after evaluating their US assets.

The first area I looked at was a vacant area to the northeast of EP-100 where

I'd worked for Strata Oil. The 272 kilometre P81 Seismic Survey had been acquired over this area by the French company, Aquitaine Petroleum, in 1981. I acquired copies of this seismic survey and although the quality was not great, the Permian limestones were clearly visible and could be confidently mapped. I mapped a large tilted fault block in the centre of this vacant area. I wondered why Aquitaine Petroleum hadn't followed up this rather obvious prospect. They drilled a well called Beekeeper-1 in 1981 to see if the Woodada Gas Field extended into their permit but when this well was dry, they lost interest and relinquished the permit. So Barrack Energy applied for this vacant area and it was granted 100% to Barrack Energy as EP-320 in June, 1986.

Strata Oil was the operator of EP-100, a large permit to the south of Mt Horner. If they weren't interested in exploring the northern Perth Basin, then Barrack Energy certainly was. So Barrack Energy negotiated a farmin agreement with the companies and individuals who made up the joint venture. The individuals included Alan Burnes and Derek Gascoyne (from the old Magnet days), John Geary (a local oil geologist) and Laurie Connell who was to become famous during the WA Inc scandals of the late 1980's. His bank, Rothwells, was cheekily referred to in business circles as the "lender of last resort". It also had the most inappropriate motto in the history of merchant banking "One day all banks will be like ours". Rothwells Bank went bankrupt in 1989!

Barrack Energy farmed into the permit in 1987, acquired over 200 kms seismic and drilled Arramal-1, an attempt to evaluate the old Arrowsmith-1 gas discovery. Arrowsmith-1 was drilled by French Petroleum in 1965 and tested 4 MMcfg/d from the Permian Carynginia Formation. Due to the poor quality of the seismic, Arramal-1 was drilled on the wrong side of the bounding fault and was a dry hole. Gairdner-1 was also drilled in this permit in 1990.

The next permit on Barrack Energy's shopping list was EP-23, a large permit to the east of Mt Horner. The permit was originally operated by WAPET who drilled Erregulla-1 in 1966 and North Erregulla-1 in 1967, both wells recovering oil on drill stem test. WAPET farmed out the permit to Abrolhos Oil in 1972 and they drilled Heaton-1 and Narlingue-1. WAPET then farmed the permit out to Mesa who drilled Erregulla-2 in 1980 and Depot Hill-1 in 1983. Western Mining Corporation ("WMC") acquired all Mesa's Australian petroleum assets (including EP-23) in 1983 and then

WAPET withdrew from permit in 1984. WMC renewed the permit in 1985 for another five years and farmed out the permit to Strata Oil and Petro Energy who drilled East Heaton-1. Barrack Energy farmed into the permit during 1987, acquired 280 kms of seismic and drilled West Erregulla-1 in 1990. This well encountered a ~30 metre gas column in the Late Permian Dongara Sandstone but the sands were tight with only 7% porosity.

The final area on Barrack Energy's shopping list was the production licences L1 and L2 covering the Dongara Gas Field which was discovered by WAPET in 1968. Production licences L1 and L2 were granted in October 1971 and production began in 1971 at 80 MMcfg/d. By 1987, production had declined to 6 MMcfg/d and WAPET was still the owner and operator. However, they hadn't explored the area for years. WAPET was made up of Chevron, Shell, Texaco and Ampolex so (excluding Ampolex) what were these giant international companies doing in the onshore Perth Basin? The Dongara gas field was virtually depleted and given their involvement in the massive North West Shelf projects, these licences had no significant value to them. So we started negotiations to purchase these licences from WAPET. The negotiations were extremely slow as every step in the process had to be referred back to the head offices of Chevron, Shell and Texaco. After several months of this, I eventually lost interest and the negotiations petered out. A few years later, WAPET came back to us enquiring whether we were still interested in purchasing the licences. By this stage, Barrack Energy was totally focussed on exploring and developing our existing Perth Basin projects so we were forced to politely decline.

Barrack Energy needed an operations manager. We hired Ken Bull who had experience with onshore operations in the Cooper-Eromanga Basin. Initially, we had an office at Barrack House with the rest of the Barrack House Group. But after we hired Peter Hall and Rob Kneale as our technical team, we needed more space and moved into a house in Alvin Street, Subiaco as our office.

Returning to Mt Horner, Barrack Energy began a technical evaluation of this oil field. With six wells and reasonable seismic coverage, there was an excellent data base to work with. Mt Horner-3 was producing about 20-30 bopd and Mt Horner-4 and -5 were shut-in. With the oil price being $US13 per barrel, these wells were not going to pay for our overheads. In May-June, 1986, Mt Horner-4 and -5 were worked over to try to establish oil production from the Basal Triassic sands but these operations were not successful. Although these original discoveries were in the Lower Triassic, the overlying Jurassic section was riddled with oil shows and the seismic interpretation indicated a simple anticline on the downthrown side of the main fault. Mt Horner-5 had a thin sand (which we called the B Sand) with excellent oil shows, high resistivities and was interpreted to be oil bearing. During the workover of this well, the B Sand was perforated and flowed at 36 bopd. Based on the seismic interpretation, the structure extended to the northwest of Mt Horner-5 so we decided to drill Mt Horner-7, 450 metres to the northeast of No 5 to evaluate the extend of the oil bearing B Sand. This was effectively a forty acre spacing.

Due to the oil price crash, exploration in Western Australia has virtually ceased. Exploration continued, albeit at low levels, in the eastern states in the Cooper, Eromanga and Surat basins where there was oil and gas production and existing pipelines. So there were no active drilling rigs in Western Australia – they were all in the eastern states. The cost of mobilising a drilling rig from the eastern states was going to add considerable costs to our drilling program on Mount Horner. Ken Bull recommended that we purchase the Bourne 5000 rig. This rig had recently drilled a well for Sydney Oil called Patience-1 in the southeast Canning Basin but after the oil price crash, the drilling company had gone bankrupt and the rig had been abandoned in the middle of the desert. The rig was now owned by the ANZ bank who had it up for sale. Given the state of the oil industry, there were no buyers. So Barrack Energy bought the rig for a song and brought it back to Perth. However, it still required a lot of costly repairs before it was finally made ready for drilling.

The 1987 APEA Conference was held at the Chevron Hotel in Surfers Paradise. Eve Howell and Dave Osborne gave a paper on the Harriet Oil Field in the offshore Carnarvon Basin and Esso gave a paper on the Flounder Field where I had worked as a wellsite geologist in 1969. After the 1986 conference in Adelaide, it was a very subdued affair. The effects of the oil price crash had sunk in and nobody was in the mood to party. Bill Nixon and Mike Voelcker tried to organise a party in their apartment on the 16th floor of a holiday complex. Mo helped Mike organise the food and drinks for the party but unfortunately, there was a series of rolling strikes and power blackouts in the Brisbane-Gold Coast area. The lifts at their apartment building were out of action and everything had to be carried up 16 flights of stairs to get to their room. It was a very quiet party but we had lots of food and drink to consume. Most of that must have ended up back at Mike's farm in Bowraville after the conference.

Mt Horner-7 was drilled in May, 1987 using the newly acquired Bourne 5000 drilling rig. I contracted Dave Osborne as the well-site geologist. It is a sad inditement of the industry at that time that someone with Dave's experience was doing well-site work but it turned out just great for Barrack Energy. When the well got to the Jurassic Cattamarra Coal Measures, the B Sand was missing and we were structurally slightly lower than Mt Horner-5. Things were not looking good. As we drilled on through the Jurassic section, I had a call from Dave that the well had intersected a sand with excellent oil shows and he recommended a drill stem test. I agreed and a drill stem test was conducted over this sand. The test recovered 39 barrels of 36° API oil. I was ecstatic. After my experience at Pancontinental Petroleum with twelve dry or non-commercial wells in its first two years, Barrack Energy had discovered commercial oil in its very first well! Mother Nature sure has a warped sense of humour but this time "we were smiling at the rocks"!

But things were even better. The oil sand in Mt Horner-7 (which we called the F Sand) was structurally lower than the same sands in Mt Horner-4 and Mt Horner-5. So why had Pacific Basin missed this oil sand in these two wells. The answer came when we got the electric logs from Mt Horner-7. The F Sand had relatively low resistivities and conventional log analysis indicated water saturations of 70%. In other words, the well should have flowed water, not oil. There was something petrophysically different about the F Sand. Pacific Basin was an American company and oil field practise in the USA was to drill to total depth, run electric logs and then test any log anomalies with shows. They rarely conduct open hole DST's. In Australia, we were trained to circulate up all drill breaks and conduct open hole drill stem tests over sands with good shows on the way down to total depth. This training, together with an experienced well-site geologist, was why Barrack Energy found the Jurassic oil in Mt Horner and Pacific Basin missed it.

So during 1988, Barrack Energy drilled Mt Horner-4A and 5A and completed these wells for production in the F Sand. At this point, it would be normal to have acquired modern seismic over the field. But the cost of acquiring onshore seismic, particularly as there were no seismic crews in Western Australia, was prohibitive. In addition, the Jurassic structure was low relief and any velocity variations in the overlying sedimentary section could impose serious errors in converting the seismic time maps to depth. So Barrack Energy decided to drill its additional development wells on a 40 acre spacing. Mt Horner-8 was drilled between Mt Horner-5 and -6 and was slightly low to prediction and although a productive well, it indicated the southwest limit of the field. Mt Horner-9 was slightly high to prediction and was the highest and most productive well on the field. Barrack Energy had five producing wells and was now in the oil business. Later, in 1989, Mt Horner-10 was drilled across Tabletop Road and was also a productive well.

In the development of the Mount Horner field, we had to prepare for the disposal of water. The formation water in the F Sand was ~ 4,000 ppm – no good for drinking but suitable for watering trees. So we decided to plant lots of trees not only to dispose of the water but also to conceal the production facilities, camp and sheds. Our first attempt failed as the rabbits ate all the saplings. In our next attempt, we had to protect each tree with wire mesh to keep the rabbits away – a very time consuming and expensive process. Eventually the trees grew around the facilities and only the beam pumps were visible from the public roads. I like to think that the Mount Horner field showed that oil production could exist with farming and the local community without causing significant environmental impact.

The Mount Horner oil was slightly biodegraded and very waxy. It had a pour point of 36°C and had to be heated before trucking to the BP refinery at Kwinana. The wax also built up in the pipelines from the wells to the storage tanks and set up hard as concrete and couldn't be cleaned out. Steel pipelines therefore became too expensive to use so we switched to polypipe which was cheap and could be disposed off after they clogged up with wax and easily replaced. We looked into using chemical additives to reduce the wax buildup but these cost more than the oil itself!

Selling the oil was also an experience as the nearest refinery was at Kwinana and operated by BP. BP imported large amount of oil from the Middle East and were not keen to buy small amounts of oil arriving on their doorsteps by truck. But they had to take our oil so the negotiations with Barrack Energy became an exercise in "plucking the goose until it started hissing". I remember negotiating with a young lady from BP who at least had the decency of smiling sweetly at me as she beat me up on the oil price.

Now that Mount Horner was producing and making money, Barrack Energy decided to sponsor one of the Dongara races. This was to demonstrate that the company was also part of the community and we also contributed to the restoration of the Old Mill in Dongara. After our sponsored race, I had to present the winning trophy and make a speech. I used the opportunity to remind the assembled crowd that Barrack Energy received only 10 cents for every litre of oil it produced. The rest of the cost of a litre of petrol was mostly government taxes! There was a huge roar of outrage from the crowd and I hoped that I had reduced the general belief that high petrol prices were entirely due to the "nasty oil companies".

After the success at Mount Horner, it wasn't all plain sailing for Barrack Energy. Our next three wells were all dry holes. Horner West-1 in L7 in July 1987, Hunt Gully-1 in L7 in November 1988 and Arramall-1 in EP-100 in December 1988.

During this time, one of the best social events on the calendar was Melbourne Cup day. Despite the low oil price, the contractors still put on a great Melbourne Cup lunch at the old Raffles Hotel. This consisted of a great buffet lunch with lots of beer and wine served by topless barmaids. After the race, there was normally a stand-up comedian and lots of "Kevin Bloody Wilson" videos. After lunch, a few of us, including Ken Bull, Bevan Cook and Ron Keenan, would retire to the bar for a few more drinks and oil field stories. It was always a ripper day followed by a taxi ride home after sunset. God I miss the "good old days"!

Barrack Energy had to acquire additional seismic in its various permits and had enough projects to mobilise a crew from Western Geophysical. The first project was in EP-96 to the west of L7. So I went to get permission from the landowner. He was not very amenable. The last seismic survey on his land had left gates open, killed some of his sheep for a barbeque and caused a small bushfire. To my horror, I realised that this was the same seismic survey acquired by Pancontinental Petroleum in 1981. I not only had to keep this a secret from the landowner but also from the Barrack Energy directors or this was going to be my shortest management contract! After considerable negotiations and many warranties, we finally got permission to acquire the seismic. We hired an experienced bird-dog to supervise the seismic survey and set down strict guidelines in the contract with Western Geophysical. At the end of the survey, I personally visited the landowner to make sure he was happy with the survey and our rehabilitation. This was probably the "squeakiest clean" seismic survey ever acquired onshore Western Australia at that time.

After this experience, Barrack Energy prepared a brochure for private landowners setting out what we were trying to achieve, how seismic worked, what was involved and what an oil development looked like if we were successful. This proved invaluable for future surveys on private land where the owners had no legal rights to stop us but still mostly resented our presence. The brochure helped to explain seismic operations and greatly assisted in negotiating access agreements.

Barrack Energy also had to acquire seismic over the tilted fault block in EP-320 which we named the Beharra Springs Prospect. The area of the planned seismic survey was mostly vacant crown land but also included nature reserves. Under the Petroleum Act, petroleum operations could be carried out in nature reserves but the environmental requirements were becoming stricter. We were required to avoid rare plants and not to cut our seismic lines down into the topsoil. In preparing existing seismic lines, previous operators had bulldozed seismic lines too deep thereby removing the topsoil. The native plants did not regenerate in the bare sand and these lines were now covered in smoke bush or weeds. Western Geophysical and Barrack Energy jointly developed operational techniques for acquiring seismic in the Perth Basin to minimise damage to the environmental. We hired a botanist to clear the seismic lines in the reserves for the presence of rare plants. The bulldozer drivers were trained to cut the seismic lines about an inch above the topsoil and the cut bush was pushed to the side of the lines. Where the seismic lines intersected access roads, they were doglegged so that they were not visible to the general public who might use the lines for access and prevent regeneration of plant growth. After the seismic survey was acquired, a backhoe was used to spread the cut bush back over the seismic lines to provide natural seeds for regrowth. This was very effective and after two to three years, the seismic lines had mostly regenerated. After five years, we could barely find them again. Our methods were inspected by the State Government Working Party on Conservation and Rehabilitation of Seismic Lines and were incorporated into the 1990 Petroleum Act guidelines for acquiring onshore seismic. In addition, Terry Grocke (seismic operations consultant), Andy Lane (seismic permitting consultant) and I published a paper "Petroleum Operations in the Environmentally Sensitive Northern Perth Basin, Western Australia" in the 1990 APEA Journal.

However mistakes still did happen. We were to acquire a seismic survey over an area covering a reserve and private land. We submitted our program to the Mines Department and carried out the required botanical surveys over the reserve. The survey was approved and the survey went ahead. However, it turned out that the rarest flora were not in the reserve. A clump of rare banksia trees was on the private land and one of our seismic lines went right through the middle of them. The landowner reported it to Conservation and Land Management (CALM) and we were in more trouble than the Russians in Afganistan. It got into the press who called us "Barrack the Big Bad Banksia Bashers". The bloke who said "any publicity is good publicity" obviously never worked for the oil industry! I got interviewed by rangers from CALM but Barrack Energy was never charged. In Barrack Energy's defence, I pointed out the proposed seismic survey was also submitted to CALM for their approval and they, being botanical experts, should have picked up the threat and insisted that the line be moved. This we could easily have done. An all round screw up by everyone!

Our most interesting area where we wanted to acquire seismic was in the Mount Lesueur reserve in EP-100. It included an area where the Jurassic Cattamarra Coal Measures outcrop indicating a structurally high area where hydrocarbons might be trapped. Barrack Energy needed to get access to acquire seismic over this structurally high area. However, the outcropping coals had created a unique ecological microenvironment where there was an abundance of rare flora. I accompanied a team of environmental officers from CALM, EPA and the Mines Department to inspect the area where the seismic survey was planned. I was amazed at the diversity of the flora in the area including a tree with blue leaves. I thought "Good grief – we're never going to get permission to shoot seismic in here". But by moving the seismic lines around and avoiding the sensitive areas, we eventually got approval to acquire a limited amount of seismic in the reserve.

In 1989, there was a proposal by Canning Resources and Hill River Power Development to develop a 2.5 million tonnes per annum coal mine in the

Mount Lesueur reserve and build a 600 MW power station nearby. Having been on my environmental field trip into Mount Lesueur, I thought "lots of luck guys!" The proposal was rejected by the EPA in 1990 and the Mount Lesueur reserve became a national park in 1992.

TRIP TO CHINA

In January 1987, I was invited through my membership of the American Association of Petroleum Geologists (AAPG) to visit the Peoples Republic of China. The Chinese Ministry of Petroleum Industry (MOPI) had announced that it was opening up ten of the southern and eastern provinces for exploration by foreign companies. MOPI invited a team of North American geologists to visit these provinces and evaluate their prospectivity and I was invited to join this team. These provinces contained nine significant sedimentary basins with working or potential petroleum systems. Barrack Energy approved my request to join the team to evaluate these opportunities.

In August 1987, I flew from Perth to Narita Airport in Tokyo and stayed overnight to join the other delegates. It was a time of protests over the expansion of the airport which was surrounded by guards with batons and rectangular shields and looking like Roman soldiers of a bygone era.

We then flew to Beijing and the following day, we met with representatives of the Chinese National Oil Development Corporation (CNODC). CNODC, a subsidiary of MOPI, had the exclusive jurisdiction over contracts for onshore exploration and development by foreign companies. These meetings were very formal and it was there that I discovered the huge Chinese bureaucracy and the complexity of the various companies that made up the national oil company. We were told that if we wanted an area, each company would have to negotiate a separate deal with the government. There was no standard Production Sharing Contract ("PSC"), as in Indonesia, to use as a guide and therefore negotiations were expected to be lengthy.

During the next three weeks we visited five basins: the Jianghan Basin in Hubei Province, the Subei Basin in Jiangsu Province, the Baise and Shiwandashan basins in Guangxi Province, and the Chuxiong Basin in Yunnan Province. Of these, the Jianghan, Subei and Baise were Tertiary non-marine basins along the eastern side of the country where there was easy access and major markets. They had lacustrine source rocks, mixed fluvio-deltaic and turbidite sandstone reservoirs, and fault controlled structures. The general geology was similar to the Jurassic section in the Perth Basin so this would be a good place to start. The Shiwandashan Basin looked poorly prospective but the Chuxiong Basin looked the most prospective with marine Triassic to Devonian sediments with good oil prone source rocks, carbonate and deltaic reservoirs, and compressional structures. It was however located in the remote Yunnan Province with very rugged to mountainous terrain.

Because of the lack of Western style accommodation, we had to share rooms. By a stroke of good fortune, my room mate was Alfred Chan **(Fig 30)**, a geophysicist from Houston. He was to become our unofficial guide, translator and facilitator for the next three weeks. The most memorable thing abour Beijing (and other large cities in China at that time) was that there were few cars and everybody rode bicycles. The one child policy was also obvious – each bicycle had dad in front, mum sidesaddle at the back and the one child in the basket at the front.

Figure 30
With my unofficial guide

You can't visit Beijing without visiting the Forbidden City and the Great Wall. The Forbidden City is truely magnificent with so many ancient buildings beautifully preserved and covering such a large area. The Great

Wall is an engineering masterpiece. It doesn't follow the topographical contours but goes up the hills and down into the valleys. The brickwork is precise and fits perfectly without mortar. Alfred pointed out that each tower is exactly far enough apart so that the entire wall can be covered by the archers in the towers.

We then drove from Beijing to Wuhan in Hubei Province. The most impressive feature of this trip was that the roads were lined with trees, not just in patches but for their entire length. When there was eventually a break in the trees, all you could see were rice paddies streaching all the way to the horizon.

In 1987, Deng Xiaoping was Chairman of the Communist Perty and he had been slowly introducing capitalism into the Chinese economy following the end of the disastrous Cultural Revolution in 1976. Deng himself spent four years working in a factory in a rural province to be "re-educated" but amazingly he somehow managed to work his way back up to the top and become paramount leader of China in 1978. The results of Deng's reforms were plain to see in the countryside. Before Deng, the state took all of each farmer's produce and then gave back to each farmer enough to live on. Under Deng, the state took two thirds of their produce with the remaining one third being enough to live on plus something left over to barter and improve their standard of living. Under the old system, the farmers planted only one crop per year as there was no incentive to do otherwise. However after Deng's changes, they planted two crops per year thereby doubling their share and allowing them to significantly improve their standard of living. You could still see some fields being ploughed with oxen but many had tractors and harvestors.

The other evidence of Deng's policies was a huge generation gap. The elderly Chinese still wore the Mao style uniforms but most of the younger generation wore Western style dress and were embracing the changes brought by Deng. This looked like trouble in the future which unfortunately came to pass two years later at Tiananmen Square

On the way to Wuhan, we had the opportunity of inspecting some outcrops. To get to the outcrop, we had to cross one of the many rice paddies in the area. Close up, these are marvels of ancient engineering. Each field is about 15 centimetres lower than the next as you move away from the river levees so the irrigation system is gravity fed. Sewerage from the various villages provides the fertilizer. The entire system requires no energy. My next thought was "how do the Chinese acquire seismic in such fragile environments". Later, we discussed this with the district Chinese geophysicists who informed us that seismic acquisition in these areas is acquired by shot hole dynamite and is entirely portable. The drill holes for the explosives are hand augered and the cables and geophones are hand carried across the rice paddies.

The main city in Hubei Province is Wuhan, at that time a city of 20 million people. Compared to sleepy old Perth, it was truely amazing. Most of the people lived in rows and rows of huge, grey apartment blocks that all looked the same. I cheekily asked Alfred "how do they find their way home after a night out at the pub?" He politely reminded me that after 4,000 years of civilisation, the Chinese could do just about anything. Being the height of summer, Wuhan was hot and humid. As the apartments were not air-conditioned, the locals all came out at night to walk around, chat to each other and generally enjoy the warm evenings. We decided to do the same and took a stoll down the main street outside our hotel. The streets were packed and all the young people would come up and talk to us to practice their English.

After Wuhan, we flew to each of the destinations on our itinerary. The first thing we learned about Chinese Airlines was that their planes were always late. Worst still, in the airports, the beer was not kept in refridgerators and was always warm. So before each flight we would get Alfred to make sure the bars put their beer in the fridge. The local Tsing Tao beer was very good but it still needed to be cold! Before the flight was finally called, the beer was beautifully cool. Even after all this, flying in China was still an adventure. On a short hop in southern China, we boarded an old Illushin-18 turbo-prop. Once on board, we were told that there was engine trouble and there would be a delay while the problem was fixed. Two technicians came out with a stepladder and a screwdriver. One of the technicians climbed up the stepladder, unscrewed the engine canopy and began fiddling about inside the engine. After about 20 minutes, he put the canopy back and they returned to base. The pilot started the engine and we took off. Great! I sincerely hoped that the technician didn't have an extra screw in his pocket!

Eating in China was also an experience. One evening we were served pigeon soup. As I started to ladel out the soup, I discovered that the pigeon was still in the pot – feathers and all. One of the American geologists, Chuck Hunnicut who was a bit of a larrikin, lifted the pigeon so that its head draped over the lip of the pot with its tongue hanging out. Everyone had a good laugh but only Chuck, Alfred and I had any of the soup. At another dinner, we had tortoise served all chopped up in its upside down shell – feet, legs and little bits of flippers. Absolutely delicious but again, not everyone wanted to try it out. At the formal banquets, there was lots of toasting with the Chinese liqueur maotai. This was ghastly stuff that I suspected they also used to send their rockets into space. Under no circumstances do you try to smell it – you just gulp it down as fast as you can, shout "Gumbai" and wait for the burning sensation moving down into your stomach.

One of the most spectacular places we visited was Guilin. Around the area of the city are some of the finest examples of karstified limestones you will ever see **(Fig 31)**. Where these are buried beneath the Early Tertiary source rocks in sedimentary basins, these karstified limestones are prolific oil producers. After inspecting them in the field, you could tell why. The limestones have caverns, giant fractures and even tunnels eroded into them. Guilin was also the only place where we saw lots of foreign tourists which was understandable given the spectacular scenery.

Figure 31
Note the huge caverns and fractures

The most interesting basin we were able to evaluate was the Sechzuan Basin in south central China. This basin has a marine Palaeozoic section similar in my opinion to the Permian Basin in North America. Unfortunately it was not available for foreign companies to explore.

After our trip across China, we eventually ended up in Guangzhou (the old Canton). This was another huge Chinese city but very modern and westernised. The hotel where we stayed was absolutely palacial compared to all our previous hotels. We had a final banquet with our Chinese hosts with lots of good food, speeches and the usual toasts with the dreaded maotais. We then crossed the border into Hong Kong and after a farewell dinner with Alfred, I flew home

Upon my return to Perth, I wrote up my notes on the field trip. Based on discussions with the local geologists, it was apparent that production rates per well in the Jianghan, Subei and Baise basins were of the order of tens to hundreds bopd. Given the expected tough terms for any Chinese PSC, these were considered to be unattractive. The Shiwandashan Basin was interpreted to be poorly prospective due to lack of oil prone source rock and effective seal. The Chuxiong Basin was interpreted as the most prospective area with marine Triassic to Devonian sediments with good oil prone source rocks, carbonate and deltaic reservoirs, and compressional structures. The rugged to mountainous terrain was expected to make seismic costs to be very high and this would have to be factored in to any negotiated PSC. I recommended that Barrack Energy purchase the data package for the Chuxiong Basin but following the October 1987 stock market crash, funds were tight and the company understandably did not follow this up.

Back to Barrack Energy

During 1987, there was a strong stock market mining boom. I had been speculating in numerous gold stocks and was looking at a tidy paper profit. After my experiences from the 1971 Poseidon mining boom, I figured this boom was approaching its peak. So in July I sold all of my gold stocks except one. I had subscribed to a new float called Centaur Mining with 20 cent shares each with a free attaching option. As soon as they listed I sold the options which paid for my original investment. Centaur Mining had discovered the Lady Bountiful gold mine in the Kalgoorlie area and their shares had risen to $5 per share. In the meantime, the stock market continued to rise and I began to wonder if I had made the correct decision. But on Monday 19 October 1987, the crash eventually came. I remember it vividly as I was driving to work and had to pull over and stop the car to be sure I heard the radio correctly. The stock market had fallen 500 points in one day and eventually fell from 2300 to 1150, a fall of 50% in only one month. Centaur Mining fell from $5 back to 20 cents per share – I couldn't believe it! But having sold all my other gold stocks at a profit, I couldn't complain. I threw the Centaur Mining share certificates into a draw and forgot about them. A few years later, I noticed that the shares had risen to $2 so I rang my broker and sold them immediately. To my utter frustration, they continued to rise and eventuall got to about $14 per share – what a rollercoaster ride! But you have to be philosophical about such things. If you play the stockmarket, you quickly learn that it is impossible to pick the peaks and troughs with any confidence.

In 1987, Minora Resources was putting together a bidding consortium for the Exmouth Gulf which had been gazetted for petroleum exploration as L87-14. Barrack Energy was invited to join the consortium. The consortium consisted of Minora 30%, Westpet 30%, Giorno 25% and Barrack 15% and its bid was successful and the area was awarded in November 1987 as Exploration Permit EP-325. Soon after the award, the Ningaloo Marine Park was declared and a small part of it extended around the northern end of Cape Range and into the Exmouth Gulf. In a major bureaucratic bungle, this part of the park encroached into EP-325. The part of the park in the gulf was a late addition otherwise the gazetted area would have stopped at the proposed park boundary. In a classic example of Murphy's Law, our best prospect, the Whalebone Prospect, was now in the marine park. This was even more frustrating as we knew there were no reefs in the area of EP-325 as the Ningaloo Reef only occurs on the Indian Ocean side of Cape Range.

The first year's work commitment was to acquire the Whalebone Marine Seismic Survey. Interpretation of the new and existing seismic showed that the Whalebone structure was complexly faulted and needed a 3D survey to mature the prospect for drilling. This certainly didn't happen so this highly prospective structure remains undrilled.

So the EP-325 joint venture acquired the 137 km Cooper Marine Seismic Survey in 1988 over other prospects in the permit. In 1989, the EP-325 joint venture drilled Rivoli-1 which was a small, non-commercial gas discovery in the Early Cretaceous Birdrong Sandstone.

The 1988 APEA Conference was held in Brisbane and at this conference, I presented a paper on the Mount Horner Oil Field. Back in Perth, I also gave a presentation on the field to the Formation Evaluation Society of WA. This centered on the oil bearing F Sand with its anomalously low resistivity, and created a lot of interest and questions.

In July 1988, an explosion on the Piper Alpha platform in the North Sea caused fires that killed 167 people and destroyed the platform. This disaster cost an estimated $US3.4 billion and resulted in dramatic changes to safety regulations for the international oil industry.

In 1988, Barrack Energy was approached by Neil Malloy who was putting together a group of Australian companies to explore for Visean (Carboniferous) reefs in the Hardeman Basin in north Texas, USA. Neil was an interesting character. He was a geologist born in Rhodesia, married to an English girl and living in Dallas, Texas. And now he wanted to operate a consortium of Australian oil companies to explore in the basin. He definitely had a serious identity problem. After evaluating the merits of this opportunity, Barrack Energy decided to join this consortium. The consortium also comprised First Australian Resources whose technical director was then Doug Jendry, and Sydney Oil Company still run by John Conolly.

The main prospects were pinnacle reefs and although small in area, some had thick oil columns. One of our wells, Parker-1 **(Fig 32)**, encountered a 30 metre oil column and tested 2,000 bopd. Unfortunately, it could only be produced at 200 bopd as required by the Texas Railroad Commission ("TRC") who regulated the oil industry in Texas. Although there was plenty of seismic data, the reefs were so small in area, that seismic interpretations were more art than science.

As an example, there was one lease where there were eleven dry holes in one square mile (250 hectares) but in the middle of this lease, there was one well which had produced one million barrels of oil. To an Australian oil geologist, this was simply amazing – "the rocks weren't just smiling, they were laughing at us". Doug Jendry suggested we write an APEA paper and stick this 250 hectare lease with its million barrel oil field in the middle of a map of the Canning Basin but it was all too depressing to even think about.

Figure 32
Parker-1 drilling in the Hardeman Basin

Why does the TRC regulate oil production in Texas you ask? The TRC started regulating oil and gas production in Texas in 1919 as it owned a lot of land with attached mineral and oil rights. Its powers increased as a result of the East Texas oil boom in the 1930's where overproduction caused oil prices to crash to 25 cents per barrel – ouch! It was legislated to conserve the petroleum resources of Texas by regulating all oil and gas production in the state.

In 1989, Neil organised a field trip to the Hardeman Basin. It was located in the northern part of Texas in the "Bible Belt". These counties were "dry" – you couldn't buy a drink in a hotel or restaurant but only in the local clubs where you had to be a member. But Neil was well organised and there was plenty of beer in an esky in the boot of the car. Transporting alcohol across county lines in the US may have been illegal but that never stopped a bunch of thirsty geologists. Hardeman County was a very depressed area of the USA. The smartest buildings in the towns were the churches. The soil was poor and the countryside flat and boring. But it was great to see all the "nodding donkeys" hooked up to their oil storage tanks.

Back in Dallas, Neil took us to the Million Dollar Saloon for a few drinks. This was a very posh girlie bar for rich businessmen - but they let us in anyway. Gorgeous scantily dressed girls came and danced on your table and you stick dollar notes in their garters. Someone bribed one of the girls to take off my glasses and stick them down the panties of one of the girls dancing on the stage. Now I'm blind as a bat without my glasses so I had to blunder my way down to the stage - all to the accompanying cheers of the assembled crowd. Fortunately one of the girls on the stage handed me back my glasses which prevented further embarrassement. All and all, it was a great night out on the town but I still haven't found and killed the bastard who bribed the table dancer. When I got back to Perth, Neil had sent a fax saying that all the girls in the Million Dollar Saloon were asking "When will the Aussies be back in town"!

Up until December, 1989 when Barrack Energy's US interests were sold, the Hardeman Basin Joint Venture drilled 26 wells for nine oil discoveries.

On the way back from one of my trips to Dallas, I detoured via Calgary to look at some prospects being farmed out by Bill Nixon. He had his own oil company Texalta Petroleum which had oil production in Alberta and Saskatchewan. As part of this evaluation, we drove to Banff and inspected some very impressive outcrops of the Upper Devonian D3 Limestone which is the prolific oil producer in Alberta **(Fig 33)**. As well as the spectacular views, the incredible vuggy and fracture porosity in the limestones can be clearly seen in these outcrops.

Figure 33
Inspecting porosity in the D3 Limestone

Back in Perth, the board decided that I should tour the Barrack Mines gold operations to see how other parts of the corporate group operated. The nearest town to the mining operations was Wiluna which was about as far from civilisation as you can get. The town was in the most unforgiving desert I had seen since my well site work in the Simpson Desert and was surrounded by old mining slag heaps (which made it considerably worse than the Simpson Desert). After checking into the motel, we headed for the local pub for a few beers. As I approached the pub, Ken Bull said to me "you can't go in there – that's the Aboriginal's bar – we go into the white man's bar". Good grief I thought – Apartheid is still alive and well and living in Western Australia! When we entered the white man's bar, we were separated from the other bar by a wall of steel mesh. It then occurred to me that Wilcannia wasn't such a bad place after all!

In October, 1988, the storage tanks at Mount Horner caught fire due to an electrical fault. The tanks were insured but Barrack Energy didn't have business interruption insurance. It took several months to replace them and get production back to normal and in the meantime, the company had no cash flow.

The APEA Conference for 1989 was held in Hobart. For me the most interesting papers were on the Saladin Oil Field in the Barrow Sub-basin and an entertaining paper by Barry Goldstein on the Canning Basin. After the conference, Mo and I hired a car and we went on a tour up the east coast of Tasmania. It had been many years since I'd toured Tasmania and that was in the northwest corner on the way to an offshore rig. We first visited Port Arthur which was very sobering – a very cruel part of Australia's history. But we enjoyed the drive up the east coast to Launceston. Tasmania is a beautidul island and we had a very pleasant and relaxing holiday. We flew back to Perth via Melbourne – this time without any stuff ups by the airline!

At the end of 1989, I was informed by the board that I was being replaced as General Manager and would become the exploration manager for Barrack Energy. I was blamed for the fire at the Mount Horner storage tanks and the lack of business insurance but they also wanted someone with more financials credentials to run the company. Although Barrack Energy was listed, it was financed by the Barrack House Group. Not being on the board, I was not aware that the Barrack House Group was in financial trouble. There were rumours in the marketplace but as we had not been told to reduce our exploration programs, we assumed that these were unfounded.

As the board of Barrack Energy no longer had any faith in my ability, I immediately resigned. The company kept me on as a consultant to hand over to David Maxwell, the new General Manager. In any event, the end was near.

During 1990, Barrack Energy had an intensive drilling program. We drilled North Yardanogo-1 in EP-320 during March and this well tested 1200 bopd from sands in the Jurassic Cattamarra Coal Measures. This was followed by Beharra Springs-1, also in EP-320 and drilled during March-May. This was my pride and joy and I'd waited four years to see it drilled. It was a tilted fault block where the fault throw was about equal to the thickness of the sealing Kockatea Shale so it was touch and go as to whether the fault would seal. It took a couple of rounds of detail seismic before the decision was finally made to drill the prospect. Beharra Springs-1 was drill stem tested at 8.5 MMcfg/d from the Late Permian and later, in June production tested at 21.9 MMcfg/d on a ½" choke. I was over the moon and "smiling at the rocks"!

The next well was West Erregulla-1 in EP-23 during May-June. This well encountered a 30 metre gas column in the Late Permian Dongara Sandstone but the sandstones were tight with less than 7% porosity. The next wells were Gairdner-1 in EP-100 during October-Novermber and South Yardanogo-1 in EP-320 during December – both dry holes.

The Barrack House Group had made some poor investments, the most damaging being a participant in the building of QV1, a large office tower in the city of Perth. During the late 1980's, recession and high interest rates meant low business activity and QV1 could not be leased. The group was being funded mostly by gold production from Barrack Mines. But when the sides of the open cut Horseshoe Lights Gold Mine collapsed, mining ceased and cash flow from Barrack Mines also dried up. The Barrack House Group became insolvent during mid 1990.

This was a disaster for Barrack Energy. The company was forced to sell a 66-2/3% interest in its recent gas discovery at Beharra Springs to Sagasco Limited in August 1990. In addition, the company itself was up for sale. I was beginning to think that I was a jinx. First Pancontinental Petroleum, now Barrack Energy. Both were technically successful companies but for economic reasons beyond their control, they were destined not to survive. To paraphrase a famous saying "to lose one company is careless but to lose two is positively irresponsible". I have looked back over my career to analyse whether I could have done things differently. In the case of

Pancontinental Petroleum, this was caused by a series of economic events totally beyond my (or the board's) control. In the case of Barrack Energy, in hindsight, I should have gone on the board. I may have been able to see the truck coming. But since the Barrack House Group controlled the board, I seriously doubt that I could have changed anything. Barrack Energy was a technically successful company with a terrific technical team and was, in my opinion, destroyed by the demise of its parent company.

Despite the disaster of Barack Energy, I still kept a strong belief in the Perth Basin. It was a proven petroleum province but was very much under-explored. In 1990, the northern Perth Basin was covered by only 6,400 kilometres of seismic or only 0.4 kilometres per square kilometre. There were only 96 exploration wells or a mere 6 wells per square kilometre. Compare this to the Cooper Basin where there was 120,000 kilometres seismic or 3.4 kilometres per square kilometres, or over eight times that of the Perth Basin. Drilling density was 16 wells per square kilometre or almost three times that in the Perth Basin.

As a footnote, Barrack Energy Limited changed name to Arrow Petroleum Limited in December 1990 and was eventually taken over by Discovery Petroleum Limited in August 1992. Eric Streitberg, the CEO of Discovery Petroleum, was the only one in the oil industry with the guts, determination and perseverance to deal with the Barrack House Group and he (quite deservedly) reaped the rewards. Discovery Petroleum drilled Mt Horner-12 and -13 and boosted production to 860 bopd in 1991. The field produced its one millionth barrel of oil in January, 1994. The Beharra Springs Gas Field began production in January 1991 at about 3 TJ/d using a temporary gas plant. After a new gas plant was commissioned in August 1992, production reached 18 TJ/d in 1993 and the field ended up producing over 77 BCF gas. It's enough to make a grown oil man cry!

In 1988, the Soviet Union abandoned its nine year war with Afghanistan and withdrew its forces. Although Mikhail Gorbachev had come to power in 1985 and had made changes to the economy, the Soviet Union's economy was in bad shape. The low oil prices of the late 1980's had crippled its oil industry and depleted its foreign exchange reserves. However, it still controlled a vast empire and wielded considerable global power and influence. In 1987, Gorbachev introduced "perestroika" a restructuring of Soviet society to be more open and democratic. By 1989, these reforms were out of control and spreading with violent protests in the Central Asian and Baltic republics. In June 1991, Boris Yeltsin became President of the Soviet Union and, following a failed attempted coup by the KGB in August 1991 the Soviet Union collapsed and became the Russian Federation. Having spent my entire life following the highs and lows of the Cold War, this was a great historic moment. After a 45 year struggle, we had finally won. However, we were all stunned by the speed of the collapse and break-up of the Soviet Union. I also thought that this little blue planet of ours might now get some peace at last. How wrong was that!

In May, 1990, we went on a very memorable gold fossicking and camping trip to Coolgardie. To do this legally, you needed permission from the company that had the mineral rights to the area. Bob Berven gave me the name of a company which allowed us to fossick for gold on one of their leases. We were joined by the Svalbe's and the Schulstad's so there were six adults and six children (five boys and one girl). We borrowed Terry Grocke's 4WD vehicle and headed off to Coolgardie. We all met in Coolgardie and headed out to the lease and set up camp. Due to the numerous deep mine shafts in the area, the kids were not allowed to wander off too far from the camp site and not at all at night. The kids all camped together and set up their own camp fire. There was plenty of dead wood scattered around so fire wood was in plentiful supply. Andy Svalbe brought along a metal detector to help find the gold nuggets that were going to make us all fabulously wealthy. However, the technology of the day was not up to the task and all we found were lots of tin cans, rusty nails and bottle tops. John Schulstad even buried his gold ring and asked Andy to find it. It didn't register on his metal detector and it took us all about ten minutes to dig around and find it again.

The return home was complicated as I had to suddenly return to Perth on business and flew back. Mo couldn't drive a manual car so Chris drove back with his provisional licence at 80 kph.

The APEA Conference for 1990 was held for the first time in Darwin. At this conference, I presented a joint paper with Terry Grocke and Andy Lane on "Petroleum Operations in the Environmentally Sensitive Northern Perth Basin, Western Australia". After the conference, we did a trip to the Kakadu National Park. Unfortunately, it was the wrong time of the year. The park was being controlled burned and was a smouldering ruin – very disappointing. But the wetland areas in the northern part of the park were fantastic. We took a tour of the rivers and were impressed by the variety of bird life and abundance of sleeping crocodiles. The boat put on a display of feeding a crocodile. They hung a piece of meat over the side and a crocodile raised itself out of the water by wagging its tail and grabbed the meat. It was the first time I had been close up to a crocodile and they have to be one of the ugliest and nastiest looking life forms on this planet. But they survived the end Cretaceous extinction event so they can't be all bad!

On the international front, in 1990 Iraq accused Kuwait of over-producting its OPEC quota thereby keeping the oil price low and costing the Iraqi economy an estimated $US14 billion a year. There were also disputes with Kuwait over its share of the giant Rumalia Oil Field which straddles the border of the two countries and the exact location of the border in the Shatt al Arab waterway entering the Persian Gulf. After much acrimonious and futile debate, Iraq invaded Kuwait in August and quickly conquered the country. The oil price shot up to $US40 per barrel. If Saddam Hussein had left Kuwait immediately after looting enough of Kuwait's gold reserves to pay for the inferred damage to Iraq's economy, he would have become the hero of the Arab world. Instead he stayed and the Western world was terrified that Iraq would move on Saudi Arabia. Their oil fields extended from the Kuwaiti border some 400 kilometres southeast to Gharwar, the worlds largest oil field. The UN Security Council set a deadline of 15 January for Iraq to withdraw from Kuwait and the US Congress and Senate narrowly voted their approval just three days before this deadline. Iraq ignored the deadline and the Gulf War began on 18 January. Following areal bombardments 40,000 troops from 30 countries invaded Kuwait. Two days later what was left of the Iraqi Army withdrew but not until after looting the gold from Kuwait's Central Bank and setting fire to over 600 oil wells. The result was like something out of "Dante's Inferno".

As well, the army placed land mines around each well head and it was estimated that it would take two to three years to extinquish them all. However, a major international effort, costing the Kuwait government a further $US1.5 billion, was undertaken starting in April and all the wells were extinquished by November only seven months later. The oil price quickly dropped back to $US20 per barrel. The most innovative of the oil fire teams were the Hungarians who used Mig 21 jet engines mounted on old T34 tanks to simply blow the fires out. How neat was that!

As well as the cost of putting out the fires, there was also the value of the oil that was burnt. It was estimated that initially five to six million barrels oil was going up in smoke each day resulting in a total loss of about one billion barrels oil allowing for the number of burning wells reducing over time as they were extinguished. At the then oil price of $US40 per barrel, this cost Kuwait an additional $US40 billion in lost revenue.

OIL POLLUTION

As well as the oil fires in Kuwait, an estimated eight million barrels of oil was also pumped into the Persian Gulf by the retreating Iraqi forces and it was feared that this would cause an environmental catastrophe and devastate the gulf for decades. The oil slick covered about 10,000 square kilometres and was about 12 cms thick. It was the worst oil spill in modern (not geological) history. However, by 1993, environmental studies showed that the gulf was back to normal and no permanent damage had been caused. The reason for this is that crude oil, contrary to popular belief, is actually biodegradable. It is estimated that over 600,000 tonnes (~ 12,000 bopd) of oil leaks naturally up faults and fractures into the oceans each year. So if oil wasn't biodegradable, it would have covered the oceans and exterminated all life within it, millions of years ago. Fortunately, over millions of years, bacteria have evolved to break down the oil that naturally leaks into the oceans and convert it into carbon-rich nutrients which actually provide food for the microscopic marine organisms. Of course, a large oil spill on the surface of the ocean will kill fish, marine mammals and birds, and this is to be avoided at all costs - but eventually the bacteria will do their work. It can be argued that the worst thing to do with a man-made oil spill is to spray detergent on it. The detergent kills the oil eating bacteria and ensures that the oil will not break down naturally. A man made oil spill should be contained and prevented from coming ashore but then left in the ocean for the bacteria to do their job and break it down.

The Torry Canyon is a classic example of what to do and what not to do with a man made oil spill. It ran aground in the English Channel in March 1967 with a cargo of 120,000 tonnes of oil (that's just over two months natural leakage into the oceans). As well as spraying detergent on the spill, British bombers dropped incendiaries and napalm on the floundering vessel. All that just made it worse. Fortunately (and I use the word loosely) most of the oil ended up close to the Brittany coast where the French dropped chalk on it sinking it to the bottom where the oil eating bacteria got to work. Brittany seems to be a dangerous place for oil tankers because in 1978, the Amoco Cadiz, carrying 220,000 tonnes of crude oil ran aground there and sank before its cargo could be unloaded. Back to the Torrey Canyon where unfortunately, some 3,000 barrels washed up on the coast of Guernsey where the locals pumped it out of the water and transferred it to a local quarry. The oil is still there today, stinking the place out and killing birds some 45 years later. All attempts to destroy the oil have failed since the oil cannot be dug out because the quarry was a German ammunition dump in WW2. Talk about Murphy's Law! In 2010, they started pumping microorganisms into the quarry to break down the oil. Let's hope they have the right ones!

CONSULTING

During late 1990, I bought my first personal computer, an Apple MacIntosh. Nicholas Lee, Jean's eldest son, came out to Perth to help me with this new technology. It fact, it was very easy to use with a "mouse" replacing the need to type in the numerous command functions. He set up various programs to catalogue my wellsite data and other geological information. Whilst he was here, we all took a break for a few days on Rottnest. Nick and I went snorkling at Little Parakeet Bay and he was amused by "Lake Bagdhad" inland from the bay since the war with Iraq had just finished.

I kept this computer for many years as I mostly used it as a word processor for my correspondence with clients and didn't need any fancy new programs. It eventually crashed with a spectacular puff of smoke. When I took it into a computer shop to get it repaired, the assistant got very excited and called in all the other computer nerds from the back room. Whilst in my possession, it had apparently become some sort of archaeological relic and was definitely the centre of attention. Alas, it was completely dead so I had to buy a new one. When I got the new computer back to my office, I was stunned by how fast it was and how much more memory it had.

The main discoveries during 1990 were both in the offshore Carnarvon basin: the Cossack Oil Field by Woodside Petroleum and the Griffin Oil Field by BHP Petroleum. The Cossack field was developed with the Wanaea field in 1995 for $740 million, six weeks ahead of schedule and $23 million under budget. A far cry from development costs these days! It was Woodside's first oil production and both fields have produced over 360 million barrels to date.

The Griffin field was developed for $600 million and began production in 1994. It ended up producing over 150 million barrels oil.

Doral Resources Limited

After the demise of Barrack Energy, I took on a number of consulting jobs for various clients. Although the oil price was low, the oil industry was still ticking over and companies, both Australian and international were stll exploring.

In mid 1990, I signed a consulting contract with Doral Resources Limited. The exploration manager of Doral Resources was a young geologist called Mark Thompson and I was to provide technical assistance to him in the running of the company's petroleum operations. Doral Resources had acquired the old Strata Oil in 1987 so here I was working for them again in a new life. The Australian oil industry is so small!

Doral Resources operated the Tubridgi Gas Field in the onshore Carnarvon Basin, just southwest of Onslow. This field was originally delineated by Jim Allender and Hugh Crocker based on the gas show in the Birdrong Sandstone in Onslow-1. They reprocessed two old seismic lines for the shallow section and these showed a strong amplitude anomaly over a broad anticline. The prospect was farmed out to Otter Exploration who drilled Tubridgi-1 to -6 during 1981-82. However, due to the remoteness of the field and the requirement that Western Australia's domestic gas supplies had to come from the North West Shelf, production didn't begin until 1991 via a pipeline to the Dampier to Bunbury natural gas pipeline. In 1989, Doral Resources purchased a majority interest in the Production Licence L9 and took over operatorship.

The main difficulty with the Tubridji field was converting the seismic time maps to depth. The gas was trapped in the Early Cretaceous Birdrong Sandstone at shallow depths of 900 metres. Some of the gas had seeped naturally up into the overlying Gearle Siltstone thereby lowering the velocities in the sediments overlying the field. This had the effect of showing a time low over the field so much so that Wyloo-1 was drilled on what was thought to be a separate structure. As if these low velocity gas zones were not enough of a problem, the overlying Tertiary limestones had high velocities which varied laterally across the field. With the assistance of Dean Powell, the seismic over the field was reprocessed and detailed velocity maps produced with the view of creating more accurate depth maps.

After a comprehensive technical evaluation, it appeared that the best locations to drill the next wells were in the centre of the time low. This was a very brave decision – I was a consultant with nothing to lose but Mark put his job on the line. Tubridgi-7 and -8 were drilled in September-October 1990 and were the highest wells on the field at that time. We all sighed with relief and "were smiling at the rocks".The field began production in 1991 after an agreement was reached with SECWA but deliveries of gas were restricted to 22 MMcfg/d due to the high nitrogen content in the gas. Doral drilled Tubridgi-9 and -10 but eventually sold its interest in L9 to Origin Energy in 1995. Origin drilled Tubridgi-11 to 18 during 1997-99 and by end 2000, cumulative production was 58 PJ. The field was sold in 2004 to BHP Petroleum who wanted to use the facilities to assist in processing and transporting the gas from its offshore Macedon Gas Field.

Doral Resources was also in a joint venture with Ampolex in an exploration permit EP-321 on the eastern flanks of the central Perth Basin. Ampolex had 60% and was the operator with Doral having 15% and Sagasco 25%. I was very impressed with Mark Thompson – for a young geologist straight out of university, he handled himself well in joint venture meetings and carefully scrutinised the joint venture accounts, picking up the odd error. The EP-321 Joint Venture drilled Barberton-1 and Yallallie-1 at the end of 1990 – both dry holes.

Doral Resources was also in a joint venture with Pan Pacific in EP-110, the exploration permit adjoining the Tubridgi production licence. Pan Pacific was the operator and drilled Abdul's Dam-1 in 1991 and Picul-1 in 1992

– both dry holes. Of these, Abdul's Dam-1 was very interesting as the Birdrong Sandstone was absent and the Triassic Locker Shale was directly overlain by the Early Cretaceous Muderong Shale. This combination set up a potential stratigraphic trap for the Birdrong Sandstone downdip from the well. A later joint venture operated by Carnarvon Petroleum drilled two wells, Sapphire-1 and -2 chasing this play but they were both unsuccessful.

It was with Doral that I had my first experience with using a computer for seismic interpretation. It was then a very tedious process compared to today's powerful workstations. You had to pin the hard copy interpreted paper seismic section to the digitising table and then calibrate the two way time scale on the left and the shot point numbers along the top. After setting everything up, you followed each mapped horizon and punched in the Two Way Times every so many shot points depending on the scale. You desperately tried not to make a mistake – otherwise you had to start all over again! But once everything was in the computer, with the press of button you could print off time and depth maps for each horizon and the relevant isochrons.

Through my consulting company, Gulliver Productions Pty Ltd, I wanted to start acquiring exploration acreage in Western Australia. I convinced Doral Resources to be the lead company and operator for the applications for new permits. Indigo Oil Pty Ltd, owned by Keith Thompson, also joined the consortium. He had moved with his family to Australia and now lived in Adelaide. The first area we bid on was in the Gregory Sub-basin of the Canning Basin. A lot of seismic had been acquired over this area by Shell and Mobil Oil in the early 1980's and a number of large anticlines had been defined. Only one, White Hills-1, had been drilled as all the permits were relinquished after the oil price crash in 1986. White Hills-1, drilled by Mobil Oil in July-December, 1982, had ~ 1,000 metres of oil shows through the Carboniferous Anderson and Fairfield formations. Like the Fitzroy Sub-basin, most of the large compressional structures were formed during the late Triassic and post-dated primary hydrocarbon migration during the Permian – the main reason why they are all dry holes. However, some of the anticlines showed structural growth and may have been affected by the mid-Carboniferous compression which affected the Amadeus Basin to the east. We applied for a large permit and EP-354 was granted in April, 1991.

The 1991 APEA Conference was in Melbourne but we did not go. It was our 20th Wedding Anniversary in June and we went to Dunk Island for a celebration. We had a great time especially as the hotels on Dunk Island were absolutely palatial compared to what was there in 1971.

The main discovery during 1991 was the 80 million barrel Wandoo Oil Field in the offshore Carnarvon Basin. The Wandoo-1 oil discovery was 2.5 kms from Enderby-1, drilled by Woodside Burmah in 1970. This well also had oil shows in *M. australis* Sand but these were ignored as no structure was visible on the old 1970's seismic dominated by multiples. I will talk about this field later.

The 1992 APEA Conference was held on 24-27 May in Perth. The conference was opened by the then Premier of Western Australia Dr Carmen Lawrence. In her address, she referred to EP-325 in the Exmouth Gulf stating that the rights of the holders of tenements granted under the Petroleum Act and the Petroleum (Submerged Lands) Act prior to the granting of marine parks and reserves will be maintained. The reality for the EP-325 Joint Venture turned out to be somewhat different. On the technical front, Mark Thompson presented his paper on the Tubridgi Gas Field at this conference.

Doral Resources was in joint ventures with WMC and Cultus in WA-128-P and NT/P 43, both in the offshore Bonaparte Basin. WA-128-P contained the Turtle oil discovery. Turtle-1 was drilled by WMC in 1984 and recovered light oil from Permian sandstones. Turtle-2 was drilled in 1989 and failed to find oil in the Permian but had excellent oil shows in the Carboniferous. Attempts to test these oil shows failed. Across the border in Northern Territory waters were the Barnett oil discoveries. Barnett-1 was drilled in 1985 and Barnett-2 in 1988 both by Australian Aquitaine Petroleum, and Barnett-3 was drilled by Santos in 1990. Although these two oil fields were very tantalising, Mother Nature put all the light oil in the poor reservoirs and all the heavy, viscous oil in the good reservoirs and they are still not commercial today. She really is a bitch but at the end of the day, "the rocks are smiling at us"!

NT/P 43 was farmed out to BHP Petroleum. The Farm-in Agreement was fine but Doral had lots of issues with the Joint Operating Agreement. The main issue was that there wasn't a Non-Consent clause. We discussed this with BHP Petroleum who politely told us that the agreement was not negotiable or the farm-in was off. We made it very clear to the Doral board that if the farm-in well was dry, that Doral should withdraw from the permit as soon as possible before incurring any on-going BHP Petroleum operating permit management costs. Billawock-1 was drilled in September 1992 and was plugged and abandoned as a dry hole. Doral immediately withdrew from the permit.

In May, 1992, Doral Resources was taken by Devex, a company controlled by George White. The company was merged with Carnarvon Petroleum (formerly Metana Petroleum) in October 1992. My time with Doral Resources was over.

Point Torment Project

At the PESA Deal Day before the 1990 APEA Conference in Darwin, Ron Prefontaine presented a farmout on EP-104 in the Canning Basin for Oil Company of Australia ("OCA"). I was intrigued by the geology of the Point Torment structural-stratigraphic trap along the Pinnacle Fault. Early Carboniferous deltaic and marginal marine sands in the Anderson Formation may potentially have trapped hydrocarbons against the Pinnacle Fault in a large three way dip closed structure. These sands were oil bearing in the West Kora-1 well to the east. This well was drilled by Esso in 1985 but the low production rates and high water cuts were not commercial at the then low oil price.

One of the amazing things about this very remote area was the large amount of seismic acquired by Esso in the mid 1980's. The area consisted of a huge tidal flat with deep tidal channels covered with bands of mangroves – not the most friendly of environments to acquire land seismic surveys. These surveys were acquired using large balloon tyres on the vibrator trucks, helicopter support and guards with high powered rifles to prevent the salt water crocodiles from eating the seismic crew! Production rates got up to three kilometres per day, not bad considering but the surveys must have cost a fortune even by Esso's standards.

I used my consulting company Gulliver Productions Pty Ltd to negotiate the right to farmout the permit on the basis of farminees paying 100% of the well cost to earn an 80% interest leaving OCA with a 20% interest in the permit. I then put together a consortium of companies including Anzoil

32%, Stirling Resources (named changed from the Magnet Group days) 10%, Basin Oil (a subsidiary of Cultus) 10%, First Australian Resources ("FAR") 8% and Austin Oil 8% to fund the drilling of the well. Gulliver retained a 2% interest, carried through the first well, in compensation for putting together the geology and hydrocarbon potential evaluation and negotiating the deal. Anzoil was appointed the operator and the Point Torment-1 well was drilled in November 1992 using OD&E Rig 27. The well was originally drilled to test the upper part of the Anderson Formation where oil was recovered in West Kora-1. This section was water bearing and the well was to have been plugged and abandoned. However, Stirling Resources and Basin Oil sole-risked the deepening of the well. The well was deepened into the lower Anderson Formation and discovered gas in deltaic sands in this section. An open hole DST over a three metres sand flowed at 4.3 MMcfg/d but pressure data indicated a depleting reservoir. However, two higher sands with good gas shows were not tested as these were penetrated before a DST could be undertaken.

Anzoil also had acreage in Vietnam and the company was taken over for these assets. The new management had no interest in the Canning Basin so Anzoil's interest in EP-104 was transferred 24% to Stirling Resources and 8% to Kiwi Australian Resources with Stirling (now with 34%) becoming the new operator. Basin Oil bought OCA's 20% interest increasing its equity to 30%. The balance of the permit was held by FAR 8%, Austin 8%, Kiwi 8%, Indigo 5.4%, Pelsoil 4.6% and Gulliver 2%. Point Torment-1 was deepened in November 1994 using Drillcorp Rig 29. The well continued to encounter gas shows through the Lower Anderson Formation but the sands were fine grained and tight, and on testing, did not flow at commercial rates. The well drilled down to the Late Devonian and encountered good oil shows at the top of these limestones. These oil shows were tested but only flowed gas at very low rates.

West Kora Project

As well as drilling Point Torment-1, the EP-104 Joint Venture also decided to re-complete the West Kora-1 well in the eastern part of the permit. The original well encountered three sands over the interval 1735-51 metres in a marginal marine section at the top of the Anderson Formation. Following a successful open hole DST, the interval was production tested at 50 barrels fluid per day of which 60% was oil. Esso put the well on production in 1989 and ceased production in 1991 when water cut had reached 85%. Cumulative production from the well was 24,000 barrels oil.

Upon reviewing the well, my interpretation was that the lower of the three sands may have been water bearing and contributing to the high water cut. In addition, there was a shallower sand at 1693-96 metres which had high resistivities and may have also been oil bearing. The well was re-entered in June 1992, the upper sand was perforated and a through tubing bridge plug installed to block off the lower perforations and minimise water flow. The West Kora pad was in the middle of the tidal mud flats and some 1.5 kilometres from the hard ground and storage tanks. A small pipeline made out of 2-7/8" tubing ran some 2.5 kilometres from the well to the storage tanks on the hard ground. The joint venture rented a beam pump from OCA and began production from the well under an Extended Production Test issued by the Mines Department. The well produced intermittently from November 1992 to January 1998 and a two block Production Licence over the West Kora structure was applied for in December 1998.

The joint venture negotiated a deal with Santos Limited who at this time was operating the Blina production licences to the east. Santos would pick up the crude oil from the West Kora storage tanks and transport it along with their own oil from the Blina area to the offloading facilities at Broome. Subject to production rates and oil prices, the tariff was around A$9 per barrel.

During the monsoon season, the West Kora pad, being on the tidal mud flats, was surrounded by water and the production operator could not drive out to the well to refuel the diesel motor that ran the beam pump. So after the first monsoon rains, he had to row out to the location in a small boat. As he approached the pad he wondered "why is the pad moving"? As he got closer he realised to his horror that the pad was covered with king brown snakes taking advantage of the only dry land in sight. As he wasn't being paid enough money to take on a nest of king brown snakes who would not take kindly to any intruders onto their small but dry sanctuary, he wisely turned the boat around and returned to shore. Life is never dull in the onshore oil industry! Production re-commenced during the following dry season.

In November, 1992, Mabel arrived in Perth on one of her many visits to Australia. Unfortunately, she got sick on the plane and was in a bad state when she arrived. Bob immediately hopped on a plane and came out to help. We all looked after her until our doctor confirmed that she had stabilised. In March, Mo accompanied her to Singapore and assisted her with her connection to London. It had been a traumatic experience for Mabel and alas she died in February 1994.

The 1993 APEA Conference was held on the Gold Coast, Queensland and it was here that I presented my fifth APEA paper "The Hydrocarbon Potential of the Palaeozoic Basins of Western Australia". In this paper I tried to dispel the industry's view that there were no source rocks in the Palaeozoic in Western Australia. I pointed out that the lack of hydrocarbons in these basins was due to later tectonic movements. This was particularly relevant to the Canning Basin where most structures that had been drilled were formed during the Late Triassic Fitzroy Movement which post-dated primary hydrocarbon migration out of the Palaeozoic source rocks.

TASMAN OIL TOOLS

In March 1991, I was approached by Peter Jones of Tasman Oil Tools to undertake an evaluation of a lease in Um al Quain in the Persian Gulf. The lease was owned by a company based in Dubai and Peter knew one of the directors who asked him for assistance to farm out the lease. So in August I left for Dubai to meet Peter and start the evaluation. I needed a visa and was told that it would be waiting for me at the airport. When I arrived, there was no visa so I was escorted to a waiting room where I expected to be put on the next plane back to Perth. Eventually Peter discovered the bureaucratic slipup and I finally entered Dubai. This city was not what I expected – it was a modern, prosperous city with luxury hotels, beautiful houses and Western-styled infrastructure. It was also full of American service personnel celebrating their Gulf War victory.

When I started my evaluation, I discovered that some of the well reports and a lot of the seismic data were missing. According to the lease owner, a lot of data had never been submitted to the government. I asked about the seismic data tapes – if the hard copies could not be found, we could process new copies. They did not know what I was talking about until I described the tapes as being large reels about so big. One of the directors said that he had a box of reels in his garage at home. Great! His garage probably reached 50°C during the summer months. The tapes were ruined. It was then that I appreciated Australia's technical data submission system controlled by the Federal and State governments. Our system wasn't perfect but it was light years ahead of this. How could countries with 75% of the world's oil reserves treat their priceless technical data in such a cavalier fashion?

At the end of a very frustrating day, we retired to one of the clubs on the shores of the Persian Gulf for a couple of well earned beers. As the evening approached, we were treated to a spectacular blood red sunset as the sun slowly set behind the thick black smoke from the raging oil fires in Kuwait some 800 kilometres to the west.

In preparing my report, I had to do with what technical data I had. There were a couple of structures, small by Middle East standards but one had the potential for reserves of about 50 million barrels. In 1991, the oil price was only $US20 per barrel so this was not as exciting as it would be today. Before finalising my report, we undertook a field trip of the area. We drove from Dubai east through Sharjah, Um al Quain, across the Omani Mountains to the Gulf of Oman. Leaving Dubai was interesting – there was no thinning out of the suburbs. Dubai instantly ceased and then we were in the desert with the sand dunes and camels. Their desert was nothing like the Australian desert which has Spinifex and shrubs dotted about to provide some green. The Arabian desert is pure sand blowing and shifting across the land – one of the most inhospitable environments on earth. We then drove to Sharjah and visited the infamous Sajaa Gas Field. This field was a huge gas and condensate field which was developed as a condensate stripping project. But instead of re-injecting the gas into the reservoir to maintain pressure, it was flared into the atmosphere. At its peak, the field was producing 60,000 barrels condensate per day and burning 600 million cubic feet of gas per day. And now that the field is depleted, Sharjah has to import natural gas for its domestic requirements. Forget about smiling – the rocks were shouting at this terrible waste of a valuable natural resource.

When we entered Um al Quain, everything changed. There were no more petroleum fields east of Sharjah. The town was a collection of poverty stricken hovels that were so depressing that we hurried through and headed for the Omani Mountains. These were not scenically spectacular but were geologically very interesting. They are made up mostly of ophiolites being at the boundary of the African and Arabian tectonic plates. When we arrived in Al-Fujayrah, we had lunch at a lovely restaurant overlooking the Gulf of Oman – a very pleasant day indeed.

I prepared a farmout report and then we all met to discuss what terms they were prepared to accept. It was then that we met our first hurdle. Since they had paid the government $US100,000 for the technical data, they wanted the farminee to reimburse the $US100,000 plus the cost of my report. We explained to them that paying the government for the data was part of the terms for getting the lease but that no farminee was going to pay this or the cost of my report. Their cost of entry into the lease would be the promote on the cost of drilling the exploration well. I had two medium sized Australian companies interested in farming into the area but they wouldn't budge from their position. We then discovered another problem – they didn't actually own the lease, they only had an option on it. At this point Peter and I agreed that there was nothing more we could do to help them when they refused to take any of our advice. So I headed back to Perth. I don't know if Tasman Oil Tools ever got paid but it was certainly one of my most frustrating projects.

In July 1991, the Greek oil tanker Kirki was on its way to the Kwinana Refinery in Perth. It lost its bow in heavy seas and was heading for the Western Australian coastline south of Jurien Bay with potentially disastrous environmental consequences. Fortunately, Ampolex was drilling Tuart-1, an exploration well in the offshore Perth Basin at the time and dispatched one of its workboats to the Kirki. It prevented the Kirki from crashing onto the coastline until government assistance eventually arrived and the Kirki was towed to Dampier where its cargo of crude oil was transferred to another tanker. The Kirki was eventually towed to Singapore where it was broken up for scrap. If Ampolex hadn't been drilling at the time, then the Kirki would have foundered on one of the many reefs along the Western Australian coastline and caused a monumental environmental disaster.

ENERGY EQUITY

In 1991, Ian Whiteley was working for Energy Equity Corporation Limited (EEC) and asked me to assist on one of their projects. EEC was then run by entrepreneur Maurice Brand and the company had acquired the company that owned the Lemetang Production Sharing Contract (PSC) in Sumarta, Indonesia. The PSC contained the Harimau Gas Field but production rates were low and the field was not making a profit. In the centre of the PSC, interpretation of existing seismic had defined a large Miocene Baturaja reef overlying a ridge of basement.

Upon reviewing the Harimau Gas Field, it appeared that it was a retrograde condensate field. The previous operator may have over-produced the field, dropping the reservoir pressure which caused the condensate to liquefy in the reservoir and restrict the flow of gas. Unfortunately the damage had been done and there was little that could be done to resurrect the field. However, by buying the company that previously owned the PSC, EEC had inherited some $US80 million of accumulated exploration tax losses. If EEC could farm out the Baturaja reef prospect and if it was successful, these losses could be recovered from future production before that production was split with the Indonesian Government.

During the term of my contract with EEC, the Baruraja prospect could not be farmed out due to the prevailing low oil prices. The prospect was eventually farmed out to Amerada Hess who drilled Singa-1 in May-June 1997. The well encountered gas in the Baturaja limestones and tested at 30.7 MMcfg/d. Reserves were estimated to be of the order of 125 BCF gas but the gas contained a high percentage of carbon dioxide and the field was not developed until 2010 with plateau production of 50 MMcfg/d. By this time, the interest holders had changed and the final ownership was Medco 74% and Lundin Petroleum 26%.

In addition to the Lemetang PSC, EEC had also purchased the Sengkang gas fields on South Sulawesi from BP. The existing three gas fields had reserves estimated to be 600 BCF but these were insufficient for an LNG export project and were of no interest to a company the size of BP. However, EEC's plan was to develop these small fields and supply gas for power generation to the towns and cities in the region.

In June 1995, EEC finalised a deal with Indonesia to build a 135 MW gas fired power station on Sulawesi. As part of the deal, EEC agreed to spend $US27.5 million over ten years on the Sengkang PSC. Then in 1996 EEC sold a 50% interest in the Sengkang PSC and power project to Tenneco for $US28 million. Tenneco also acquired a 9.9% shareholding in EEC.

Both companies then arranged project financing of $US232 million for the construction of the Sengkang gas and power project.

In Australia EEC, through its 100% subsidiary Australian Gasfields Limited, progressively acquired 100% of exploration permit ATP-172P containing the Gilmore Gas Field. This field is located in the Adavale Basin in western Queensland and was discovered by Phillips Petroleum in 1964. Gilmore-1 tested up to 7.9 MMcfg/d from Devonian sediments in a reverse faulted anticline. As there was no market for the gas at that time and Phillips did not find any oil, they eventually relinquished the permit. In 1981, the Magnet Group was granted ATP-172P and they acquired a seismic survey over the Gilmore structure. Nomeco acquired the permit in 1985 and drilled Phfarlet-1 in 1986 looking for oil in the shallower Jurassic Hutton Sandstone. These sandstones were water wet but Nomeco ran 9-5/8" casing to 1853 metres and suspended the well capable for re-entry and deepening to the Devonian gas bearing sandstones.

In December, 1991, EEC signed a $400 million contract to supply Queensland Electricity Commission with electricity for 20 years. It intended to generate this electricity from natural gas supplied from the Gilmore Gas Field and from gas discoveries in the Cooper-Eromanga Basin via a farmout from MIM Petrolem. In 1992, EEC deepened Phfarlet-1 to test the Devonian reservoirs. They also drilled Phfarlet-2 in the same year. Both these wells encountered gas bearing sands but the wells failed to flow gas at commercial rates.

Rick de Boer and I wrote a report for EEC on the field in October 1993. Rick followed this up with an APPEA paper in 1996. Our conclusions were that most of the gas was in the Log Creek Formation, a deltaic unit containing clayey sandstones with poor reservoir quality. Immediately overlying this formation was the Lissoy Sandstone, a thin, basal transgressive unit fining upward into the Cooladdi Dolomite seal. In our opinion, most of the gas production came from the thin, permeable Lissoy Sandstone but that this unit was prone to mud invasion and formation damage. In the 1960's, drilling rigs had less powerful mud pumps and mud invasion was not as severe as today's modern drilling rigs. This is a possible explanation of why Gilmore-1, -3 and -4, drilled in the 1960's flowed gas at rates up to 7.9 MMcfg/d but the Phfarlet wells drilled in 1992 did not flow at commercial rates.

During 1994, Rick and I expanded our study from the Gilmore Gas Field into the Adavale Basin in general. Energy Equity operated ATP-601-P and ATP-572-P which covered a significant part of the basin. The Gilmore structure was a fault controlled compressional anticline but the gas was trapped in the fault independent part of the closure. The Cooladdi Dolomite was a good vertical seal but the overlying Buckabie Red Beds were probably not and more importantly, very unlikely to provide a lateral seal for the bounding faults. It turned out that Gilmore was the only structure with fault independent closure and in all the other structures the Lissoy Sandstone was faulted against the Buckabie Red Beds which did not provide a lateral seal. So there wasn't any follow up with this play. So we looked at the potential to the east in the Devonian Bury Limestone. This was sealed by the Boree Salt so if there were deeper source rocks in the basin, it could have developed into a new petroleum system. However, we could not map any drillable prospects from the existing seismic. We wrote a detailed but negative report on the Adavale Basin for Energy Equity in April 1995.

In early 1996, I prepared a technical report on the Dingo Gas Field in RL-2 in the Amadeus Basin. Dingo-1 and -2 were drilled by Pancontinental Petroleum in the early 1980's and Dingo-3 and -4 were drilled by Magellan Petroleum Australia in 1990-91. The first three wells were all gas bearing but Dingo-4 was a dry hole. Energy Equity had plans to develop the field and build a 42 kilometre pipeline to its Yalara LNG plant and a planned 8.7MW power plant. The power plant was built at a cost $13 million and officially opened March 1997. EEC re-entered Dingo-1 but found that the casing had collapsed from salt flow and the well had to be abandoned. Without Dingo-1 as a producer, the project proved uneconomic and was abandoned. EEC subsidiary Australian Gasfields Limited is now owned by Central Energy Power Pty Ltd which operates the power station with gas supplied from the Palm Valley gas field. It wasn't until 2013 that Magellan signed a long term gas contract to supply 30 BCF gas from the Dingo Gas Field to NT Power and Water Corporation over 20 years. In 2014, Central Petroleum bought Magellan's onshore assets for $35 million and in March 2015, successfully commissioned the Dingo Gas Plant and Pipeline project. So 34 years after its discovery, the Dingo Gas Field was finally in production.

Energy Equity was taken over and name changed to Energy World Corporation Limited in Janauary 2002. Energy World Corporation Limited was de-listed in October 2011.

SAGASCO HOLDINGS LIMITED

In late 1992, I was contracted by Grant Samuel & Associates to prepare a review and valuation of the exploration assets of Sagasco Holdings Limited which was under takeover offer from Santos Limited. The hardest part of this project was to value the remaining exploration potential of their Cooper-Eromanga Basin assets. There were 500 oil and 300 gas leads and prospects in PEL's 5 & 6 alone. I had to add up the estimated potential resources of each lead and prospect and then apply the average probability of success. It turned out that gas prospects had a higher success rate than oil prospects. I also selected the odd prospect to check the resource estimates by the operator which also happened to be Santos. I concluded that there was a 90% chance that an additional 34 million barrels oil and 425 BCF gas remained to be discovered in PEL's 5 & 6 in South Australia. The Grant Samuel accountants used these numbers to calculate their net present value. I then repeated this process for ATP 259P and ATP 488P in Queensland. The overall value obtained by Grant Samuel & Associates was higher than the Santos offer and the takeover bid failed.

One year later, Sagasco was under attack again, this time by Boral Limited. Again I was contracted to provide a review and valuation of Sagasco's exploration assets but this time excluding the Cooper-Eromanga Basin tenements. The Boral bid was eventually successful and Sagasco became a subsiary of Boral Limited. In February 2000, Boral Limited demerged its energy business as Origin Energy Limited.

NERDLICH

In late 1992, I met an American oil man called Tom Knowlton. He was very eccentric with long hair in a pony tail and smoked large Cuban cigars. He claimed to have access to secret US military technology which allowed him to predict the location of large oil fields. He claimed that large oil deposits being resistive, deflected the earth's surface electrical currents and this disturbed the earth's magnetic field. While all this was scientifically plausible, I doubted if such minute disturbances could be accurately measured, even by the US military. He claimed that some large oil fields were located in the Canning and Officer basins and wanted to contract me to obtain the acreage and work up the geology. As well as attractive consulting rates, he offered me a 2% royalty on any production discovered by Nerdlich in the areas I worked up for him. It was a terrific offer almost too good to be true, so I agreed to the contract.

My first task was to evaluate a well Nerdlich had drilled in 1990 and which had hit basement at unexpectedly shallow depths. The well was Anna-1 drilled in the Spencer Gulf. He didn't believe the well bottomed in basement and wanted me to check it out. So I flew to Adelaide and visited the Mines Department core store to examine the cuttings. The well intersected Quaternary sands and then went into the pinkest granite I'd ever seen. On returning to Perth, I looked at Nerdlich's report on Spencer Gulf and the Anna Prospect. It was very confusing. Firstly, the aeromagnetic data over the gulf clearly indicated shallow basement but the seismic indicated a thick sedimentary section folded into a gentle anticline. Upon closer examination, the processing of the seismic was suspect as the interval velocities between the main reflectors had unrealistic values. The Anna-1 well clearly proved that no thick sediments had been deposited in Spencer Gulf.

During the next gazettal in October 1992, I assisted Nerdlich in applying for three large permits in the southern Canning Basin in December 1992. Since the Canning Basin wasn't the most sought after acreage in the early 1990's, I recommended modest bids to secure this acreage. However, Tom insisted on enormous bids. For each area, the bids were First Year aeromagnetic survey and seismic reprocessing, Second Year 100 kilometre seismic survey, Third Year one exploration well, Fourth Year 100 kilometre seismic survey, and Fifth Year one exploration well. Nerdlich also submitted financial accounts indicating the company had net assets of $US408 million. The permits were granted as EP-374, EP-375 and EP-376 in April 1993.

Nerdlich also applied for a permit in the Officer Basin in South Australia in February, 1993. However, because there were other bids for acreage in the same area, the South Australian Department of Mines and Energy invited all the bidding groups to reapply in September 1994 on a competive work program bidding system.

I was also negotiating my consulting agreement with Nerdlich. His solicitors, Greaves, Wannan & Williams, sent me the first draft in January 1993. John Ralph of Ralph & Edwards assisted me but while the agreement was mostly OK, we could not accept the Governing Law being California. I'd have been better off shooting myself in the head than fighting a law suit in California! So the negotiations dragged on.

My first task was to reprocess the existing seismic in the Canning Basin permits. The reprocessing project was tendered in March, 1993 and awarded to Digital Exploration in May 1994. I was busy arranging copying of the seismic tapes and getting the project underway using Dean Powell to supervise the reprocessing.

In April, 1993, World Geoscience tendered to acquire the aeromagnetic survey. It was to be acquired in two stages, the first being the reprocessing of existing data in the permits to the northwest costing $40,000. The second stage was to fly 15,500 square kilometres over the permits to the southeast at an estimated cost of $250,000. The second stage was to be flown in June 1993 but was not flown until August 1994.

In September 1993, a cheque from Nerdlich for my last month's consulting invoice bounced. I didn't think this was a very likely event for a company supposedly worth over $US400 million. What's that wonderful saying "If something sounds too good to be true, it usually isn't!" I contacted Tom about the bad cheque and he claimed it was a clerical error. But I put a halt on the seismic reprocessing until I could determine the real state of Nerdlich's finances.

By May, 1994, it was clear that Nerdlich was not the company it claimed to be and didn't have the finances to fund the large work programs in its Canning Basin permits. I arranged a meeting with the Mines Department to apply to extend the permit work programs for six months. The Mines Department wanted to cancel all three permits for non-performance so I had a lot of negotiating to do. I finally arranged for Nerdlich to keep EP-376 (in my opinion the best of the permits), relinquish the other two permits and even get refunded the $60,000 security bonds. Nerdlich would have to acquire the aeromagnetic survey, carry out the seismic reprocessing and drill one well in EP-376 by May 1995. I thought I'd made the best of a bad situation but Tom was furious and terminated my services immediately.

In EP-376, I had defined a large reverse faulted anticline called the Patience Prospect. The folded Early Palaeozoic sediments were overlain by flat Permian sediments so the structure was not a Late Triassic Fitzroy Movement structure. If Ordovician Acacia Sands were present at this location, there was a good chance of find a significant oil accumulation in the structure. Patience-1 had been drilled by Sydney Oil in 1986 but the well bottomed in the Devonian Tandalgoo Sandstone and did not reach the deeper Ordovician sediments.

Somehow Nerdlich managed to hold onto all three permits for the next six years when the company drilled Patence-2 during 2000 using OD&E Rig 16. The well encountered oil and gas shows in the Bongabinni and upper Goldwyer formations but these were never tested. Alas, the Acacia Sandstone was absent. The well reached a total depth of 4184 metres and was suspended. Later electric log interpretations indicated 3.6m net gas pay in the upper Goldwyer Formation and 7.6m net gas pay in fine grained sands in Bongabinni Formation with porosities of 15-16%.

Soon after, Nerdlich left Australia leaving many contractors unpaid. Rig 16 was left in the desert until it was recovered and made fit for drilling again in 2003.

Ashland Oil

In late 1992, I was approached by David Wood who was the exploration manager for the US company Ashland Exploration. They had farmed into an exploration permit EP-341 on the Peedamullah Shelf along the southeastern margin of the offshore Carnarvon Basin. They needed a geologist to represent them in Australia and asked me if I would take on this role. As my time with Doral Resources had finished, I readily agreed and we signed a consulting agreement.

Ashland was an interesting company – it had two main assets: oil production in Nigeria and shale gas production in the Appalachian Mountains in the USA. Dave Wood had done his time in Nigeria and related the most harrowing stories about operating there. On arriving at Lagos Airport, he would be met at the airport to be escorted to the Ashland office. On clearing customs, there would be 4 or 5 Nigerians holding up signs saying "David Wood". He would have a secret password to determine who his real contact was and who were there to kidnap him. He was then driven to Ashland's office in a bullet-proof limousine. Not my idea of a fun international posting!

Ashland's shale gas production was even more amazing as I had never heard of such a thing. It came from Devonian marine shales at very shallow depths. The flow rates were so low that each well had to be drilled and completed for no more than $US60,000 or the well would not pay out. But Ashland drilled thousands of these wells and the production and cash flow were amazing. Twenty years later and using horizontal drilling and fraccing techniques, the shale gas industry was to revive the US economy and cut their petroleum imports. Ashland's gas wells were in the now famous Marcellus Shale gas play.

Back in Australia the first order of business was to tidy up the farm-in agreement and negotiate the Joint Operating Agreement (JOA) for EP-341. The original permit holder was Peedamullah Petroleum Pty Ltd, a wholly owned subsidiary of Discovery Petroleum who was carried 30%. The farminee companies were Command Petroleum (who was to be the operator) 20%; Mount Isa Mines Limited 20%; Louisiana Land and Exploration (LL&E) 20%; and Ashland Exploration 10%. In 1987, the Australian assets of Nomeco had been vended into Command Petroleum and John Doran was the Managing Director. Command was eventually taken over by Cairn Energy, a UK company in a $365 million deal and the company was de-listed in December 1996.

Because Ashland was the last company to join the consortium, it couldn't acquire its desired interest of 20% via the farm-in so it purchased 10% from Discovery Petroleum. Being a purchase of property, the State Government sent them an $80,000 bill for stamp duty. I remember trying to explain to Ashland's financial controller the concept of stamp duty but he was totally outraged. He replied that such a tax was monumentally unjust and said that we should rise up, like America did in 1776, against our cruel masters. Not a bad idea except that "we" voted "our cruel masters" into power so a revolutionary war was definitely out of the question.

Command Petroleum was located in Sydney and on one of my trips over from Perth I flew in a Boeing 747 on the domestic leg of an international flight. Being in business class, this was not only very pleasant but I also ended up sitting next to James Bolan. I had a very enjoyable flight chatting with him and he, like most overseas visitors, was amazed at the sheer size of the country and the enormous amount of empty desert between Perth and the eastern states. It also turned out that he never liked one of my favourate TV programs "When the boat comes in".

Negotiating the JOA was a tedious but necessary process. The negotiations came to an impasse over the issue of "pre-emptive rights". LL&E and Ashland wanted pre-emptive rights and Discovery Petroleum was equally opposed to the concept on the basis that they impede a company's ability to trade in the permit. I suggested a "first right of refusal" as a possible compromise. Under pre-emptive rights, a potential buyer can spend a lot of time and money evaluating a tenement, make an offer and then have one of the existing joint venture companies pre-empt and take over the purchase. So most companies will not even bother to look at a sale where there are pre-emptive rights in the JOA. Under a first right of refusal, the seller agrees on a price, offers the interest to the other companies in the joint venture and then, if these companies do not accept the selling price, then the seller is free to market the interest unconditionally at that price. All the companies agreed to this clause and the JOA was finally signed.

New seismic was acquired and a number of small prospects defined. The size of the prospects were disappointing, particularly to a large company like LL&E, but the joint venture agreed to drill two wells. Crackling-1 was drilled in April 1994 and intersected oil shows over the interval 340-349 metres in the tight Mardie Greensand but alas, the porous Birdrong Sandstone was water wet.

Santa Cruz-1 was drilled in April, 1994 and encountered oil staining and bright yellow fluorescence in the Birdrong Sandstone over the interval 432-445 metres. These shows were considered to be non-commercial and the well was plugged and abandoned without testing. Ashland considered sole risking a test but, not being set up to operate offshore operations, this really wasn't feasible.

At the same time, Ampolex was looking for partners for an application in the inner Dampier Sub-basin next to its operated Wandoo Oil Field. The Wandoo Oil Field was discovered in WA-202-P by Arco Australia in partnership with Ampolex in 1991. The oil was biodegraded (API 19.5°) with a high viscosity (14.4 centipoise) and Arco, believing that the oil would not flow, withdrew from the joint venture before the well was production tested. Ampolex ended up with 100% of the permit and later tested the well at 4,600 bopd. The rocks definitely had a good laugh at Arco's expense!

An application area to the northeast of the Wandoo Field was gazetted by the Federal Government and Ampolex decided to bid on this area. Before the closing date, their bidding partner withdrew and they were looking for a company to take up the interest of this withdrawing company. After evaluating the Ampolex leads and discussing these with David Wood, Ashland joined the bid with Ampolex. The new permit was awarded as WA-243-P.

In the meantime, Dave Wood was interested in the Wandoo Oil Field. Ampolex had a 100% interest in this field and was looking at having to fund a very expensive development on its own. Dave thought there was an opportunity for Ashland to negotiate the purchase of an interest in this field and assist Ampolex in the cost of the development. I was not involved in these negotiations but I know that Dave was close to a deal when Ashland's head office decided against investing in the project. Dave was naturally very upset and Ashland missed out on acquiring an interest in what was to become a very profitable oilfield. Ampolex went on to develop Wandoo on its own and the field came on stream in late 1993 at 10-20,000 bopd from the Hakuru jackup rig acting as a temporary production facility. This was changed out to the 80,000 tonne Wandoo B Platform built for $100 million at the Henderson yards south of Perth and installed in 1997. Production rose to 50,000 bopd before its natural decline and the field is still producing today with cumulative production of over 80 million barrels.

Ampolex acquired a closely spaced 2D seismic survey in WA-243-P. The seismic quality was poor due to the weak reflectivity at the top of the Barrow Group sands and the abundance of multiples as a result of the shallow depth of the reservoir being ~600 metres. Ampolex's interpretation of the seismic delineated a Wandoo look-a-like structure they called Kanji. Using paper sections, I came up with a similar looking structure at Kanji. Kanji-1 was drilled during February-March 1994 using the jack-up "Ron Tappmyer" but was a dry hole. The most interesting story about Kanji was the site survey carried out prior to the drilling of the well. A video from the ROV showed bubbles of gas coming from the sea floor. This video was carefully stored in case the joint venture was later accused of having a leaking well with a poor cement job. The video is however testament to the prolific Carnarvon Basin and the large volumes of oil and gas being generated, an infinitesimal part of which is currently and naturally leaking from the sea floor! Two other prospects, Tingle-1 and Hyperno-1 were also drilled by the joint venture using the "Ron Tappmyer" in 1995 but both were dry holes.

At the request of Ashland, I prepared a summary of Australia's oil exploration, production and consumption statistics. At this time Australia's oil consumption was 650,000 bopd and production was slightly lower at 600,000 bopd. I calculated that over the past seventeen years, our industry had found some 1.7 billion barrels of oil, a discovery rate averaging 100 million barrels per year. I also determined that our finding costs were approximately $4-50 per barrel with development costs of a similar $4-50 per barrel. Given that oil prices had varied from $15 to $30 per barrel over this time, this was quite a respectable effort.

Part of my contract with Ashland was to review all farmout opportunities on the North West Shelf. This gave me the opportunity to update myself on the new geological thinking on the area since my days with Woodside. I got to review farmouts in the Timor Sea, Browse and Bonaparte basins and the Zone of Cooperation with Indonesia. I also prepared a report on the Browse Basin outlining the potential for oil accumulations on the southeastern flank of this basin. This potential was later confirmed by the future Gwydion and Cornea oil discoveries but the commerciality of this trend has yet to be realised.

One of the most frustrating of these farmout opportunities was WA-209-P. This joint venture consisted of BHP Petroleum (operator), Phillips Petroleum and a Canadian company Norcen International and they were farming out the permit for the drilling of an exploration well on a prospect called Stag. The Stag Prospect was mostly a Permian structure with a small Cretaceous drape. It was high risk but the Cretaceous drape was very interesting, particularly after the success at Wandoo. However, there was only about 12-13 metres of relief on the Cretaceous structure and estimated potential reserves, even if the well was successful, would have been non-commercial. On my recommendation, Ashland turned down this farm-in opportunity. The permit was farmed out to Apache Energy, Santos and an American company Globex Far East who funded the drilling of the

Stag-1 well which was drilled in 1993. The well intersected a 15 metre oil column in the *M. australis* Sand, pretty much as predicted. Unfortunately for Ashland (and my pride), the 15 metre oil column was part of a larger 60 million barrel stratigraphic trap. Bloody hell, those damn "rocks were smiling at us" again! The Stag oil field was developed for $200 million, began production in 1998 and reached peak production of 30,000 bopd in 2000. In early 2015, Apache sold the bulk of their Australian assets (including Stag) for $US2.1 billion to the Macquarie and Brookfield consortium which became Quadrant which in turn was taken over by Santos for $US 2.15 billion plus contingent payments. The field itself was purchased by Jadestone Energy in 2016 and was still producing at 3,600 bopd.

During 1994, the main petroleum discoveries on the North West Shelf were the Laminaria and Elang oil fields. Laminaria-1 was drilled in AC/P 8 by Woodside and encountered a 102 metre oil column in the Middle Jurassic sands of the Elang Formation. The well was tested at 7500 bopd. For me this was an amazing discovery. In my BOCAL days, I had worked on the Timor Sea and we had mapped the Laminaria structure. However being located in 350 metre water depths, the drilling technology to drill the prospect was not available in 1972. Based on our maps, the structure was heavily faulted and based on Woodside's later seismic interpretation published in the 1996 APPEA Journal, some of these faults went almost to the sea bed. Such recent faulting would normally indicate a high risk of leakage and downgrade the prospect. I remember seeing Henry Askin at a later APPEA conference and he remarked "I can't believe Laminaria-1 came in with a 100 metre oil column – there shouldn't be a drop of oil in that structure". I had to agree with him but then I remembered – "the rocks are smiling at us".

Elang-1 was drilled in ZOCA 91-12 by Petroz NL and intersected a 176 metre gross oil column in the Jurassic Montara Sands and Plover Formation. The well was tested at 5,800 bopd. ZOCA 91-12 has an interesting history. Prior to the establishment of the Zone of Cooperation (ZOC) with Indonesia, the area was covered by WA-74-P granted under the Australian Petroleum (Submerged Lands) Act. WA-74-P was awarded in 1977 to Pelsart Oil and was farmed out in 1979 to a group including Mesa Australia Limited who became operator. Pelsart was acquired in 1986 by

Petroz NL who assumed operatorship. A large area of the northern part of the Timor Sea was disputed by Indonesia following its annexation of East Timor in 1976. After lengthy negotiations, the Timor Gap Treaty was signed by Australia and Indonesia in December, 1989 and ratified by both governments in February, 1992. As part of this treaty, the northern part of WA-74-P was excised from the permit. The WA-74-P joint venture was offered the right to match the highest bidder on a nominated block in the ZOC and ZOCA 91-12 was awarded to the joint venture in February, 1992.

WMC Resources Limited (who bought Mesa Australia) sued the Commonwealth for compensation for part of its WA-74-P petroleum asset being nationalised. In February, 1998, the High Court ruled (four to two) that no compensation was payable as there was no acquisition of property involved. The main defence of the Commonwealth was that the PSLA did no more than create an exclusive right to explore for petroleum in a designated area. In other words, an exploration permit is not property. Justice Kirby, one of the decenting judges, concluded that if permits could be cancelled at will, without just terms, "it is unthinkable that permittees would assume the extremely heavy obligations, involving long term investment and commitment, associated with such permits". And yet the Australian petroleum industry continues to do so some 20 years later.

The 1994 APEA Conference was held in Sydney in March. This was the last time the conference was held in Sydney as the focus of the oil industry had shifted to Perth and there were no longer enough companies based there to support future conferences. In my opinion, the most interesting talk at the conference was by Elwood Horstman on the effect of inertinite on Rock Eval analyses. Inertinite does not generated hydrocarbons but its carbon content is included in the calculation of Total Organic Carbon and has the effect of underestimating hydrogen indices.

In the same year, BHP Petroleum who once owned 40% of the shares of Woodside Petroleum Limited sold the last of its shareholding in Woodside Petroleum for $4-76 per share. Some fourteen years later, Woodside's share price would reach a peak of over $65 per share in 2008.

It was also in August 1994 that another great Australian oil company, Bridge Oil Limited, was taken over by a US company Parker & Parsley for its US assets. In March 1996, Parker & Parsley sold most of its Australian assets to

Santos for $200 million. These included a 25% interest in the Patchewarra Central Block; a 15% interest in WA-214-P including East Spar; and a 15% interest in the Airlie oil producing fields. The sale excluded its 22.5% interest in the Bayu gas discovery in ZOCA 91-13. The takeover turned out to be a terrible deal for the Bridge Oil shareholders as the Australian stock market was giving no value to the company's US assets. It was a great deal for Parker & Parsley as they got what they wanted, the US assets for a very low net price.

The next Western Australian Basins Symposium was held in Perth in August 1994 and I presented a paper on the onshore Carnarvon Basin which mostly dealt with the Palaeozoic potential in the Gascoyne and Merlinleigh sub-basins. Additional age dating information had become available since my 1993 APPEA paper on the Palaeozoic Basins of Western Australia. The age of the Tumblagooda Sandstone which was thought to be entirely Silurian, was pushed down into the Ordovician and late Cambrian based on conodonts. The main thrust of my paper was the hydrocarbon potential of the Late Devonian in the Gascoyne Sub-basin. Based on only five wells, this potential petroleum system contains reservoir, seal and source and is virtually unexplored.

In July, 1995, Discovery Petroleum merged with Minora Resources in a friendly takeover. Ironically, it was Minora's interest in the Kakap gas project in Indonesia that made Discovery Petroleum itself a takeover target. It was then taken over by Premier Oil Australia Pty Ltd, a subsidiary of the Premier Oil Plc, and delisted in January, 1997. In February, 1997, the Australian assets of Discovery Petroleum were bought by AWE Limited.

The year 1995 was a good year for gas discoveries on the North West Shelf. In January 1995, WMC announced the development of the East Spar Gas Field at an estimated cost of $250 million. This field was discovered in 1993 and was developed via a pipeline to Varanus Island and then into the $400 million Goldfields pipeline owned 62.5% by WMC. Production from East Spar commenced in 1996 and the field is now depleted.

The giant Perseus Gas Field was also discovered in 1995 with Perseus-1 intersecting a 112 metres gas column. Perseus is a giant structural-stratigraphic trap with reserves estimated in excess of 6 TCF gas plus over 150 million barrels condensate. The presence of this field was first indicated by production from the NRA22 development well on the Rankin

Platform. This well had been producing since 1991 without significant pressure drop in the reservoir. The NRA22 well was re-evaluated and following interpretation of the 3D seismic acquired in 1992, the trap was delineated.

The giant Bayu-Undan Gas Field was discovered in 1995 with the drilling of Bayu-1 which intersected a 155 metre gas column. Bayu-1 was drilled by Phillips Petroleum on that part of the structure that extended into ZOCA 91-13. It was its last commitment well before withdrawing from Australia. Very embarrassing for Phillips but the rocks had a good laugh! They had already closed their Perth office and had to set up all over again. Undan-1 was drilled by BHP Petroleum in August 1995 on that part of the structure in that extended into ZOCA 91-12. After further appraisal drilling by both operators, the total structure was estimated to contain 3 TCF gas plus 300 million barrels condensate. A joint development of the field, then estimated at $2 billion, was complicated by disagreement over the development plan. BHP Petroleum wanted to build a concrete gravity structure on an artificial island and use a Compact LNG Processor. However Phillips preferred the traditional and less risky 500 kilometre pipeline to an onshore LNG Plant near Darwin. The issue was resolved when BHP Petroleum sold its interest in the field (plus the Elang-Kakatua oil fields) to Phillips Petroleum in April 1999 for $110 million.

In August, 2000, Phillips also made a cash offer for all the shares of Petroz NL who had an 8.25% interest in Bayu-Undan and a 15% interest in the Elang-Kakatua oil fields. However, in November 2000, Eni made a higher offer of $127 million (56c/share) for Petroz forcing Phillips to increase their offer to $158 million or 70c/share. Eni had an 11.8% interest in Petroz and initially prevented a compulsory takeover of the company. They eventually agreed to the takeover and Petroz was delisted in November 2001. By February 2003, Phillips had acquired over 90% of the shares in Petroz but a group of dissident shareholders prevented the final acquisition until November, 2006.

Eventually in 1998, Ashland's petroleum division was merged with Marathon Oil. Ashland's Nigerian production was sold to Total and its Australian assets were sold to Tap Oil and my contract finished. I very much enjoyed working with Ashland and particularly with Dave Wood – it was definitely one of my better contracts.

WILDCAT AUSTRALIA

During the early 1990's, exploration was moving into the shallow waters along the inner Dampier Sub-basin and Peedamullah Shelf. Not only was the water shallow but so were the objective Early Cretaceous sands at between 600 and 1,000 metres. Using a conventional jackup rig, the cost of mobilising the rig was greater than the actual cost of drilling the well. For example, Santa Cruz-1 took only five days to drill but four days to demobilise. Jim Carver formed a company called Slimdrill Pty Ltd with the idea of building a small rig that would fit on a barge. The barge mounted rig would drill a slim hole expendible well and considerably reduce exploration drilling costs. If a discovery was made, then the costs of drilling the conventional wells for appraisal drilling could be justified. Jim hired an engineer called Ray Rampling to design and build the Slimdrill Rig. Prior to this, Slimdrill had made enquiries to the Western Australian Mines Department as to whether such a barge mounted rig would be allowed to drill in these environmentally sensitive shallow waters. The Mines Department had no objectives in principle but told Slimdrill they would have to prove the rig's capabilities by having it drill some onshore wells. A company called Wildcat Australia Pty Ltd was formed to acquire onshore exploration acreage and work up prospects for drilling with the Slimdrill rig. I was contracted in January 1993 by Wildcat Australia to do just that.

The onshore Canning Basin was the ideal place to test the Slimdrill Rig. There were many thousands of kilometres of existing seismic data acquired during the early 1980's oil boom, and plenty of prospects to drill. Wildcat Australia would apply for permits where there were undrilled prospects and avoid having to acquire expensive onshore seismic.

Wildcat Australia applied for drilling reservation over an interesting prospect which we called Maroon. Maroon was an anticlinal rollover at the top Devonian on the downthrown side of the Pinnacle Fault. Drilling Reservation DR-2 was awarded to Wildcat Australia and soon after, the company was approached by a Canadian company called Bow Valley who wanted to farm into the tenement. The exploration manager for Bow Valley was Les Beddoes who I had worked with in Esso Australia's Sydney office in the late 1960's.

Bow Valley Australia Pty Ltd farmed into DR 2 to earn a 65% interest by drilling one well leaving Wildcat Australia Pty Ltd with a 35% interest in the tenement. Bow Valley changed the prospect name to Chestnut-1 which was drilled in July-August, 1994. This well encountered a 300 unit gas show in a 10 metre sand at 1350 metres in the Late Devonian. The well then intersected a 30 metre sand in the Late Devonian at 1800 metres with a 450 unit gas show. This sand was drill stem tested over the interval 1795-1842 metres and flowed gas to surface at a low rate. The sands were gas bearing but tight.

Bow Valley also drilled another well called Wattle-1 in the Canning Basin but not in Wildcat's acreage. In my opinion, this was one of the most significant wells drilled in this very difficult basin at that time. Bow Valley 40% and Cultus 10% farmed into the EP-129 Joint Venture (collectively 50%) to drill a well on the flanks of the Fitzroy Sub-basin. At this time, the EP-129 Joint Venture consisted of Petroleum Securities, Laurel Bay Petroleum, Minora Resources, Terratek Drilling Tools, Command Petroleum and Vamgas Limited.

Wattle-1ST1 was drilled in September-October 1994 and encountered good oil shows in the Early Carboniferous. The well was drill stem tested from 2529-2550 metres over an Early Carboniferous calcarenite and recovered two barrels 35° API oil and 2.8 barrels oil cut mud. The well was not drilled on a structure but on a stratigraphic trap within a wedge of sediments thickening into the Fitzroy Sub-basin from the Pinnacle Fault. This discovery with an equivalent flow rate of about 30 bopd was not commercial at the then oil price. But the real value of this oil discovery was the potential it revealed for stratigraphic traps in the Early Carboniferous along the northern margins of the Fitzroy Sub-basin.

The next task was to finance the drilling of these wells. Through their accountant, Wildcat Australia formed a very complicated, tax effective trust to raise high risk exploration funds from high income individuals. Although complex, the tax effective scheme had been approved by the Australian Taxation Office. These exploration funds were managed by another company called Wildcat Management Pty Ltd. An accountant Rob Collins and I were appointed directors to provide independence and control of the management of the funds from the individuals to Wildcat Australia for the drilling of the wells.

Wildcat Australia applied for and was granted Exploration Permit EP-371 in the Fitzroy Sub-basin. The company drilled Nemile-1 in June, 1994 using Drillcorp Rig 29 as the Slimdrill rig was not quite ready for drilling. The well had oil shows in fine sands within the Permian Noonkanbah Formation but these were not significant and the well was plugged and abandoned. Metters-1 was drilled in October-November 1994 using the Slimdrill HTA 3000 Rig 1 but it too was a dry hole.

Wildcat Australia also applied for a permit over the Calamia Prospect in the southern Canning Basin in November 1993 but no wells were drilled in this permit. It also applied for a Drilling Reservation over old Barrack Energy EP-330 where they had identified the West Moorak Prospect.

Wildcat Australia tried to drill this prospect but attempts to find water for the drilling operations failed and the operations were terminated. Eventually, further investment was not forthcoming and Wildcat Australia and Wildcat Management were eventually wound up.

AMADEUS ENERGY

In 1993, I was asked by Geoffrey Towner to look at the Savory Basin, a new basin that had been discovered from field mapping by the WA Geological Survey. Geoff and the colourful senator Ross Lightfoot were directors of a private company West Australian and Northern Territory Oil Search Pty Ltd (cheekily abbreviated to Wantos) and wanted to take out a petroleum permit over the entire basin. A quick review of Bulletin 141 by Ian Williams revealed that the Savory Basin was indeed part of the Officer Basin and contained a Late Proterozoic section with reservoir, seal and structure. Source rocks were the main risk but if the Savory Basin was part of the Centralian Superbasin, then source rocks could be expected in the older units beneath the salt. This part of the section did not outcrop and had not been reached in any of the wells drilled in the basin.

Wantos applied for permit over the Savory Basin in April, 1993. Exploration permit EP-380 was awarded to Wantos in June 1993 – it consisted of the maximum 200 graticular blocks and covered an impressive area of 15,762 sq km. Wantos also applied for two permits in the Western Officer Basin in June 1993 but these were never taken up.

In August 1993, I undertook my first field trip to the Savory Basin with geologists Ian Williams and Andrew Svalbe from the Geological Survey of Western Australia who were continuing their mapping of the area. I met up with them in Newman and they showed me around the most important outcrops in the basin. The area was very barren but the outcrops were good **(Fig 34)** and we camped in Savory Creek which was very picturesque **(FIG 35)**.

Figure 34
The barren landscape of the Savory Basin

Figure 35
A shady spot to camp along Savory Creek

I followed this up with another trip in September with Geoff Towner. This involved a flight to Newman (at the ungodly hour of 6-00am), hire 4WD vehicle and drive east to the Jigalong Aboriginal Community. Even though we were just passing through, I applied for and was granted a "Permit to Enter a Reserve" by the Aboriginal Lands Trust to enter the Jigalong Reserve. As we approached the Jigalong Community, the area was strewn with abandoned cars and the community itself was a collection of dilapidated buildings with smashed windows and rubbish littering the streets – I thought I had suddenly been teleported into some third world country in the middle of central Africa. I handed our permit to the Jigalong administrator and he was very impressed that we had taken the trouble to get a permit – obviously most people didn't! I guess some people didn't like the idea of having to get a permit to visit parts of their own country.

On 11th December, 1994, my mother died at the age of only 73. She had had dementia for the past four years and had been in care at Sunset Hospital. It is a terrible disease. Towards the end, I used to visit her once a week but she didn't know who I was and stared blankly at me. All I could do was hold her hand and talk to her hoping that somewhere in there, she understood. The hospital itself was beautifully located at Dalkeith overlooking the Swan River. But a lot of elderly patients had dementia and I would pass them on my way to visit mum. They would be seated in chairs staring blankly into space and I would shudder, thinking "I hope I don't end up like that".

In January, 1995, the State Energy Commission of WA was broken up by the State Government into Western Power Corporation, Alinta Gas Corporation and the Office of Energy, the new industry regulator. This was designed to increase competition in the energy sector for WA.

The 1995 APEA Conference was held on 2-5 April in Adelaide. At this conference, there were three interesting papers on the Timor Sea: one on the Elang Oil Field, one on hydrocarbon-related diagenesis in the Vulcan Sub-basin, and Gorter and Kirk's on the Kimmeridgian. After the conference, we met up with the Thompson's and another family for a houseboat trip on the Murray River. The Unforgettable Houseboats were definitely the most luxurious houseboats we'd ever seen. This was a first for us and we were very impressed by the way they were fitted out and the quality of the furnishings. We took the boat up the Murray as far as we could, taking into account the time to return to Mannum. Each evening, we stopped by running the boat onto the bank and tieing it up to the nearest trees. We got pretty good at spotting the ideal location being one with two large trees close to each other. During the day, the slow pace of the boat allowed you to admire the scenery and the incredible diversity of bird life on the river. One afternoon after we'd tied up, I went for a swim in the Murray. It was freezing cold but I gritted my teeth and pretended that it was lovely and warm. The other families had two daughters each and two of them believed me and jumped into the Murray. You could have heard the shrieks in Adelaide! All in all, it was one of our more relaxing holidays in Australia.

In mid 1995, I had a meeting with John Kopcheff. He was planning a survey by Knight Industries over some of Victoria Petroleum's permits in the onshore Carnarvon and southern Canning basins and asked if Amadeus Petroleum would like to extend the survey into EP-380 and split the costs. Amadeus Petroleum agreed and the survey took place in September 1995. We first surveyed the onshore Carnarvon Basin EP-41 originally held by Wapet, then Ampolex and now Victoria Petroleum, and visited the CRA Kintyre Uranium deposit which was undeveloped due to the State Government's policy on uranium mining. Our next visit was to the town of Wittenoon which had been abandoned due to the dangers from the blue asbestos which was mined near the town. It was a beautiful town set in a valley in the Pilbara ranges, with magnificent trees and colourful hills surrounding the town. As we walked down the main street, we could see the blue asbestos embedded in the bitumen on the road. No wonder the place became so deadly!

Our next field trip to the Savory Basin was an archaeological survey carried out in August, 1997 over our proposed drill sites before our drilling campaign could commence. The archaeologist supervising the survey was Christine Martin and I attended a meeting with Mines Department in July to ensure we would cover all aspects of clearing any sites of significance. During the survey, we camped overnight by a magnificent outcrop of glacial deposits called tillites in the Boondawari Formation **(Fig 36).**

Figure 36
Tillite deposited by retreating ice sheets

I remember waxing lyrical about the vast ice sheets that once covered this land some 800 million years ago. Perhaps one too many red wines around the campfire! After the survey, Christine Martin prepared a detailed report clearing the drill sites with some minor areas to be avoided around the Boondawari site.

On 25 September, 1997, I submitted an application for a 60 block area over the Proterozoic Tarcunyah Group to the north of EP-380. It was granted as 10/97-8 subject to reaching native title agreements with the Birriliburu and Martu claimants. Negotiations were progressing with the Birriliburu but broke down with the Martu who were making very tough demands including up front payments before negotiations even started. Amadeus Petroleum would not agree to these demands and the permit was returned to the government.

The drilling campaign in the Savory Basin began in October, 1997 using Western Deephole Drilling Rig 8. All three wells were to be continuously cored. Mundajini-1 was drilled in October 1997 to a total depth of 600 metres; Boondawari-1 was drilled in October-November, 1997 to a total depth of 1,364 metres; and Akubra-1 was drilled in November 1997 to a total depth of 181 metres. There was a very minor oil show in Mundajini-1 but the most interesting result was that the main objective Spearhole Sandstone in Mundajini-1 had up to 15% porosity and permeabilities of up to 532 millidarcies. Not bad for 800 million year old rocks.

I reviewed the results of the three wells. It was clear that the Spearhole Sandstone was a good reservoir and was sealed by thick red mudstones in the Mundajini Formation. Based on outcrop mapping, there were structures in the permit area but without expensive seismic, it was impossible to predict the crest of the structures at the objective Spearhole Sandstone level in the subsurface. Lack of potential source rocks was also a problem and if these were present in EP-380, they should be further west in the main Officer Basin.

In March 1998, the state government sold the Dampier to Bunbury natural gas pipeline and the AlintaGas gas transmission business to the Epic Energy consortium for $2.4 billion. This was way above the state government's expectations and nobody could take the smile off the then premier Richard Court's face!

EP-380 was renewed for a further five year term on 14 July, 1998.

My next field trip was in September, 1998 to map the various outcrops in the eastern part of the Savory Basin. Because of the vast distances involved, the survey was carried out using a helicopter. Since helicopters keep their engines running during short stops on the ground, the main safety rule when you work with helicopters is "always approach a helicopter from the front – never from behind". This hopefully will prevent you from walking into the tail rotor.

In March, 1999, there was an amazing proposal from a company called Pangea to dump nuclear waste in the Savory Basin. Pangea was planning to use the Synroc technology, invented by Ted Ringwood of the Australian National University, which locks the nuclear waste into synthetic crystalline rocks prior to disposal. This plan was supported by Senator Ross Lightfoot but the Federal Government Resource Minister Nick Minchin, was not in favour and neither were State Premier Richard Court and State Energy Minister Colin Barnett. Why Pangea picked a sedimentary basin to dispose of nuclear waste, I'll never understand. Any nuclear waste escaping from the synrocks, would end up in the permeable Spearhole Sandstone and (eventually – after thousands of years) leak all around the edges of the basin. They should have proposed to dispose of it in the Yilgarn Block where it could have been effectively and safely sealed in tight metamorphic rocks for at least the next million years. In any event, the proposal would certainly never have been approved.

In September, 1999, Amadeus Petroleum farmed into permits PEL's 2, 4, 5 & 267 in the Sydney Basin. The permits were held by Sydney Gas for coal seam gas exploration and Amadeus farmed in for the conventional rights. I was contracted to work up two prospects for drilling. It had been many years since I worked on the Sydney Basin so on 12th September, I flew to Sydney and visited the NSW Department of Mines and Energy the following day to research the latest geological studies on the basin. The most useful document was an excellent summary of the basin by Chris Herbert. The Sydney Basin has good source rocks in the Permian, good seal, plenty of structures but, as mentioned previously, has very poor reservoirs. The low permeabilities in the sandstones were caused by volcanoclastic detritus being eroded off the New England High and washed into the basin from the north and possibly from the offshore areas to the east. The southern and southeastern parts of the basin had better reservoir as they were eroded from the granites and metamorphic rocks along the Tasman Geosyncline.

PEL 2 was located in the southwestern part of the basin in very rugged terrain. However, the Camden and Camden Extension seismic surveys had been acquired by AGL in 1987 and 1988. Based on the interpretation of these surveys, I delineated two prospects, Glenmore and Fairlight, for drilling. They were poorly defined and there were associated faults but they were in an area where better reservoir could be expected eroding from the metamorphic rocks along Tasman Geosyncline.

In November, 1999, Amadeus Petroleum held its Annual General Meeting in Sydney. I attended this meeting and gave a technical presentation on their prospects in the EP-380 and their new permits in the Sydney Basin. I then travelled to Camden to visit the area to review the terrain and talk to the landowner regarding access to drill a well on his property. I was amazed at the ruggedness of the terraine and marvelled at how the seismic vibrator trucks acquired the survey without toppling over. But the landowner remembered the surveys and described the trucks driving up and down the steep slopes and crossing the many creeks. I sat down with him and we negotiated an acceptable access agreement.

Amadeus Petroleum contracted Mitchell Drilling to drill the wells using one of their mineral drilling rigs. The wells were continuously cored with 63mm HQ-3 from the 114mm intermediate casing to total depth. The drilling costs were considerably cheaper than the costs using a normal petroleum drilling rig. I contacted Doug Short to act as wellsite geologist. I hadn't seen Doug Short since our wellsite adventures with Delhi in the Cooper Basin in 1971 and it was good to catch up with him and chat over old times.

Amadeus Petroleum drilled Glenmore-1 in PEL 2 in February-March, 2000. The well encountered good oil and gas shows in the Early Triassic Lower Bulgo Sandstone and an open hole drill stem test was conducted. I was present for the testing and it was a miserable, cold day with a steady drizzle of rain **(Fig 37)**. The test began with a strong blow and, as expected, eventually died. The reservoirs were simply not good enough for liquid hydrocarbons to be expected to reach the surface. As we pulled the test string out of the hole, we waited anxiously to see what was in the pipe – oil or water? DST-1 recovered 12 barrels of fresh water and our mood quickly changed to that of the weather that continued to drizzle on our sorry, pathetic well-site.

Figure 37
Glenmore-1 well site on a miserable day

The well was plugged and abandoned.

Fairfield-1 was drilled immediately after Glenmore-1 in March-April, 2000 but didn't encounter any significant shows and was also plugged and abandoned. After the drilling of these two wells, Amadeus Petroleum withdrew from the farmout agreement. Sydney Gas continued with its coal seam gas drilling eventually reaching production of 4.5 PJ per year from its Camden Project by May 2001. Production problems and land access issues plagued the project and in January 2009, the company was taken over by AGL and delisted in April.

In February, 2001, I was invited to visit the Department of Mines Core Store to examine the cores from the Mundajini-1 and Boondawari-1 wells. Geologists from the Geological Survey were describing these cores and they were all laid out for inspection. It gave me a great opportunity to look at the complex environments under which the Spearhole Sandstone was deposited. I was also very impressed with the Core Store which had been completely restructured and modernised. All the cores were neatly stored and efficiently catalogued.

In January, 2002, Amadeus Petroleum changed its name to Amadeus Energy Limited.

Back in the west, EP-380 was renewed again in April 2003. Following the renewal of the permit, Amadeus Energy decided to conduct a geochemical survey over the mapped surface anticline that was to be the next drill target. The geochemical survey was carried out during 24-25 August, 2004 using Petrofocus operated by John Conolly. Following a geochemical survey for Empire Oil, we drove from Exmouth to Newman across the Pilbara. The scenery was spectacular with the outcropping hills of the Banded Iron Formation. We stopped at Tom Price for a coffee and were both surprised by the small size of the town given the major iron ore mining projects in the area.

Once we got to Newman, we checked into the motel and set up the chromatograph and associated equipment. Newman was a large and booming town, and totally different to Tom Price. The next day, we drove to the airport to meet the helicopter. At the airport, we were met by Max, our helicopter pilot. We loaded our equipment and headed off to the Savory Basin, some 200 kilometres to the east. It was going to be a two hour flight and with the hot, desert conditions, the helicopter was bouncing all over the sky. After about an hour of this, Max said "to hell with this – I'm going down lower and surf the dunes" Not knowing what this meant, we said nothing. We soon discovered that "surfing the dunes" meant the helicopter flying about 10 metres above the ground and hopping over the dunes at a similar height. Max assured us that natural hot air up-drafts lifted the helicopter over the dunes. But since the skids underneath the helicopter were clipping the vegetation on top of the dunes, we were not convinced. John and I were white-knuckled in the back of the helicopter trying to remember if our life insurance policies were paid up!

Eventually we got to the Canning Stock Route and began the survey. The survey consisted of the helicopter flying about one kilometre, landing, John would exit the survey, collect a sample and pop back into the helicopter and we would fly to the next sample point. Max decided that this was too slow so he took the door off the helicopter to make it quicker for John to jump in and out. He put the door under a tree so that it would reflect the sunlight and be seen from the air. So, away we went to complete the survey. Once the survey was completed, we returned to find the door of the helicopter. In 2004, GPS was not as accurate as it is today so when we arrived back at the location, we couldn't find the door. Max was very concerned – if he returned to Newman without the door (and perhaps also one of his passengers), he would lose his licence. Eventually, after circling for at least 15-20 minutes, we eventually found the door, landed and Max fitted the door back on the helicopter and we all "safely" returned to Newman. After this hilarious episode and "surfing the dunes", Max earned our enduring affection and the permanent nickname of "Mad Max".

The results of the geochemical survey were extremely negative with values so low that they indicated a complete lack of mature source rocks. I recommended that the permit be relinquished. Amadeus Energy held on to the permit for a few more years and it eventually expired in July 2009. That was basically the end of my contract with Amadeus Energy.

As a last hurrah, in August 2007, I was asked by Amadeus Energy to review three licences in the Weald Basin, onshore UK held by Magellan Petroleum. PEDL's 135, 136 & 137 covered some 500 square kilometres of the southern margins of the UK. The Weald Basin contains the giant 450 million barrel Wytch Farm Oil Field which extends from the onshore under Poole Harbour. The oil is trapped in Jurassic and Triassic sandstones sourced by Lower Jurassic shales. Magellan's licences had a number of prospects the most attractive of which was the Collendean Prospect, a tilted fault block updip from the Collendean Farm-1 well. This well, drilled by Esso in 1964, had excellent oil shows in the porous Jurassic Portlandian Sandstone and Greater Oolite. This prospect had a prospective resource of over 20 million barrels oil and I gave Amadeus a strong recommendation to farmin subject to negotiating more favourable terms than those being asked by Magellan. Unfortunately, Magellan would not renegotiate their farmin terms and Amadeus walked away from the opportunity.

In October, 2012, there was a reverse takeover offer by an American company Lonestar Resources. This was supported by a majority of the shareholders and resulted in a change of management at Amadeus Energy. The company changed its name to Lonestar Resources Limited in January 2013.

CLIMATE CHANGE

In 1995, the Kyoto Protocol claimed that carbon dioxide was a greenhouse gas and was causing the earth to warm up. I was somewhat confused by this as I thought everyone knew that the earth was warming up because we were coming out of "the Little Ice Age". There must be new scientific information so I started to do some research. And sure enough, there was new scientific information – the Greenland ice cores. These had been drilled on the permanent ice shelf in Greenland and air bubbles trapped in the ice had been analysed using mass spectrometers, scanning electron microscopes and gas chromatographs among other technologies. The contents of minor gases (such as carbon dioxide, nitrous oxide and sulphur dioxide) can be measured and based on oxygen isotope analyses, the average temperatures for each year can also been calculated. This gives us a clear picture of the earth's climate going back hundreds of thousands of years.

From the human civilisation point of view, the end of the last ice age can be clearly seen with the temperatures in the northern hemisphere rising a whopping 18°C from 12,000 to 9,000 years ago. The carbon dioxide content in the atmosphere also rose from 180 ppm (parts per million) to 280 ppm as it was driven out of the warming oceans. As the ice sheets gradually melted, it is estimated that the sea level of the oceans rose ~ 130 metres. Fortunately, there were no coastal cities back then, just nomadic hunters and gatherers who simply picked up their meagre belongings and moved to higher ground.

However, for the past 9,000 years, the carbon dioxide content in the atmosphere stayed constant at 260-280 ppm despite a number of warming and cooling cycles being identified **(Fig 38)**. These were a severe cold period around 2,000 BCE, a very warm period around 1,200 BCE, a warm period around 100 BCE (the Roman Warm Period), a cold period around 500 AD, a warm period around 1,000 AD (the Medieval Warm Period) and of course the Little Ice Age from 1,500 to 1,750 AD. The carbon dioxide content in the atmosphere has increased to 400 ppm over the last 100 years clearly caused by the industrialisation of human civilisation and the burning of fossil fuels.

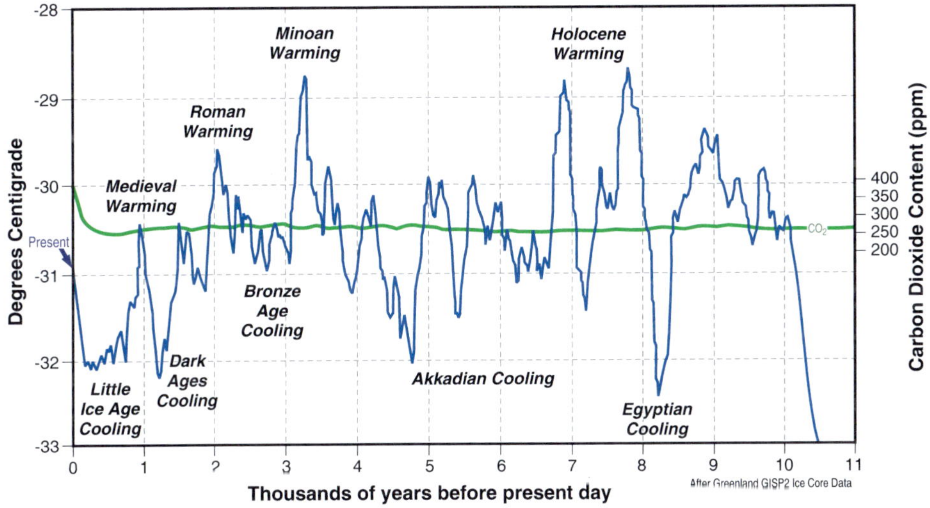

Figure 38
Global warming and cooling cycles based on ice cores

The current confusion comes from the fact that the carbon dioxide started to increase at the same time as the earth started to warm around 1,800 AD, the start of the industrial revolution. However, if increases in the carbon dioxide content in the atmosphere are the cause of our current global warming, what caused the warming periods in 1,200 BCE, 100 BCE and 1,000 AD where there were no increases in carbon dioxide and certainly no industrial societies?

The really scary thing about all this is that we don't really know what causes these warming and cooling cycles – and if we don't know what causes them, we certainly can't predict them. There is an impressive theory that they are caused by sunspot activity on the sun and we do know that during the Little Ice Age, sunspot activity on the sun was abnormally low. This theory was also supported by Michael Economides, Professory of Petroleum Engineering at the University of Houston at a talk he gave to the APPEA Conference in Adelaide in 2007. He considered that the main drivers of global warming are small increases in the sun's temperature according to Stephan Boltzmann Law and indicated by increased solar activity. In simple terms, this law states that the amount of energy radiated from a hot body is directly proportional to the fourth power of its thermodynamic temperature. So even a very slight increase or decrease in a body's temperature can cause a disproportional increase or decrease in the energy released. If this theory is true, it would not be very helpful in controlling global warming – we obviously can't control changes to the sun's temperature! But at least we might be able to predict the sunspot cycles.

In any event, global warming is not such a bad thing. History tells us that global warming during the Roman and Medieval warm periods was beneficial – crops yields increased and living conditions were relatively better. However global cooling is definitely bad as it causes lower evaporation from the oceans and increases the chances of drought conditions developing. The severe droughts caused by the 2000 BCE cold period contributed to the collapse of the Aakadian Empire along the Tigres-Euphrates river system and the collapse of the Harrapan Civilisation along the Indus River. Severe droughts almost destroyed the Egyptian Civilisation with civil uprisings ending the Old Kingdom but fortunately the rulers of the Middle Kingdom managed to restore order and learned from the mistakes of the Old Kingdom. They stopped building massive pyramids and began building more grain storage!

The cold period around 500 AD is believed to have caused mass migrations of people to the south including the barbarian invasions that destroyed the Roman Empire. The Little Ice Age was also time of crop failures in Europe along with mass starvations, food riots and increased diseases such as the Black Death. History tells us that life is not good during global cooling periods.

At the peak of the last major Ice Age, around 20-25,000 years ago, the carbon dioxide fell to a low of 180 ppm. This is because carbon dioxide is more soluble in the oceans at lower temperatures than higher temperatures. The process of photosynthesis ceases to function below 150ppm so this planet came within 30ppm of all plant life becoming extinct – no trees, no plants and no crops. This is a truely terrifying scenario. Now that human activity has raised the carbon dioxide content in the atmosphere to 400ppm, this possibility will be considerably reduced during the peak of the next major Ice Age. Don't panic – this won't happen for another 50,000 years.

SOCDET

In mid 1994, I was approached by Charlie Morgan to provide technical assistance for the Sydney Oil Company Drilling and Exploration Trust (SOCDET). SOCDET was at that time a listed trust that had been suspended from the ASX. Charlie had recently purchased the management company SOCDET Management Pty Ltd and he sold me a one third interest in SOCDET Management Pty Ltd in June, 1994. Our first task was to organise a reconstruction and recapitalisation of SOCDET. Once this was completed, SOCDET raised about one million dollars by issuing 15 million capital units and, in November, 1994, the ASX removed its suspension.

The only asset of SOCDET was a 2% interest in WA-208-P in the offshore Carnarvon Basin. This permit had a well commitment coming up and the operator MIM Petroleum wanted to drill a prospect that most of the joint venture hated. The prospect was a structure along the Legendre Trend where I had worked in my old Woodside-Burmah days. I thought the prospect looked pretty good with reservoir, seal and plenty of source rocks and oil migration out of the Lewis Trough. SOCDET needed to raise some money and needed some drilling activity. SOCDET also needed more equity in WA-208-P. We negotiated to purchase Command's 22% interest in the permit. Command's management didn't like the prospect but they didn't want to withdraw from the permit completely. Charlie devised a cunning plan where Command would sell their interest in the permit and SOCDET would give Command the option to buy back in for 11% by carrying SOCDET for its 22% share of the first appraisal well if the exploration well was successful. This satisfied their dilemma so they agreed and SOCDET now had a 24% interest in WA-208-P. Prior to entering the permit commitment year, Minora Resources and Phillips Petroleum withdrew from the permit. Much to the annoyance of the operator MIM Petroleum, SOCDET took up its pro-rate share of their interests and now had a 45% interest in WA-208-P. This permit had a very expensive offshore well commitment and SOCDET had very little money in its bank account. Charlie was getting a little nervous and asked "Bevan - are you sure you know what you're doing?" A very reasonable question but I couldn't tell him that SOCDET was playing the oil industry equivalent of a game of "Russian roulette". So I assured him that the WA-208-P permit was prime offshore Carnarvon Basin acreage and, as long as we weren't greedy, we'd be able to farm the permit out.

Both Ampolex and Woodside were interested in farming into WA-208-P. I offered them both a very soft "third for a quarter" deal – they would fund SOCDET's 45% share of the well cost to earn a 33.75% interest in the permit. Given more time and a stronger bank balance, we could probably have got a better deal but we didn't have the time and we didn't have any money. Once the cash call for the drilling of the well was made, we could be put in default and lose our entire interest (and Ampolex and Woodside knew it!). Woodside came back wanting a cap on the well cost but SOCDET couldn't agree to any cap on the well cost – we simply didn't have the money even for that! Ampolex agreed to the terms without conditions and got the deal. The well Spinner-1 was drilled in March-April, 1995 but was plugged and abandoned as a dry hole. SOCDET could not afford to participate in the ongoing exploration costs of an offshore permit, so we sold its remaining 11.25% interest in WA-208-P to Woodside for $600,000. Not a bad deal – SOCDET had just made $600,000 cash without spending a cent on exploration!

Prior to selling out of WA-208-P, SOCDET earned a 3.75% interest in

WA-254-P by participating in Bellerophon-1 in WA-254-P operated by Apache Energy. This prospect extended across the permit boundary with WA-208-P and the well was drilled in June-July, 1995. SOCDET also participated in the Helicon-1 well in WA-254-P but had farmed down to a 1.0714% interest. The well was drilled in July, 1996 but was a dry hole. In August 1997, SOCDET sold its interest in WA-254-P to Tap Oil. While all this was going on, SOCDET was able to raise a further $2.2 million by share placements.

As part of the deal on WA-208-P, SOCDET purchased Command Petroleum's 12.8-13.3% interests EP 129 and L6 and L8 in the onshore Canning Basin. Once SOCDET was re-listed, SOCDET participated in the drilling of Sundown-3H then operated by Santos. This well was drilled during October-November, 1995 and was the first horizontal well in the basin. The well flowed 1,174 barrels oil per day but by March 1996, the water cut increased, and the well settled down to oil production similar to vertical wells. The casing could not be set far enough along the horizontal section to get past the water zone and this may have contributed to the early water cut.

After this I spent some time looking at other potential prospects in EP-129. I was intrigued by the dolomites of the Yellow Drum Formation in the Blina Field. It was porous and permeable and probably contributed most of the oil production from Blina. To the north of Blina, this dolomite pinched out to create a potential stratigraphic trap. In addition, two wells drilled in the area, Janpan-1 and Janpan North-1, had excellent oil shows in this dolomite. This prospect needed more seismic to better define a drilling location but we couldn't convince the joint venture to do this

Instead the EP-129 joint venture drilled Millard-1 in October 1996 but this well was also a dry hole. After all this disappointment, SOCDET sold its interests in EP-129, L6 and L8 to Shebandowan Investments for $175,000.

SOCDET farmed into ATP 565 P in the Surat Basin by contributing 14.2% to the drilling costs of Green Swamp-1 and Riverview-1 to earn a 10% interest. These wells were drilled by Mosaic Oil in December 1995 but both were dry holes.

Also during 1995, Ross Macleod introduced SOCDET into an international opportunity in the Philippines. Fletcher Challange was selling its interest in GSEC 77 in the Palawan Basin. The original permit GSEC 49 was awarded to a Petrocorp Exploration operated consortium in January 1989. Petrocorp was taken over by Fletcher Challenger and the permit changed to GSEC 77. Between 1991 and 1993, the GSEC 77 joint venture drilled Calauit-1, Calauit South-1, Busuanga South-1 and Coron-1 and acquired 1,000 kilometres of 2D seismic.

I had a quick review of the permit and thought that the basin had a working petroleum system with good potential for further discoveries. It was also a huge permit covering over 14,000 square kilometres so Charlie and I travelled to Manila to review the technical data and negotiate terms for purchasing their interest. GSEC 77 turned out better than I expected. There were a number of un-drilled Miocene reefs plus, the Calauit and Calauit South oil discoveries and a giant exploration prospect called Coron North. The Coron North structure covered an area of 17 square kilometres with vertical closure of 270 metres. Potential resource estimates ranged from 280 million up to a billion barrels oil. It was however located in 350 metres water depth so the well cost would be very expensive. Offsetting this, the Philippines had the most attractive PSC terms in South East Asia and the economics of any oil discoveries in the country would have been excellent. SOCDET purchased a 20% interest in GSEC 77 for $200,000. The sale was approved by the Department of Energy in December 1995,

After ongoing technical studies of the Palawan Basin, SOCDET farmed into SC 14 to earn a 30% interest by acquiring a 3D seismic survey with option to drill a well. SC 14 contained the Nido - Matinloc reef trend plus the Galoc and West Linapacan oil discoveries. SOCDET also farmed into SC 6B containing the Cadlao Oil Field to earn a 30% interest. It now had a significant acreage position in this very prospective basin. SOCDET also acquired a 5% interest in GSEC 78 in the Visayan Basin in the Philippines. The operator was Stirling Petroleum and the trust participated in an application in the Cagayan Basin to the north of Luzon and GSEC 84 was awarded in March, 1996. It was also operated by Stirling Petroleum and contained the large undrilled Fuga Island Prospect, a reverse faulted anticline.

Back in Australia, SOCDET purchased a 100% interest in EP 366 from Empire Oil in December, 1995. This permit is located in the Fitzroy Sub-basin in the Canning Basin and contains the Selene Prospect, an anticline updip from Cycas-1 where oil shows were encountered in the Carboniferous.

SOCDET also purchased a 100% interest in EP 371 from Wildcat Australia in September, 1995. This permit is also located in the Fitzroy Sub-basin of the Canning Basin and contains the Audrey Prospect. The trust then exchanged a 30% interest in EP-371 to Stirling Resources for a 30% interest in EP-390, on the Broome Arch of the Canning Basin. SOCDET also acquired a 30% interest from Stirling Resources in EP-391 containing the Mahe anticlinal structure.

With all our common interests in the Canning Basin, SOCDET and Stirling Resources decided to join forces and present our Canning Basin permits and prospects at the North American Prospect Expo (NAPE) in Houston during January to February, 1996. NAPE was held every year in Houston and various American oil companies would set up booths to present their prospects for drilling by companies looking for drilling opportunities. We prepared a very glossy brochure showing the Mahe, Praslin, Admiral and Chingal prospects in EP-390 & EP-391 for Stirling; and the Audrey, South Scarpia and Selene prospects in EP-371 & EP-366 for SOCDET.

I travelled with Terry Neilsen from Stirling Resources to Houston to present these prospects. We attended NAPE and set up meetings after the conference but to no avail. It was all too frontier for the Americans and we couldn't attract a single farminee. As an example of how risk averse the Americans were, this fellow walked into our booth and I asked him if his company was interested in international exploration. He replied "Son – anything outside of Wyoming is international to me!" After this very disappointing result, SOCDET sold its interests in EP-366, EP-371 and EP-391 to Stirling Resources for a 7.5% interest in GSEC 78 (taking SOCDET's interest to 12.5%) and a 5% interest in GSEC 84 (taking SOCDET's interest to 10%).

On the personal front, in June 1996, we sold our house in Carine and moved into an apartment in Observation Rise along the Scarborough beachfront. Alex was still at university and moved in with us but it was a three bedroom apartment so there was plenty of room. Mo and I then went off to attend the APPEA Conference in Darwin in June 1996. APEA had changed its name to the Association of Petroleum Production and Exploration Association now abbreviated to APPEA. After my recommendation to Ashland Oil, the most interesting paper for me at the conference was by Crowley and Collins on the Stag Oil Field. Rick de Boer also gave his paper on the Gilmore Gas Field at this conference.

In September, 1996, Ampolex was taken over by Mobil Oil, mainly for their interest in the Gorgon Gas Field. This was a terrible event for the Australian oil industry as Ampolex was a very successful, independent Australian oil company with a very good technical team and an excellent acreage position in the offshore Carnarvon Basin. In my opinion, it could have become a multi billion dollar company particularly as it also had an 11.6% interest in the 250 million barrel Kutubu oilfield in PNG as well as other interests in the Moran and Gobe oil fields on trend. It also owned a PSC in Java where it had defined and was preparing to drill a large compressional anticline. This was one of the untested reverse faulted anticlines I had mentioned in my report to Pertamina (p103).

Drilling operations had been held up by a dispute with Pertamina over its equity in the PSC and this was unresolved by the time of the Mobil takeover. The structure was eventually drilled by ExxonMobil in 2001 and resulted in the discovery of the giant Cepu Oil Field with estimated reserves of 250 million barrels oil. The dispute with Pertamina continued until 2006 when equities in the PSC were finally agreed at Pertamina 45%, ExxonMobil 45% (courtesy of Ampolex) and the Central and East Java local governments 10%. The Cepu PSC reached peak production of 200,000 bopd from Cepu and additional oil discoveries in the PSC in 2017.

While all this was going on, John Howard became Prime Minister of Australia in November, 1996 and I drank my 1976 bottle of Penfolds Grange Hermitage to celebrate. It was absolutely beautiful and a fitting celebration to the man who was to turn out to be, in my opinion, Australia's best prime minister.

In early 1997, Charlie and I were invited to visit PGS's office in Stavanger. We decided to also visit the Maersk office in Copenhagen to make farmout presentations on our acreage in the Philippines. We flew to London and visited a financial institution to promote the assets of SOCDET. The next day we flew to Stavanger. Although it was winter, I remember that it was a clear day and as the plane came into land, it flew right up the fjord and gave me a great view of a classic glacial valley. Stavanger was a very pretty city with very steep rooftops to handle the heavy winter snows. We visited the PGS office, they showed us their latest seismic technology and we discussed the pros and cons of multi-client seismic surveys.

The next day, we flew to Copenhagen to meet with Maersk. It was bloody freezing – the temperature was -11°C. It was so cold that the harbour had frozen over and as the various cargo ships ploughed through the water, there were great chunks of ice blocks bobbing about in the harbour. To make matters worse, the wind blew off the North Sea and cut through you like a knife! The meeting with Maersk was polite but not very enthusiastic so we weren't holding our breath about getting a deal with them.

During 1997, SOCDET applied 100% for SPR94 at the southern end of Palawan Island just north of Sabah. The existing seismic showed a giant Miocene reef with what appeared to be a hydrocarbon/water contact. The prospect needed additional seismic to properly define it for drilling and SOCDET contracted the AGSO Rig Seismic to acquire this seismic. The vessel was being mobilised to Korea for Beach Petroleum and had to go virtually past our SPR94 on its way. In 1996, Beach had amazingly been able to negotiate a 28,000 sq km PSC off the east coast of North Korea covering a basin with over 5000 metres of sedimentary section. The Rig Seismic vessel was originally chartered in 1988 by the old BMR to acquire seismic over Australia's continental margins outside the areas currently being explored by the Australian oil industry. These areas extended east to Lord Howe and Norfolk islands and included the offshore part of Australian Antarctic Territory. Under the International the Law of the Sea, Australia had to carry out scientific investigations over these remote areas in order to claim them as part of Australia's legal resources. Dave Falvey was supervising this project and in my opinion the Rig Seismic did a technically and financially sound job on behalf of the Australian taxpayers. The BMR became AGSO and when not acquiring seismic for the continental margins project, it began contracting out the Rig Seismic for private work, including SOCDET's and Beach Petroleum's surveys. The petroleum industry objected and the new Liberal government, after lobbying from APPEA, eventually sold off the Rig Seismic.

SOCDET acquired 500 kilometres over the reef prospect and other leads in SPR94 in May 1997. The Rig Seismic then went to Beach who acquired 1000 kms seismic over their North Korean PSC. The most frustrating part of our survey was the clearing of fish traps which the Filipino fishermen set in the open ocean. They are mostly floating wooden cages but you don't want to run over them with expensive seismic cables, as well as creating antagonistic relations with the local fishing communities. The AGSO Rig Seismic also acquired 153 kilometres of 2D seismic around Fuga Island in GSEC 84.

During 1995, SOCDET units varied from 2 to 5 cents per unit but during the first half of 1996, the units gradually rose in recognition of its strategic acreage position in the Philippines. In February 1996, SOCDET announced a $22 million exploration program in the Philippines. In October, 1996, SOCDET reached agreement with Novus Petroleum Limited to issue them 53.8 million shares at 16 cents per share to raise $8.6 million and give them control of the trust. SOCDET also undertook a 3 for 4 rights issue to existing shareholders at the same price to raise a further $4.6 million. This was approved by the unitholders in January 1997 and allowed SOCDET to participate in the acquisition of the 3D seismic surveys, the testing of the Calauit Field, and securing a further 16.7% interest in GSEC 77 plus a 30% interest in SC 14 containing the Nido, Matinloc and West Linapacan oil fields. The SOCDET units also increased in price to 24 cents per unit by the end of 1996.

SOCDET negotiated with both Geco Prakla and PGS Nopec to acquire the 3D seismic survey and managed to get the cost down to $US4.35 million for 1,000 square kilometres on a multi-client basis. The successful contractor was the Geco-Prakla Seisventurer and the NW Palawan 3D seismic Survey was acquired in September and processed in October, 1997.

In April, 1997, Mo and I attended the APPEA Conference in Melbourne. There were a number of interesting papers at this conference. AGSO gave a paper outlining the six main Palaeozoic petroleum systems in northwestern Australia. There was a paper on the Permo-Triassic boundary using both palynology and carbon isotope ratios. BHP Petroleum gave a paper on the Gwydian oil discovery on the southeastern flanks of the Browse Basin.

After the conference, we decided to take a short holiday along the southwest coastline of Victoria. We hired a car and drove down to Torquay and then on to the famous Bell's Beach. The southwest coastline of Victoria is famous for its stunning views including the impressive "Seven Sisters". It is also a geologist's paradise with the road winding its way along endless cliffs through the Tertiary sediments of the Otway Basin. We were blown away by the towns of Lorne and Apollo Bay which were so picturesque and delightful to visit. We drove as far west as Warnambool and from there, connected north to the main freeway back to Melbourne and flew back to Perth.

In July, 1997, we returned to Mauritius for a holiday. After our fond memories, including my tequila adventures during our visit in 1975, we were looking forward to seeing the island again. It hadn't changed much except for a lot more tourists. We stayed at the Tres Biche Resort and had a chalet with spectacular views overlooking the bay and reef. We visited the market in Port Louis and drooled over all the varied and beautiful spices there. As before, we couldn't buy any to bring home.

On the international front, in July 1997, Hong Kong reverted to China after 156 years of British rule. It was also the beginning of the Asian Financial Crisis. It started in Thailand and spread to Indonesia and South Korea. The crisis had major repercussions for our northern neighbours. In May, 1998, the devaluation of the Indonesian rupiah caused widespread rioting and forced the resignation of President Suharto after being in power for 30 years.

Back to Australia, Chevron had discovered that during the production of the Saladin Oil Field, the existing wells were only draining the Barrow Group sands and the oil/water contact in the Mardie Greensand had not moved. Although it was a poor reservoir, it did have thin fine grained sandstone interbeds and there was too much oil-in-place to be ignored. So during 1997, they drilled a number of horizontal but sinusoidal development wells through the Mardie Greensand where they achieved flow rates of up to 4,000 barrels oil per day.

It was also in 1997, that the Carnarvon Basin overtook the Gippsland Basin as Australia's main oil producing basin. Boosted by new production from the Wanaea, Cossack, Wandoo and Stag oil fields, oil production from the Carnarvon Basin rose to over 300,000 bopd.

GSEC 77 had an exploration well commitment so the joint venture decided to undertake an extended well test ("EWT") of the Calauit-1B well. This well, drilled in 1991, had intersected a 113 metre gross oil column in the Miocene Nido Limestone and was therefore ideal for an EWT to monitor the well performance to determine optimum production rates. It would also help with the evaluation of the exploration potential of GSEC 77.

Therefore, the joint venture applied to exchange the well commitment for an Extended Well Test (EWT) of the Calauit-1B well. This was approved by the Philippine Department of Energy.

Prior to the testing of Calauit-1B, SOCDET farmed into Dragon Oil, Basic Petroleum, Palawan Oil & Gas, and Trans Asia to increase its equity in GSEC 77 to 43.427% by carrying the parties for Calauit-1B EWT. The Calauit-1B EWT was operated by PNOC using Schlumberger as the Project Manager. The drilling rig "Pioneer Driller" was contracted to re-enter the well and conduct the EWT. Reservoir studies predicted a flow of 10,000 bopd with 5.5" tubing. The Calauit-1B EWT was conducted at the end of 1997. Initial production was 6,500 bopd with no water, and was gradually increased to 9,535 bopd. Unfortunately water cut also increased and the EWT was abandoned at 1,600 bopd with 9,200 b/d of water. Those sneaky "rocks were smiling at us" again.Novus Petroleum changed SOCDET to a limited company and relisted it as Nido Petroleum Limited on 30 July, 1999. Nido Petroleum went on to participate with a 22.28% interest in the development of the Galoc Field which began oil production in 2008. The company eventually farmed out and drilled the Coron North Prospect (renamed Gindara-1) in May-June, 2011. Unfortunately, Gindara-1 was plugged and abandoned as a dry hole.

THE SPACELIFT STORY

After the sale of SOCDET to Novus Petroleum and its conversion from a trust to Nido Petroleum, there was no further use for SOCDET Management Pty Ltd which Charlie and I still owned. It still had some funds in it but rather than wind it up, Charlie came up with another entrepreneurial proposition. He had been approached by Yuri Sokol who had migrated to Australia in 1979 and become an Australian citizen. He had contacts in the Russian aerospace and defence sectors. Dr Sokol wanted to use Russian rockets to send commercial satellites into space and use Australia as the launch site. I said to Charlie "This is the craziest idea I've ever heard of but let's give it a go anyway" So we decided to use the funds to see what could be done.

Spacelift Australia Limited was registered in August 1999. The company was to utilise the highly reliable Russian START family launch vehicles modified from their Intercontinental Ballistic Missiles (SS-25's or Topol's). Some 2,000 of these vehicles were available due to SALT disarmament agreement between the Russians and the Americans. In addition, a treaty between USA and the Russian Federation, executed in 1993, confirmed that the START family systems could be used commercially. If they weren't used commercially, they would have to be systematically destroyed. They had also been successfully launched 420 times (72 times in front of US observers) without a single failure.

The market niche for the Russian launch vehicles was to send small payloads (under 800 kg) into Low Earth Orbit (LEO). Market research indicated that this market was estimated to be worth $8 billion per annum. The marketing was also enhanced by the failure of other LEO communication systems such as Iridium and Orbcomm. Dr Sokol believed Australia was the perfect location for this project. Although Australia had strong ties with the US, it was not a member of NATO. It also had a good climate and the old Woomera site still had infrastructure that was no longer being used. In December 1999, a technical team from STC Complex-MIHT and Puskovie Uslugi (STC/PU), Russian Government Entities, visited Australia and validated Woomera as the preferred launch site in Australia.

Negotiations in 2000 led to an agreement between SOCDET Management (owned by Charlie and myself), and Spacelift Australia Limited owned by Yuri Sokol and Dilankex Pty Ltd, also owned by Yuri Sokol. I should point out that Charlie did all the negotiating – I was merely a passive investor. Dilankex had an agreement with STC/PU to provide the launch services to Dilankex. Dilankex would assign these rights to Spacelift and in return Spacelift would establish a commercial space launching business in Australia. SOCDET Management would attract investors who would fund the establishment of the commercial space launching business in Australia. Spacelift would need to deal with twenty three Federal and State agencies for the approval and establishment of Woomera as the launch site. Once the approvals were in place, contracts could be negotiated and signed, the Russian rockets brought to Australia and transported to Woomera and the launch of these rockets and their payloads into LEO undertaken. Senator Nick Minchin, Federal Minister for Industry, Science and Resources, and the Premier of South Australia, John Olsen, were supportive.

Unfortunately, the project stalled due to the number of approvals required and the number of government departments to deal with. The project eventually died from government bureaucracy and inertia, and a potential billion dollar industry was lost to Australia.

In October, 2015, Australia had to use facilities in French Guiana, South America to launch a satellite for our National Broadband Network. I immediately thought of Spacelift and what a damn shame it was that our government didn't grab that opportunity all those years ago. This satellite alone would have saved us $500 million! To make matters worse, in March 2018, there was a proposal by the Ukraine to use northern WA as a major launch site for commercial launches on behalf of our Asian partners Japan, Singapore, South Korea and Indonesia. Charlie was certainly "ahead of his time"!

The PNG – Queensland Pipeline

In early 1997, there was a proposal to build a 3000 km pipeline from the gas fields in Papua New Guinea to Gladstone in Queensland to supply the increasing domestic gas market in the eastern states. Gas supplies from the Surat and Bowen basins, and the Queensland side of the Cooper Basin had been declining and power generating companies were becoming concerned. The pipeline was to start at the Exxon operated Hides gas fields, then run down to the Chevron (then) operated Kutubu fields and extend down to the Iehi and Barikewa gas fields and then to the coast. From there it would go to an offshore platform, the Kumul Terminal, for the liquids to be stripped out and shipped to Asian markets. From the platform, the gas would continue along the sea bed to the tip of Cape York and then down to Gladstone. It was a massive project estimated even then to cost $US3 billion. This eventually rose to $US6.5 billion by 2002. I always thought that this was a crazy idea. For $US6.5 billion, the industry could have built a pipeline across Australia connecting the North West Shelf gas fields to Moomba and then onto the eastern states via the existing pipelines. This pipeline would have only been 2,400 kms long, much less complicated, and certainly a lot cheaper, to build. In addition, Australia would be using its own gas and not importing it from another country.

By 2007, LNG prices had risen to a point where ExxonMobil proposed that an LNG project would be a better economic prospect than the pipeline to Queensland. Front end engineering design began and a final investment decision was made in 2009. The final cost for the PNG LNG Project was revised to $US19 billion in 2012 and the project was finally completed in 2014. The first cargo of LNG from the project left for Japan in May, 2014.

The Coal Seam Gas Boom in Queensland

The gas supply shortage for the domestic Queensland gas market forecast in 1997 was to be initially solved by the coal seam gas exploration boom. I completely missed this boom and it hurts my pride to write about it here. It began in earnest about 1996 and over the next decade, the exploration effort was to discover over 4 TCF of 2P coal seam gas reserves. Most of this gas was discovered by Origin Energy and Santos who had the lion's share of the acreage in the Bowen-Surat basins. The rest was discovered by small exploration companies and led to a spate of takeovers which greatly increased the wealth of their shareholders. In June 2008, Shell signed a $776 million alliance with Arrow Energy for its CSG assets; in October 2008, British Gas ("BG") bought Queensland Gas for $5.6 billion; in June 2009, BG bought Pure Energy for $1 billion; in July 2010, Shell and Petro-China bought Arrow Energy for $3.45 billion; and in June 2011, Santos bought Eastern Star Gas for $924 million.

All these takeovers were undertaken to establish gas reserves to supply initially four LNG export projects; the BG Curtis Island LNG Project, the Santos-Petronas Gladstone LNG Project; the Origin Energy-ConocoPhillips Australia Pacific LNG Project and the Shell-PetroChina Arrow LNG Project. The latter of these projects never eventuated but in early 2016, Shell bought BG itself for a trifling $US53 billion. According to a BG spokesperson, the lack of collaboration between the Queensland CSG LNG projects was "all the more staggering in the light of high construction costs" and contributed to the large cost over-runs. The total cost of all three LNG projects ended up costing almost $80 billion and with the then low LNG prices, the economic return on these projects was expected to be considerably reduced.

BG was first cab off the rank and loaded its first cargo of LNG in December 2014 with Santos following in October 2015 and Origin Energy in January 2016. According to Citi at the time, the planned production rates from the three LNG plants would consume two and a half times the domestic demand for the entire east coast. Although electricity demand in the eastern states had been falling since 2011, due mainly due to increased energy efficiency and solar panels, the gas diverted to these LNG plants caused serious cost increases for power generating companies in eastern Australia starting in 2016.

WEST OIL

After the success of SOCDET, Carmichael's suggested that Charlie and I float a new company to take advantage of the slowly rising oil price and the ongoing successes in the Timor Sea. Between 1994 and 1997, there had been the following discoveries in the Timor Sea: Elang, Laminaria, Kakatua, Bayu-Undan, Elang West, Corallina, Juhal, Buffalo, Kakatua North, Tenacious and Sunset. West Oil NL was incorporated on 24 January 1997. The company acquired a 50% interest in AC/P 22 and a 33-1/3% interest in ZOCA 96-16, both in the Timor Sea. AC/P 22 contained the Puffin-1 and -2 oil discoveries made by Arco Australia in 1974. Puffin-2 tested at 4,608 bopd from a sandstone in the late Cretaceous Puffin Sandstone. Puffin-1 recovered oil from a separate sand within the Puffin Formation. ZOCA 96-16 contained the Minotaur-1 well drilled by Marathon in 1994. This well had good oil and gas shows in the Jurassic Elang and Plover formations. Based on existing seismic, there could have been a structure updip from Minotaur-1 but additional seismic would have to be acquired to confirm this.

West Oil also applied for an area in the offshore Perth Basin south of Rottnest Island. This area contained a number of wells with good oil shows and was awarded as WA-272-P on 12 August 1997 for a six year term. West Oil had a 50% interest and was the Operator.

West Oil also farmed into EP-342 (Leatherback block) to earn a 10% interest by contributing to the costs of drilling another well on the Leatherback structure. Leatherback-1 was drilled by LASMO in 1993 and tested at 2,332 bopd from a 7 metre oil column in the Triassic Mungaroo Formation. Post well interpretation of the seismic indicated that the well may have intersected a fault and there may have been a significant oil column on the upthrown side of the fault. Leatherback-2 was drilled in December, 1997 but was plugged and abandoned as a dry hole. West Oil withdrew from the permit in 2000.

West Oil also tried to farm into WA-234-P. This permit covered the West Barrow-1 well with its associated oil and gas shows in the Barrow Group. Based on this well, the Federal Government put the area up as a "Cash Bonus Bid" as opposed to the usual work program bidding system, a decision criticised by most of the oil industry. Ampolex was the only bidder and put up a $1 million cash bid. Most of the oil industry thought that this was too much but I thought it was an absolute steal. For $1 million, they got title to the permit area for 6 years and could explore the area at their own pace without being pressured by government deadlines. I was later proved to be correct on this one!

In 1990, I had worked on the West Barrow-1 and -2 wells with Ross Macleod who was then contracted to Hallmark Minerals, the last permit holder of the old West Barrow permit WA-64-P. The permit was about to be cancelled by the Federal Government in November 1989, so in January 1990, Petroz, BP Petroleum Development, Petrofina Exploration Australia and Stirling Petroleum NL withdrew from the permit and assigned all their interests to Hallmark Minerals who became the sole permit holder. Discussions were being held with Japex and Marathon to farm-in and jointly renew the permit in a desperate attempt to save the permit from cancellation. We even had access to an old Reading & Bates report dated October 1982, estimating the potential oil reserves in the Barrow Group sands in West Barrow to be 16 million barrels recoverable.

West Barrow-1/1A was drilled in the first half of 1982 – it was a very expensive well due to drilling and mechanical problems. The well hit the top of the Barrow Group at 2504 metres where there was a drill break and a 40 unit gas show with up to C4 hydrocarbons. No oil shows were recorded on the mud log but there was fluorescence and cut in SWC's. The 9-5/8" casing was set at 2507 metres. Despite these difficulties, log analysis indicated possible moveable hydrocarbons down to 2513 metres. Moveable hydrocarbons could also be interpreted in deeper sands at 3228-32m and 3290-3320m. In the upper of these sands, 2% oil was recorded in the mud when drilled and in the lower sand, there was a drill break with associated oil shows and the well kicked.

West Barrow-2 was drilled in mid 1985 at a location interpreted from seismic to be up-dip from West Barrow-1. Although up-dip in time, it was down-dip in depth and the Barrow Group sands were water wet. By correcting for the obvious velocity gradient, a depth structure could be mapped at the top Barrow Group up-dip from West Barrow-1 to the south.

By 1997, Ampolex had been taken over by Mobil Oil who were only interested in their 14.28% equity in the Gorgon project. Mobil was farming out its 100% interest in WA-234-P. West Oil negotiated with Mobil Oil to earn a 10% interest in WA-234-P by paying 20% of the cost of Woolybutt-1. Unfortunately for West Oil, Woolybutt-1 was drilled in April, 1997 before the prospectus could be finalised and an unconditional agreement with Mobil Oil executed. Based on the shows in West Barrow-1, I always thought that Woolybutt-1 would discover condensate rich gas but instead, it tested 7,900 bopd. The oil business can be somewhat like fishing – there are lots of stories about the "one that got away" and Woolybutt got away from West Oil big time! Despite being correct about Ampolex getting WA-234-P "for a steal", I still went out and got seriously pissed. The Woolybutt Oil Field was eventually abandoned in 2012 after producing over 35 million barrels of oil.

Due diligence for the West Oil Prospectus started in May and continued to July, 1997. Meetings were held with Edwards Thompson on the solicitors report and with Norwest Energy on their operated ZOCA 96-16. Robsearch Australia did the Independent Geologist's report for the Prospectus and after many broker presentations in Sydney and Melbourne during July, West Oil eventually listed on the Australian Stock Exchange on 24 July 1997. The oil price was then $US22/bbl but by end 1998, the oil price reached its next cyclical low of $US10-80 per barrel. ZOCA 96-16 had been awarded on 14 November 1996 for a 30 year period to a joint venture consisting of Moondance Energy Pty Ltd (a subsidiary of Norwest Energy) 33-1/3% and operator, ZOCA Pty Ltd (a subsidiary of West Oil) 33-1/3% and Indo Pacific Energy 33-1/3%. The first two years obligatory work commitment was to reprocess and interpret the seismic in the permit. Robsearch Australia was contracted to interpret the seismic and the results of their mapping were presented to the joint venture on 9 September, 1997. The main outcome of the interpretation was the delineation of the Coleraine Prospect, a large fault block on trend with the recent Elang and Bayu-Undan discoveries.

On Thursday 11 September, 1997, West Oil shares rose from 35c to 47c. On Friday, 12 September, West Oil was queried by ASX. As West Oil was not involved in any operations, there was no obvious reason why our shares had risen. The only explanation was the definition of the Coleraine Prospect had leaked to market which had reacted to the news despite the fact that the prospect was not farmed out and would not be drilled for another three years. On Monday, 15 September, West Oil made an ASX announcement on ZOCA 96-16 and the Coleraine Prospect. Then the Australian Securities Commission threatened civil proceedings against West Oil for breaching ASX Listing Rule 3.1. I was stunned by all this – none of the directors had traded in West Oil shares on 11 September and no ASX release had been made as I considered (incorrectly as it turned out) that the seismic mapping of a prospect was not material to the company, at least not until it was farmed out and ready to drill. After much legal negotiations with the ASIC (the nightmares of the United Canso lawsuit came flooding back), West Oil finally settled with the ASIC by signing a deed over the non-compliance with full disclosure rules and the payment of a $5,000 fine. The board quite rightly blamed me for this fiasco and I was very badly shaken up by the whole affair. It was very clear to me that my interpretation of what's material for disclosure by an exploration company and the ASIC's interpretation of what's material were in fact galaxies apart. At the end of 1997, I resigned as Managing Director becoming Technical Director and Charlie Morgan was appointed Managing Director.

In September 1997, Charlie and Alan Stein from Ikoda went to Cape Town to meet with Sasoil to discuss farming into their acreage off southern Africa where they had made several small gas discoveries. I joined them later to review the prospects and evaluate their potential commerciality. I remember flying over the Mozambique coastline and looking down at endless white beaches very similar to the Western Australian coastline but without any towns or developments. Cape Town was a very nice city dominated by Tabletop Mountain and surrounded by beautiful beaches. Our meetings with Sasoil were very interesting as they kept asking us what our market capitalisation was and how many employees we had. When Charlie said "three", that was just about the end of negotiations. Obviously Sasoil didn't deal with small companies. While returning to our hotel, I watched the clouds roll over the top of Tabletop Mountain and drop down like a waterfall. About half way down, the clouds reached a certain temperature and simply disappeared – very spectacular! Back in our hotel, Charlie was reading the financial section of the local newspaper and observed that there were virtually no small resource companies listed on the South African

Stock Exchange – their mining industry was dominated by a few giant companies who controlled virtually all their mineral resources. This was in total contrast to the Australian mining industry where there were hundreds of companies ranging from giant BHP all the way down to the "penny dreadfuls". I may be biased, but in my opinion, our system is far superior being diverse, vibrant and so exciting. Our system also resulted in more exploration and, "surprise, surprise", more mines.

Following its listing on the ASX, West Oil progressively purchased a 100% in AC/RL-1 (Talbot Oil Field) individually from Santos, Beach Petroleum and Norcen Energy Resources, the previous joint venture parties. After reprocessing the existing 3D seismic, West Oil renewed AC/RL-1. in February, 2000. In November 1997, I flew to Sydney to meet with John Carmody of Mosaic Oil and Dave Bennett of Indo Pacific Petroleum to discuss a joint bid on an offshore gazettal block in the Timor Sea, adjoining AC/RL-1.

It was in on 25th November 1997 that I attended my first Lilac Hill cricket match as a guest of Empire Oil. They had a marquee which they shared with Tollhurst stockbrokers. It was a terrific day with lots of food, drink and good company. The match was between the ACB XI and South Africa. South Africa won by 31 runs.

AC/P 22 had been awarded on 30 May, 1997 for six year period to a joint venture consisting of Cartier Oil Pty Ltd (a subsidiary of West Oil) 50% and operator, and Westranch Holdings Pty Ltd (a subsidiary of Norwest Energy) 50%. The obligatory work commitment was to acquire a 3D seismic survey over the Puffin oil discoveries in Year One and drill an appraisal well in the Puffin area in Year Three. Strictly speaking, appraisal wells are not counted as wells in the work program bidding system since "wells" means "exploration wells". But I firmly believed that, due to the velocity variations and problems with depth conversion, any well in the Puffin area was really an exploration well. In fact, given its appraisal drilling history, any appraisal well in the Timor Sea is an exploration well! However, I put "appraisal well" in the application to make sure there was no future argument with the government.

An amusing story after the award of AC/P22 was West Oil received a letter from some very serious lawyers in New York. The letter threatened legal action for using the name Cartier without the permission of the very famous Cartier jewellers. Of course, West Oil had taken the name Cartier from the Ashmore Cartier Islands in the Timor Sea but we were not going to argue with them over such a trivial matter. If they had contacted West Oil directly, we would have agreed to change the name without all those hefty NY legal expenses they must have incurred. We changed the name to Ashmore Oil Pty Ltd.

PGS, a seismic acquisition company, had decided to acquire a multi-client 3D seismic survey over part of the Timor Sea. By this method, PGS would acquire the seismic and then have the ability to sell it to as many companies they could find, thereby lowering the price. West Oil agreed to purchase 300 square kilometres of the Onnia 3D Seismic Survey if it extended the survey into the southeast corner of AC/P 22 to cover the Puffin oil discoveries. The agreed cost was $1.3 million instead of about $3.5 million for a proprietary survey. Before committing to the survey, PGS had to convince the government to increase the 2 year confidentiality limit for seismic surveys to 8 years for multi-client seismic surveys. Otherwise no company would purchase a multi-client survey if it was going to be released by the government after only 2 years for free. After considerable negotiations, the government agreed to this proposal and the 8,500 sq km Onnia 3D Seismic Survey was acquired from March-July 1998. For the AC/P 22 joint venture, the 300 square kilometres of the survey satisfied the first year's commitments for the permit.

The processing of the Onnia survey was completed by the end of 1998 but not without a lot of testing. I attended the first meeting on the initial processing tests. PGS gave the presentation and asked for questions. Now I'm not a processing expert but there was something seriously wrong with the initial processing tests. The top Permian limestone reflector, a booming event on the 2D seismic, was not visible on the 3D data. None of the larger companies seemed at all concerned – I guess because they were going to reprocess the data themselves anyway. Fortunately, Phil Woods of Cultus Petroleum was at the meeting. Cultus had acquired its interests in the Timor Sea in 1993 by buying Esso Timor Sea Limited which owned 18.75% of the Jabiru, Challis and Cassini oil fields plus a 9.38% interest in AC/P4 for $30 million. Phil came to the rescue and asked the obvious question. It turned out that in trying to remove noise, PGS had removed some of the real data as well. After further testing, the problem was fixed and top Permian was at least mappable. I'll never forget receiving West Oil's part of the processed survey because technology was certainly advancing rapidly. Instead of boxes full of reels, all I got were four exabyte tapes. I felt cheated! I wanted more than this for West Oil's $1.3 million.

In the meantime, Charlie had been negotiating with Stuart Cairns to takeover Frontier Petroleum who owned Cambridge Petroleum Royalties. A takeover would provide Frontier shareholders with some exploration blue sky upside and it would have provided West Oil with some cash flow. A formal offer was made to the Frontier directors in September, 1998 on the basis of 5 West Oil shares for 9 Frontier shares. Although the offer was supported by Stuart Cairns, it was rejected by the other directors and the proposal failed.

On the home front, we bought a new and expensive couch for our apartment. Alas when it was delivered to the apartment, it wouldn't fit in the lift. We had to hire a crane to lift the couch over the balcony and into apartment. The couch was on sale at half price so it was still worth the cost of $1300 for the crane. When we sold the apartment, we had to include the couch in the price. The new owner rented the apartment to Danny Green and we saw the infamous couch on TV while he was being interviewed.

At the same time, BHP Petroleum had put its Timor Sea assets up for sale. This included the producing Jabiru and Challis oil fields. West Oil had joined with Premier Oil in a combined study of their tenements with the view to bid $35 million to be funded 80% Premier and 20% West Oil. However, Premier withdrew at last minute and West Oil didn't have access to sufficient finance to bid on its own. BHP Petroleum sold their Timor Sea assets to Gulf Canada in June, 1998.

In January, 1998 I went to the North American Prospect Expo ("NAPE") in Houston to try to farm out the drilling of our Puffin prospects. I took my son Alex with me to help man the booth – he was in his third year of his geology degree. We flew to Los Angeles and onto Las Vegas. We were both impressed by Las Vegas – it had grown so much since my last visit and Alex had never seen it. It is a surreal place, ugly in the daylight but truly magnificent by night. The next day we flew over the Grand Canyon and landed in Falstaff. It was a beautiful day without a cloud in the sky so the views were magnificent. Being January, there was snow on the ground so it was very chilly walking around. Neither of us had seen the Grant Canyon so it was quite a treat. After this adventure, we headed on to Houston and the NAPE conference from 27-29 January.

The conference was huge and was significantly larger than the one in 1996. The international section had also grown but attracting farminees was going to be difficult given the low oil price. In the evenings, dinner was always an adventure for Alex. He was not quite 21 (birthday in May) and was not allowed to be served alcohol in the US. Being tall, sometimes he got away with being served but mostly he had to show his ID and drink soft drinks. After Houston, we flew onto Washington where I wanted to visit the Smithsonian Institute and look at lots of fossils! Washington was an interesting city as it didn't have any high rise office towers like most other US cities. We stayed in a very convenient hotel but not in the best part of the city. It was a short bus ride down to the area where most of the museums were situated. The next day we got on the bus (I made sure we had the correct change!) and as we travelled down to the museums, Alex nudged me and whispered "Dad, do you realise that we're the only white people on this bus?" I looked around and sure enough, he was right – and it took a lot of self control to stop myself from laughing out loud. After we got to the museums, we went to the Smithsonian first and it did not let my expectations down. It was just great, lots of fossils to look at and I was in Palaeontology Heaven. Alex was not so interested and didn't understand what I was so excited about. After the Smithsonian, we went to the Air and Space Museum. This museum was huge – you go in the front door and start with the Wright Brothers and end up at the exit looking at a Saturn rocket. It was so huge that we had to come back the next day to finish the tour. For me, the best was a Messerschmitt 262 hanging from the ceiling – the first jet plane ever built – and somehow it survived the war. After the Air and Space Museum, we did some tourist things – look at the White House, Lincoln Memorial etc. After Washington, we flew to London. At that time, round the world tickets were cheaper than flying from Australia to the US and back. At last, Alex could legally have a beer – which he did!

The APPEA Conference was held on 8-11 March, 1998 in Canberra. At that time Chris was working at the Academy of Science and we had dinner with him during the conference. At the conference, Woodside gave a paper on the giant Perseus Gas Field describing this complex structural-stratigraphic trap. John Gorter presented a fascinating paper on buried impact craters and their petroleum potential. Malcolm King gave a paper on Shell's recent well, Looma-1, in the southern Canning Basin. This was a most frustrating well with good oil shows in both the Ordovician Nita Formation and Acacia Sandstone. The Acacia Sandstone had porosities of up to 12% and oil saturations up to 70% but testing indicated tight reservoirs in both units.

In June, 1998, Mo and I went on an Asian cruise on the "Gemini" out of Singapore for our wedding anniversary. We'd been on one of these cruises with the Bervens in 1995 and had such a good time we thought we'd do it again. Following this, I attended the Timor Sea SEAAOC Conference in Darwin for West Oil.

During the 1998 Federal gazettal of offshore acreage, Charlie had this great idea that West Oil should apply for every gazettal block on the North West Shelf. So we did – 14 applications in all! West Oil had already been involved in one successful application in the Timor Sea. AC/P 26 was awarded in February to Mosaic Oil 35% and operator, Indo Pacific Energy 35% and West Oil 30%. Out of all these applications, the company was awarded 100% of AC/P 28 in the northern Browse Basin in June, 1998 and 100% of WA-284-P in the Bonaparte Basin in August, 1998.

In AC/P 28, West Oil purchased 454 square kilometres of the 10,000 sq km North Browse TQ3D multi-client seismic survey acquired by Geco-Prakla in 1999. Interpretation of this seismic data defined the Galileo and Arlo prospects which were rollovers of the Puffin sands on the flanks of the Discorbis horst block. Discorbis-1 had been drilled by BHP Petroleum but the well encountered an Early Triassic shaley section without reservoir quality sands beneath the Intra-Valanginian Unconformity. The well did, however, drill through the Puffin sands in the graben overlying the Triassic horst block and these had exhibited good oil shows in the cuttings.

Interpretation of the 2D seismic in AC/P 28 also defined an interesting prospect which we called Aristarchus, in the BHP tradition of naming prospects after astronomers. Aristarchus was the Greek astronomer who determined that the Earth revolved around the Sun (and not the other way round) in 300 BCE, somewhat earlier than Galileo. The Aristarchus Prospect is a stratigraphic trap set up by the pinchout of the *M. australis* Sand between Asterias-1 where it is 30 metres thick and Copernicus-1 where it is absent. Top and base seals are provided by the marine shales of the Early Cretaceous Echuca Shoals Formation which are also interpreted to be possible source rocks. The prospect has a potential recoverable resource of 400 million barrels. It covered a large area and in the late 1990's, it was cheaper to drill than acquire 3D seismic.

West Oil was unable to farm out any of the prospects in the permit and AC/P 28 was cancelled by the government in 2003. The Arlo Prospect was drilled by Apache Energy in 2009 and although it had oil shows in the Puffin sands, these were not commercial and the well was plugged and abandoned. The Aristarchus Prospect, however, has not yet been drilled and still awaits a future intrepid oil explorer.

In September 1998, there was an explosion at Longford gas processing plant in Victoria which stopped all petroleum production from the Gippsland Basin. Fortunately for the Victorian industrial companies, gas production was restored after two weeks but oil production did not recommence until January 1999.

In WA-284-P, there was only one well in the permit, Rambler-1, drilled by Sagasco Resources Limited from October 1993 to January 1994. However, when I tried to get the well data from the government, I was told the well was not open file as the five year confidentiality period had not expired. I argued that this was patently absurd as the acreage had been relinquished and the well was no longer of any interest to the companies who drilled it, and that therefore the well should be open file. But to no avail – government policy prevailed over common sense and the well remained confidential. After much scrounging, I eventually got a copy of the well completion report. Rambler-1 had numerous oil and gas shows through the Early Cretaceous Darwin and Flamingo formations which were over-pressured and very difficult to drill – hence the long time taken to drill the well. Rambler-1 was extensively tested but without commercial success. It was one of those "Catch 22" situations – if you went too far into the basin, the source rocks were mature and generating hydrocarbons but the reservoirs were tight (the Rambler-1 location) but on the edge of the basin, the reservoirs were great but the source rocks were immature and not generating, and wells like Tamar-1, drilled by Getty Oil in 1979, were dry. "These rocks were definitely smiling at us".

Newfield Exploration had recently bought Gulf Canada's producing Timor Sea assets. They also operated the permit adjacent to WA-284-P so

West Oil managed to farm out the permit to them. Newfield Exploration acquired 252 sq km 3D seismic with the option to drill a well. But after drilling Wambenger-1 which was a dry hole, in their own permit, they did not exercise their option to drill a well in WA-284-P. After that, West Oil applied to cancel the permit.

At the end of 1998, Mobil was selling its Timor Sea assets which consisted of 6.25% interests in AC/L 1, 2 & 3 (the producing Jabiru, Challis and Cassini oil fields) plus interests in AC/L4 (the Montara oil and gas discovery, and AC/RL2 (the Oliver gas discovery). West Oil's evaluation indicated a price of about $3 million for which the company had arranged financing. We submitted our tender but heard nothing for several months. When I contacted them, they told me that they had decided not to proceed with the sale and had terminated the tender process. They just forgot to tell the bidding companies.

West Oil decided to attend the next NAPE Conference in Houston in February 1999. We had plenty of prospects to farm out and my private company Gulliver Productions had a few more. While chatting to Craig Marshall, I discovered that Empire Oil had a number of farmouts and was also thinking of attending NAPE. So we all decided to pool our resources and get a large booth at NAPE and present all our prospects with an Australian flavour. After we had set up the booth, Craig had brought along a number of inflatable kangaroo dolls which we prominently displayed around our booth. The blowing up of these kangaroos was quite exhausting and most undignified **(Fig 39)**. Everyone at the conference wanted to take one home so at the end of the conference, Craig raffled them all off but he rigged the raffle so that they all went to the best looking women at the conference.

While we were in Houston, we ended up in a restaurant recommended by John Todd. Craig gave me the wine list and said "you choose the wine". The wine list consisted of mostly American and European wines, the better of which were outrageously expensive. However, in the Australian section, they had good vintage Penfolds 389 at a very reasonable price. It was one of my favourite wines and, given that Empire Oil's most prospective permit was EP-389, I ordered a bottle of that. Of course it was followed by many more bottles and we all had a very merry evening.

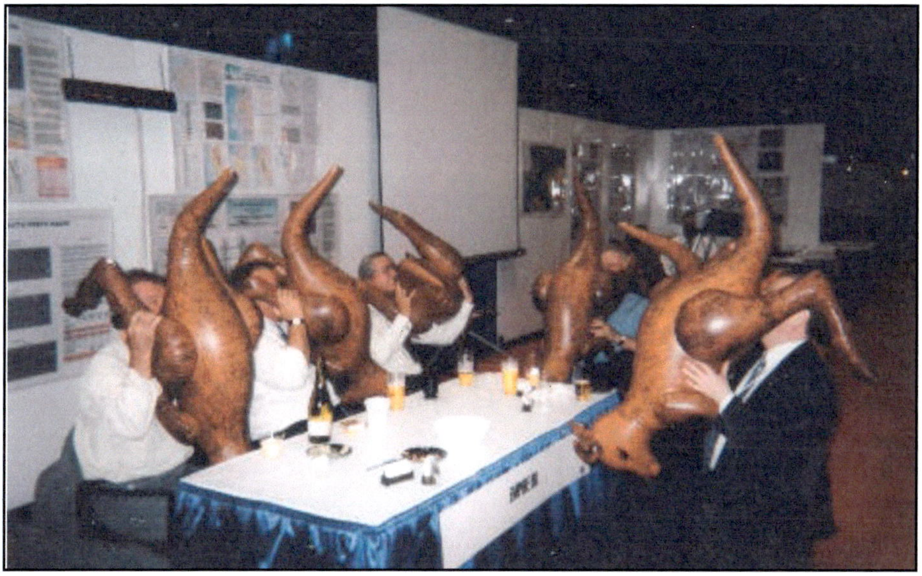

Figure 39
What are those perverts doing!

FUSION OIL & GAS

West Oil was using the consulting group Ikoda to carry out seismic interpretations and prospect evaluations in its operated permits. Alan Stein, the manager of Ikoda, put a proposal to West Oil for West Oil and a company, Fusion Investments Limited ("FIL"), to jointly apply for exploration acreage in West Africa. West Oil would put up the initial funds and FIL would provide the technical expertise. Charlie and Alan went to Africa in September, 1997 to negotiate for the grant of exploration tenements. In West Africa, the permitting system was to firstly be granted a Technical Cooperation Agreement ("TCA") allowing companies a period of time to evaluate a large area after which a smaller concession area of their choosing could be applied for with a firm exploration program. West Oil 50% and FIL 50% signed six month TCA with Senegal-Guinee Bissau Joint Authority (AGC) in 1998. This TCA gave West Oil and FIL the right to apply for a 1,000 square kilometre concession within the TCA. West Oil and FIL also signed a TCA with Gambia and a three year Petroleum Agreement with Ghana.

During early 1999, West Oil and FIL formed a new company called Fusion Oil & Gas ("Fusion") to hold their combined West African interests. By this stage, Fusion had a 10% interest in Production Sharing Contract (PSC) A and a 20% interest in PSC B in Mauritania with Hardman Resources; a 100% interest in the TCA in Gambia; a 100% interest in the TCA with the AGC; a 90% interest in the North Tano Licence in Ghana and a 100% interest in a TCA in Liberia. West Oil owned 20% of the shares in Fusion. Following the signing of an agreement with Woodside Petroleum to farm into Hardman's Mauritanian PSC's, Fusion raised £15 million with an initial public offering of the issue an additional 30 million shares and listed on the AIM stock exchange in London in September 2000. The directors of West Oil agreed to distribute the company's now 8% shareholding in Fusion *in specie* to the West Oil shareholders.

Another major event in the international petroleum industry took place in early 1999. Esso, now renamed Exxon, bought Mobil Oil for $US125 billion and the combined company became ExxonMobil. And now there were only "Five Sisters"!

During 1998, seventeen countries out of the twenty eight members of the European Union decided to join together to embrace a common currency called the Euro. The Euro was introduced on 1st January, 1999. The British decided not to join and kept their pound sterling. This turned out to be a very good decision as it gave them greater flexibility during the uncertain financial conditions after the 2008 global financial crisis.

In February, 1999, Mo and I went on a cruise on the QE2. It was always a dream of Mo's to take a cruise on the QE2 so when we found out that it was going to be in Fremantle, we booked the leg to Hong Kong. We left Fremantle on 15th and cruised up to Bali. In reality, the QE2 was showing her age and was not the luxury cruise liner she once was - but she was still very elegant. We arrived in Bali and went ashore by tender. Indonesia was in the throws of the Asian Financial Crisis and the docks were crammed with locals desperately trying to make a meagre living. We pushed our way through the crowds and found a nice beach to spend a few hours before returning to the ship. Our next port of call was Manila which was more relaxed and we were able to do some sightseeing and shopping without getting mobbed. The most interesting passenger on the ship was an old lady who was on her own and spent her time permanently cruising on the

QE2. It reminded me of that joke about cruising in your old age – you get all your accommodation, food, travel and medical supplied free and when you die, they simply throw you overboard! From Manila, we sailed north to Hong Kong where we did the cable car up to Mount Victoria and a bit of shoping before flying back to Perth.

In April 1999, Apache bought out British Borneo's 8.4% interest in the Harriet Joint Venture and 10% interest in the East Spar Joint Venture for $US80 million. On the international front, BP purchased Atlantic Richfield for a cool $US26.8 billion.

The 1999 APPEA Conference was held in Perth from 18th to 21st April. It was an opportunity to hold technical and operating committee meetings on AC/P22 and ZOCA 96-16 where we were preparing to drill a well on each permit. It was a busy time and I didn't get to see any papers and had to read the relevant papers in the journal after the conference.

On the personal front, we sold our beach house in Falcon in October 1999. The boys had left home and we didn't need such a large beach house. I also had a large tax bill to pay! But we still wanted a beach house so we bought a smaller place at 9A Galtea Road. John Schulstad had also bought himself a beach house at Seabird north of Perth. It was right on the beachfront, just perfect for swimming, fishing or just relaxing. He invited us up for New Year's Eve at the end of 1999.

After New Year's Eve, I was soon off to Houston to join Ivan Burgess at the NAPE Conference held in late January to early February, 2000. This was a busy time for us and we gave presentations on AC/P 22 and ZOCA 96-16 to Phillips, Anadarko, Walter Oil & Gas, Burlington Resources, Occidental, Anadarko Petroleum, Kerr McGee, Pennz Energy, Tesoro, Newfield Exploration and other US independents. We were both impressed by the foyer at the Anadarko office. There was a very large sign instructing us to "Check your Firearms at Reception"

Attendance at the 1999 and 2000 NAPE conferences turned out to be very beneficial for West Oil. West Oil farmed out AC/P 22 to Alberta Energy who would earn a 60% interest in the permit by carrying West Oil and Norwest Energy 20% each for the drilling of one well. Interpretation of the Onnia 3D seismic had defined three prospects of which updip Puffin-1 was the lowest risk. Puffin-1 was drilled by Arco Australia in 1972 and encountered a two metre oil column at the top of the Late Cretaceous Puffin Formation sands. The next well, Puffin-5, was to be located updip from Puffin-1 to intersect an interpreted 30 metre oil column in the Puffin Formation sands.

Norwest Energy also farmed out ZOCA 96-16 to Phillips Petroleum who would earn a 66% interest by drilling one exploration well with West Oil and Indo Pacific Energy each being carried 10% and Norwest Energy staying at 14% by partly contributing to the well. Phillips Petroleum became the operator. In October, 1999, the ZOCA 96-16 Joint Venture acquired a 1500 km 2D seismic survey to satisfy the Year 3 permit work commitments.

In April 2000, one of the largest gas fields on the North West Shelf was discovered. Jansz-1 intersected 29 metres of net gas sand in an Oxfordian shallow marine sandstone. This was followed up by Io-1 in January 2001 and this well encountered 44 metres of net gas sand in the same reservoir. The field is a giant stratigraphic trap on the western limb of the Kangaroo Syncline and covers an area of some 2,000 square kilometres. While attending a future presentation on the Jansz Gas Field at the 2003 APPEA Conference, I immediately thought of the Troll Field in the North Sea from my Burmah Oil North Sea days. Except the reserves for the Io-Jansz Gas Field were over 20 TCF gas making it 50% larger than Troll in barrels of oil equivalent terms.

The APPEA Conference was held during May 2000 in Brisbane. Woodside presented an excellent paper on the Sunrise-Troubadour gas fields which took me back to my days at BOCAL. After the conference, we joined up the Gorter's and the Schroder's for a holiday on Stradbroke Island. The Schroder's had organised the rental of a house for a few days and it was very comfortable for the six of us. Its only drawback was that it was high up and well back from the coastline and beaches.

During August, 2000, Charlie and I made presentations to Unocal to farm into West Oil's other Timor Sea permits. Their technical team was very interested in the opportunities and negotiations were well advanced until Unocal head office decided not to proceed.

Prior to the drilling of Puffin-5, I wanted to be present for the testing of the Puffin sands. I had not been on an offshore rig since Swan-2 in 1979 and safety issues had certainly improved over the intervening 20 years. I had to undergo helicopter training to get my HUET certificate before I could go offshore. This training took place at the swimming pool at Challenge Stadium where a group of us undertook simulations of helicopter crashes in the ocean. This involved being seated in a cage with seat belts fastened and then being tipped upside down and dumped into the swimming pool. It's amazing how disoriented you get the first time this happens. It took a few simulations before I got hang of it and made it safely out the window. After the training, I was totally exhausted but as I walked back to my car I laughed to myself "God I love this oil industry. Where else would you actually pay for someone to try and drown you"!

Puffin-5 was drilled during June, 2000 by the Ocean General and encountered a 9.4 metre oil column in very porous Puffin Formation sands. A core was cut through the sands and it consisted of 100% sand with 28% porosity and multidarcy permeabilties. Electric log analysis indicated oil saturation of 85% due to the excellent reservoir quality and that the oil-water contact was the same as Puffin-1. Of course a 9.4 metre oil column was not what was predicted from the seismic and was definitely not what we were hoping for – those "rocks were smiling at us" again. We needed at least a 20 metre oil column to give us sufficient reserves for an economic development. However, seismic depth conversion is not an exact science and is improved with every well drilled, so there was still hope. At least Puffin-5 proved that there was an anticlinal oil field in the Puffin-1 and Puffin-5 area – we just didn't know how big it was.

It was decided to test the oil column in Puffin-5 to get reservoir and production information. Ivan Burgess and I went up to the rig for the test. The last time I'd been to the Timor Sea was Swan-2 in 1979 and I'd forgotten how long it took to get there. We flew to Darwin staying overnight before flying to Troughton Island and helicopter to the rig. Troughton Island was a flat, barren desolate place and the wind howled across the island at about 70 kms/hour. I think I preferred my Swan-2 route via the Kalumburu Mission. The helicopter flight was interesting especially after the HUET training. They packed us in like sardines so if the helicopter had actually crashed into the sea, I doubt if many of us would have made it out and survived. To make matters worse, the bloke sitting next to me was ~ 160 kgs and would have struggled to get through the window. So I made my plans that if the helicopter did crash, I would head for the window on the other side! We eventually (and safely) landed on the Ocean General and settled into our rooms. The test of the oil column in Puffin-5 commenced the next day and built up to about 8,000 bopd before having to be abandoned due to sand production clogging up the test separator. This was very disappointing as we were unable to get any meaningful production data from the test.

In October-November, 2000, the Coleraine-1 well was drilled in ZOCA 96-16 using the Ocean General. While drilling through the top hole section, the well encountered oil shows in Early Tertiary limestones. This caused quite a panic as there was no intermediate casing in the hole and we were drilling with seawater. Fortunately, the limestones were tight and the well didn't flow in on us. The intermediate casing was quickly run and the well drilled ahead. At 2,391 metres, the well intersected the Darwin Radiolarite which was also oil bearing. The oil accumulation was trapped in low permeable clayey glauconitic sands and was not economic offshore. The well continued down to the main objective Jurassic Elang and Plover sands where it encountered a 164 metre residual gas column overlying at least a 32 metre residual oil column. This was a crushing blow! There was once a huge field at Coleraine but the oil and gas had all leaked out up the faults and was gone. The faults controlling Coleraine had the same orientation as Bayu-Undan but they didn't seal for us – it was all very depressing. Those damn "rocks were smiling at us" again!

On a happier note, I had attended many Lilac Hill cricket matches over the years and they were always a lot of fun **(Fig 40).** The most memorable match was against the West Indies in November 2000. The organisers picked a date that clashed with the Melbourne Cup race and many fans were very upset at this timing. Even the gods were angry – heavy rain washed out the match!

There were Operating Committee Meetings in January 2001 to discuss the disappointing results at Coleraine and to try to understand why the faults leaked, and again in November to outline future exploration strategies for the permit. There was one further meeting with the Joint Authority in Jakarta in March, 2002 but West Oil eventually withdrew from ZOCA 96-16 in early 2003.

On 27 January, 2001, Mo's sister Jean and her partner Jon arrived in Perth to get married and have a holiday. The wedding was on 31st and was held around the pool at Observation Rise. During February, we went down to

Figure 40
With Ross Macleod at Lilac Hill

the beachhouse for a few days and then went across to Rottnest staying at the Rottnest Lodge and touring round the island including our favourite spot, Little Parakeet Bay. On returning to Perth, we then travelled down to Bunbury staying at the Forrest Lodge and touring around the area. They then caught the India-Pacific train to Sydney and Mo and I joined them there to finish their holiday. Mo and I did the Harbour Bridge climb which was a great experience. It was a fine day and the view from the top was truly breathtaking.

Alberta Energy drilled Puffin-6 in January-February, 2001 in an attempt to get updip from Puffin-5 but unfortunately, the well came in low and was a dry hole. In 2002, Pan Canadian took over Alberta Energy and the combined entity became EnCana. EnCana withdrew from AC/P 22 in May 2002, with West Oil and Norwest Energy reverting to their 50% interests in the permit.

As a result of presentations at the NAPE conferences, Anadarko farmed into AC/P 26 in the Timor Sea. This permit had been awarded to Mosaic Oil, Indo-Pacific Energy and West Oil. Mosaic was the operator and had purchased the Talbot 3D seismic. Their interpretation of the 3D and 2D seismic had defined the Anson North Prospect updip from the Anson-1 well where oil shows had been encountered. West Oil had free carry for

12.5% in AC/P 26 by Anadarko drilling Anson North-1. Anadarko also farmed into AC/P 27 and AC/P 25, permits adjoining AC/P 26. Charlie Morgan negotiated a complicated deal with Daytona Energy which was listed in Canada and had interests in AC/P 25 and AC/P 27. In the deal, West Oil had options to earn 5% in AC/P 27 and 10% in AC/P 25. Daytona got 14.5 million shares and 6 million options in West Oil plus $100,000 cash. West Oil invested $600,000 in Daytona and was issued 4 million Daytona shares and warrants. West Oil now had interests in all three wells being drilled by Anadarko.

The 2001 APPEA Conference was held in Hobart from 7-11 April. This was the last time the conference was held in Hobart. The APPEA Conference had grown too large and the convention centre in Hobart was no longer large enough to accommodate the numbers of delegates now attending the conference. The talking point during the conference was the Australian Government's rejection of Shell's A$10 billion takeover offer for Woodside Petroleum at A$14-80 per share. This turned out to be great for the Woodside shareholders as Woodside shares eventually rose to $65 per share in 2008.

After the conference, Mo and I hired a campervan to tour the west side of the island, having done the east coast after the 1989 conference. We drove southwest to the town of Southport which was a very small town in a very pretty location. From there we drove northwest across the island to the Gordon River and on to Strahan at the north end of Macquarie Harbour. The drive went through endless kilometres of wilderness with lots of hills which severely tested the horsepower of the campervan. From Strahan we drove north to Wynyard turning west to Stanley. The north coast of Tasmania was very picturesque but our best experience was in the little town of Penguin. We pulled up to a caravan park and for the princely sum of $13 for the night I backed the campervan into the allotted site. We opened the curtains at the back of the van to be greeted with a panoramic view of Bass Strait. Out came the beer, champagne, nuts, olives and dips, and to this day, this has to have been the cheapest accommodation with million dollar views that we have ever stayed in. The owner of the park told us that we should also drive down into the farm lands to the south. The countryside was green and stunning but Mo had the best time as they were selling raspberries by the roadside about every kilometre – Mo just loves raspberries! From there we drove onto Launceston and flew back to Perth.

During 2001, Chevron merged with Texaco. And then there were only "Four Sisters" and only two of them were American! Now all it would take for the old Rockefeller Standard Oil to be reunited after 90 years would be for Chevron and ExxonMobil to merge!

Mo and I went on an overseas holiday departing June, 2001 for London. We hired a car and drove down to Bornmouth to visit Rosemary who was dying of cancer. It was a timely visit as she died the following August. We then drove across the country to Pembroke to catch the ferry across to Rosslare in Ireland. It was just after the time when Mad Cow Disease had ravaged the UK and bans were still in place. When we arrived at Rosslare we had to drive the car through a bath of disinfectant before we were allowed to enter Ireland. From there we drove to Glengarriff to visit Bob before returning to London. From London, we flew to Athens for an Aegean cruise. Our trip got off to a bad start with Mo's purse being stolen during breakfast at the hotel. We had to fill out police report in a very run down police station but of course, the thieves were never caught. They managed to use the credit card to buy some items before the cancellation went through. The worst part was that they were able to withdraw some cash from a bank. We always keep our pin numbers separate from our cards so there had to be some collusion for this to occur.

After all this frustration, we eventually joined the cruise on 29 June. Our first stop was Istambul and I was very impressed. The city has a terrific location and although we only had one day, we managed to visit the Blue Mosque, the Topkapi Museum and the Church of Sophia and still had time to buy a rug. From Istambul, we sailed south to Kusadasi. I wanted to visit the ruins of Ephesus but I got sick and couldn't make it. From there, we sailed to Rhodes and although I was still sick, Mo managed to drag me around the town and view the sites.

My favourite stop was Santorini which was the location of a massive volcanic eruption around 1,500 BC. The resulting tsunamis caused massive damage to the ancient cities on Crete, destroyed the Minoan fleet and led to the decline and eventual demise of that Bronze age civilisation. As the ship sailed into the collapsed cauldera, you could see the layers of volcanic ash and a new volcanic island forming in the centre. After docking, we took the cable car up to the top and the view was truely magnificent. There were two other cruise ships in the harbour and they all looked like toy boats in a bathtub. From Santorini, the cruise returned to Athens and we flew home.

Soon after returning to Perth, on 17 July, I had an audit by the Australian Taxation Office. I expected this to be a daunting process but it turned out to be mostly accounting for the GST and correcting my invoices to conform to the stardard tax invoice.

Twin Towers

On the morning of Tuesday 11 September 2001, two hijacked planes crashed into the twin towers of the World Trade Centre. It was the evening in Australia and we were watching television when the program suddenly switched to live coverage of the attacks. Initially we thought it was an advertisement for a new disaster movie but quickly realised that this was for real! At this stage, only one plane had hit the north tower but then we watched in real time as the second plane hit the south tower. It seemed to me that the pilot may have had second thoughts about meeting Allah as the plane appeared to swerve just before hitting the tower. For about an hour and a half, we watched as the fires gradually took over the upper parts of the towers and desperate people lept from the windows to avoid the flames. Then the unthinkable happened - the Twin Towers collapsed! We thought that tens of thousands of people would have been killed but due to the efficient emergency procedures and the calmness of the evacuees, only 3000 people were killed. But it was still the greatest pre-emptive military strike in modern history since Pearl Harbour. We learnt later that a third plane had crashed into the Pentagon and a fourth plane, believed to be heading for the White House, crashed in a field after some brave passengers tried to take it back. The oil price crashed from $US30 per barrel to $US22 per barrel before recovering.

The implications of this pre-emptive strike were enormous and the world has changed forever. Initially, the first task was to find out how this meticulously planned attack was not uncovered by the intelligence services of the USA. As it turned out, there were some obvious warnings. In July, the FBI office in Phoenix was warned of an unusual number of Middle Eastern men enrolling in US flight schools, one of whom told an instructor that he didn't need to learn how to land!. In August, the FBI arrested Zacarias Moussaoui after a tip-off from suspicious flight school trainers. One of the agents warned that he may have been planning to "fly something into the World Trade Centre". Any alarm bells ringing yet?

Of the 19 terrorists who boarded the four planes that fateful morning, fifteen were from Saudi Arabia, two from UAE and one each from Egypt and Lebanon. So what did the Americans do? They invaded Afghanistan in order to kill Osama bin Laden and the other leaders of Al Qaeda! Al Qaeda eventually left and the Americans ended up fighting the Taliban who had little to do with the attacks on the Twin Towers. They committed billions of dollars and the lives of thousands of American and allied soldiers to a war they couldn't win. Good grief - if they just wanted to kill Osama bin Laden, surely a better option would have been an elite Special Forces team which they eventually did in 2011. Today, almost twenty years later, the war is still dragging on against the Taliban.

The Good Oil Conference was held on 17-18 September, 2001 but following the events earlier in the month, there were very few delegates. Nobody was keen on flying, especially those from overseas.

West Oil also farmed into EP-410 in the onshore Carnarvon Basin to drill Carlston-1 on an interpreted Devonian prospect. The company also farmed into Strike Oil's Altostratus Prospect in TP/17 in the offshore Carnarvon Basin to earn a 10% interest in the permit. As a public relations exercise, West Oil made up T-shirts for its shareholders with the prospects to be drilled in our future drilling program printed on the back. These were the Carlston, Anson North, Sebring, Sleeper and Altostratus prospects. Altostratus-1 and Carlston-1 were drilled during January- February 2002 but both were dry holes.

The West Oil office was in Kandahar House in Ord Street, West Perth. It had a very large balcony which was ideal for entertaining. Mo and I organised the West Oil Christmas party and set it up on the balcony. It was a huge success so we did it again the following year.

During December, 2001 and January, 2002, Anadarko drilled Anson North-1 in AC/P 26, Sleeper-1 in AC/P 27 and Sebring-1 in AC/P 25. All three wells drilled by Anadarko were trouble free and under budget but were all dry holes. One amusing situation occurred during the drilling of one of the wells operated by Anadarko. The daily drilling reports were being emailed to Anadarko but were not being received immediately but some hours later. We couldn't understand this until it was pointed out that the word "Sleeper" may be triggering internal security alerts in the US following the attacks on the Twin Towers in September. Presumably, the e-mails to Anadarko were being intercepted by US security until they eventually realised that the word "Sleeper" actually referred to a lonely exploration well being drilled in a very large sea on the other side of the planet and therefore posed no immediate threat to the USA!

In May-June, 2002, we took a holiday to Vienna. It was my best man Al Raam's 60th birthday and he invited us to attend. It was our first visit to Vienna and I marvelled at the magnificent historic buildings. The city was mostly destroyed by the Russians in WW2 and most of the buildings had been rebuilt with immaculate precision to their original state. Al Raam's birthday was a blast with a cruise on the Danube, lots of food and drinks and of course the obligatory speeches. My speech rambled on about the great times we had at university and the waterskiing adventures in Port Hacking. After his birthday, we headed off by train to Paris. The train trip to Paris was very entertaining. We all tend to think of Europe as being overcrowded and in the cities it certainly is. But the train journey travelled through endless green countyside with agriculture, rivers and the odd village. We spent two nights in Paris and spent a whole day touring the city. Our hotel room was the size of a toilet but the hotel was very central. We saw the fabulous Sacre Coure, walked down the Champs-Elysees and did a river cruise along the Seine. We then took the train through the Channel Tunnel to London. I returned to Perth but Mo stayed on a few extra days to visit her sister Jean in Lancashire.

West Oil also had an 18% interest in AC/P 32 acquired through the Daytona deal. The other companies in the permit were Norwest Energy 34% and operator, Eagle Bay 30% and Daytona 18%. In September, 2002, OPIC, the Chinese Petroleum Corporation of Taiwan farmed in to earn a 25% interest in the permit by funding 36% of the cost of drilling one exploration well. There was a very elaborate signing ceremony with the minister and the press on 2 September. The only problem with this deal was that the smaller companies couldn't afford to fund the remaining 64% of the cost of the well. The drilling of the well was put off in order to farm out these costs but West Oil withdrew from the permit soon after.

STAR PATH

The Star Path Project was an opportunity for West Oil to invest in an exploration program targeting low risk gas prospects onshore Texas. The project covered 300 square kilometres in the Rio Grande Embayment of southern Texas on the border with Mexico. Targeted reservoirs were Eocene deltaic sediments of the Wilcox, Reklaw and Queen City formations at depths of 1,500 to 5,500 metres. The Star Path participants had underwritten a 320 square kilometre 3D seismic survey and interpretation of this seismic had delineated 43 prospects of various sizes and depths. In June, 2001, Charlie had negotiated at deal for West Oil to purchase Star Path Exploration's 10.5% interest (at the usual 75% net revenue interest) in the Star Path Project including the oil and gas leases. The deal was subject to 60 days due diligence and I travelled to Houston in August 2001 to review the data. The trap styles were faulted anticlines, fault blocks and downthrown fault traps. The 3D seismic was excellent and the prospects were well defined. The lowest risk prospects were the shallower Queen City prospects with fine grained, shaley sandstones deposited in coastal barrier, strandplain and delta front environments. Porosities were good but permeabilities were low and wells had initial flow rates of 2-4 MMcf/d with high decline rates. Potential recoverable reserves were also low, of the order of 3 to 15 BCF gas. The higher risk prospects were the deep (greater than 4,500 metres) Wilcox prospects with fine grained, quartzose sandstones deposited in deltaic and slope fan environments. Porosities were good and permeabilities better with wells having initial flow rates of 2-15 MMcf/d. The section is overpressured so well costs would be expensive even allowing for the greater depths. Potential recoverable reserves were also higher, of the order of 20 to 70 BCF.

There were four initial prospects to drill, Castor, Capella, Europa and North Perseid, and West Oil needed to raise $3 million to participate in the project. The Capella and Europa prospects were the lower risk, low reserves Queen City prospects but could potentially have generated sufficient cash to fund the drilling of the deeper, higher risk Castor and North Perseid prospects with potential reserves of 68 and 65 BCFgas. Gas prices in the US had risen to $8 per Mcf in January, 2001 and were still hovering around $5 to $6 per Mcf at the time of these negotiations.

In the end, West Oil was unable to raise the money to fund this opportunity and the deal with Star Path Exploration lapsed. After this, US gas prices went on a rollercoaster ride, falling to $2 per Mcf in 2002, rising to $8 per Mcf in 2003, falling to $4-50 per Mcf in 2004, rising to $13 per Mcf in 2006, falling to $5 per Mcf in 2007, rising to $12 per Mcf in 2008 before falling to $3-5 per Mcf between 2009 and 2011. Due to the shale gas revolution in the US, the oversupply of gas caused gas prices to drop to only $1-50 per Mcf during 2016. Imagine trying to develop a gas field in the US with these fluctuations in the price of your product!

LA BRER TAR PITS

On my return to Australia, I had to fly back via Los Angeles airport. Being an ex palaeontologist, I'd always wanted to visit the La Brer Tar Pits. This is a lake of oil just outside LA where over the past tens of thousands of years, a variety of mammals, birds, reptiles, insects and plants had got trapped in the sticky tar and perished. So I stayed overnight in LA and the next day, hopped in a taxi and went there. The La Brer Tar Pits are actually located at the end of Wiltshire Boulevard and as soon as I got out of the taxi, I was hit by the smell of oil. So I wandered in and there it was – a large lake of oil with the odd bubble of gas rising to the surface. The lake itself is fed by oil migrating up faults from the underlying Salt Lake Oil Field and was instrumental in the field being discovered in 1902. By 1917, 450 wells had been drilled and the field eventually produced over 50 million barrels oil.

Around the edge of the lake, there were numerous shafts where palaeontologists were digging down and collecting all the animals that had been preserved in the tar. Oil is a very good preservative and prevents the oxidation and decay by bacteria of the animal remains. The deeper you go, the older the fossils are, so the palaeontologists have obtained a record of all the animals that lived in western North America over the past ~ 40,000 years.

Further on from the lake was the George C Page Museum. George Page was an entrepreneur who got his start in life packaging and distributing California oranges. He made his fortune in real estate development and owned the land around the tar pits. He financed the building of the museum which was completed in 1977 and presented it to the city of Los Angeles. The museum itself was absolutely stunning. It was so well laid out and beautifully presented. It was also good to see that it was full of school kids hopefully learning from the experience. The fossilised mammals included mammoths, bears, wolves, sloths, saber tooth cats, horses, camels and lions. Most of these became extinct at the end of the last Ice Age although horses were reintroduced by the Spanish during their 16th century conquests. But the birds were the best! Such was the preserving qualities of the tar, that some still had their feathers attached. They had tried to feast on the dying animals stuck in the tar but ended up sharing the same fate.

Indonesia

During 2001 to 2002, West Oil, through John Ferry, a business acquaintance of Charlie's, began negotiating for acreage in Indonesia. There were many small oil fields and many more prospects to be drilled on the island of Sumatra. Due to the then low oil prices, these had been abandoned and ignored by the larger oil companies and the licences had been picked up by local businesses. These licences were now being offered for farmout and most of the ones I evaluated were very attractive. I travelled to Jakarta in March 2002 and met with two directors of companies who owned some of these fields. However, when we began negotiations with the owners, a depressing picture emerged. The owners had borrowed money from banks to develop the existing oil fields in their licences. So far so good but Shock! Horror! Instead of developing the oil fields, they spent the money on themselves and living the good life. So they not only wanted us to develop their oil fields but also repay to the banks their associated debts. Given the very tough terms for PSC's in Indonesia, all these promising opportunities suddenly turned out to be uneconomic. With all their connections with government officials, we expected that the licences were unlikely, at least in the short term, to be cancelled. I returned to Perth very frustrated and disappointed. As a footnote, oil production in Indonesia has continued to decline from its peak of 1.65 million bopd in 1991 to 825,000 bopd in 2015 and oil reserves declined from 5.9 billion barrels in 1990 to 3.2 billion barrels in 2015. With domestic consumption continuing to rise, the country became a net importer of oil in 2004.

The West Australian Basins Symposium was held in Perth in October, 2002. Ian Longley presented a great paper on Woodside's perspective on the prospectivity of the North West Shelf. There were biostratigraphic papers on Cretaceous nannofossils and foraminifera; Triassic conodonts; Late Triassic palynozones and Permian correlations based on a compilation of many fossil groups. There were papers on various oil and gas fields in the offshore Carnarvon Basin, the Blina Oil Field in the Canning Basin and the Cliff Head Oil Field in the offshore Perth Basin. It was a very enjoyable and instructive conference. It was also the last really technical conference on the WA basins for quite a while. The technical content of APPEA conferences had been steadily declining and the next WABS conference in 2006 was combined with the AAPG and the papers were never published.

During January 2003, John Roberts, the owner of Multiplex Constructions, attempted to get ownership of the West Tryal Rocks Gas Field which was discovered by Wapet in 1973 and had been held by Chevron under the retention licence system ever since. Multiplex had tried to buy the retention lease off Chevron for $70 million but the offer had been declined. Multiplex claimed that the field was commercial and put up a proposal to the government to develop the field by building another pipeline to Geraldton for a gas fired power station or alumina smelter. However, the government rejected the proposal and WA-5-R renewed for another 5 years.

In February, 2003, West Oil and Norwest Energy farmed out AC/P 22 to the US based companies Century Exploration and Concordia Resources. These companies would fund the reprocessing of the Onnia 3D Seismic Survey for the option to earn an 80% interest in the permit by funding the cost of drilling one well. The Farm-in Agreement was signed with Longview, a company jointly owned by Century Exploration and Concordia Resources.

The 2003 APPEA Conference was held in Melbourne in May.We decided to have holiday before the conference instead of after when we were suffering from overeating and drinking. We flew to Avalon and stayed overnight in a motel near the airport. The next day, we drove to Queenscliff and caught the ferry to Sorrento. From there we drove across Mornington Peninsula to Stony Point and caught the ferry to Philip Island. It was a beautiful sunny day and we had a very pleasant lunch overlooking the bay. We then drove to Melbourne for the conference. For me, the most informative papers were on the Mutineer-Exeter and Legendre oil fields and the Jansz Gas Field. Having worked on the Legendre-1 oil discovery during my Woodside days, I was amazed at the stratigraphic complexity of the oil bearing Late Jurassic sands.

West Oil's exploration strategy was to accumulate acreage, acquire 3D seismic to define drillable prospects and then farmout the drilling of those prospects to larger companies. This strategy was ultimately unsuccessful mainly due to wild swings in the oil price. During the time from West Oil listing in 1997, the oil price fell to $US10-80 in November 1998, then rose to $37-50 in August 2000 and then fell to $17 in November 2001. In addition, there was also a change in the oil industry itself with companies reducing exploration expenditure and preferring to grow by mergers and acquisitions. West Oil eventually ran out of money and the ability to raise additional high risk funding for exploration. I resigned from West Oil in April 2003 to allow for the company to change its focus. West Oil called an Extraordinary General Meeting for December, 2003 to obtain shareholder approval for the consolidation of the existing shares and change in the corporate direction of the company. West Oil became Stirling Products Limited in February, 2004.

Puffin Oil Field

Following the demise of West Oil, a company called AED Development acquired AC/P22 in April 2005 and listed on ASX on 3 May 2005. An independent report by Gaffney, Cline & Associates estimated recoverable reserves of 7.6 to 9.2 million barrels in the Puffin-1/-5 structure and 3.4 to 4.9 million barrels in the Puffin-2 accumulation. By this time the oil price had risen to US$60 per barrel and based on these reserves, a field development commenced in February 2006. Using the "Stena Clyde", Puffin-9 was drilled some 1.2 kms south of Puffin-2. Puffin-7 was then drilled to the northeast of Puffin-5 and then horizontally drilled into the K1a reservoir. The well was completed and tested in May at 9,700 bopd. In May 2006, production licence AC/L6, consisting of three blocks covering the two fields, was awarded to AED. During this time, the oil price had fallen to $US26 per barrel before rising to $US50 per barrel.

In February 2007, Puffin-8 was drilled horizontally into the K1a reservoir, completed and tested at 11,660 bopd. Based on these results, AED announced a contingent resource of 40 million barrels oil in the Puffin fields in July 2007. Production commenced in November 2007 from Puffin-7 at 16,000 bopd and Puffin-8 at 12,000 bopd using the FPSO Front Puffin. In March, 2008, Sinopec International Petroleum Exploration and

Production Company purchased a 60% interest AC/P 22, AC/L6 and AC/RL-2 for a staggering $US561 million and became the operator of all the joint ventures. Soon after, early water cut had reduced production which, despite drilling additional wells, had fallen to 3,000 bopd by the end of 2008.

In my opinion, potential reserves could have been up to about 20 million barrels but it was difficult to get to the 40-100 million barrels stated by AED. I believed that the Puffin fields should have been produced slowly to avoid early water cut. The Puffin sands had no shale breaks and the vertical permeability was probably similar to the excellent horizontal permeability. If the wells were produced at too high a rate, early water cut was always possible. The field ceased production in May 2009 after recovering a total of only 2.2 million barrels oil. AED went into administration in August 2011 with assets of $5.4 million but owing $46 million.

GULLIVER PRODUCTIONS PTY LTD

From 1998 to 2003, I used my private company Gulliver Productions Pty Ltd to take on a number of projects in Australia and overseas.

EMPIRE OIL PERTH BASIN

In mid 1998, Gulliver Productions Pty Ltd ("Gulliver") was contracted by Empire Oil to provide an independent review of the seismic data in EP-389 in the onshore Perth Basin. In September, 1997, CalEnergy had farmed into Empire Oil's EP-389 to spend $14 million on drilling the Gingin-3 well to earn a 50% interest in the permit. Gingin-3 was drilled in 1998 at a location that was interpreted to be updip from the Gingin-1 gas discovery drilled by Wapet in 1965. It turned out to be a dry hole. My seismic interpretation was completed in December and revealed that a cross fault cutting the main bounding fault had resulted in the well being located on the wrong side of this bounding fault. I also recommended that they drill a short sidetrack to the east to get onto the right side of the fault. Fortunately, for reasons that will become clearer later, they did not do this.

In March 1999, Gulliver was contracted by Empire Oil to prepare a report on and assist in farming out the Gingin West Prospect in EP-389 in central part of the Perth Basin. The main objectives in this prospect were the Jurassic Cattamarra Coal Measures which were up to 2,000 metres thick and contained interbedded reservoir quality sands and shales to act as seals. Based on reports by previous operators, the associated coals and organic rich lacustrine shales were excellent oil and gas prone source rocks. Gas discoveries at Gingin-1, Gingin-2 and Bootine-1 confirmed a working petroleum system in the permit.

I remapped the existing seismic over the prospect and prepared a report on the hydrocarbon potential of the prospect. My mapping was very similar to Empire's own mapping and confirmed the robustness of the structure. The structure was also updip from the Bootine-1 well which tested 2.3 MMcfg/d from a seven metre gas column in the Cattamarra Coal Measures. As mapped, the prospect had potential recoverable resources of 47 BCF gas and 45 BCF gas respectively from two sands with good sealing shales and coals.

Over the next four months, I contacted a number of companies and sent copies of my report to those who expressed interest. However, despite three commercial gas fields and one commercial oil field being discovered in the Perth Basin, it had a poor reputation in the industry that was difficult to dispel. Of course, an oil price at $US10 per barrel didn't help the cause either. In the end, I had no more success than Empire Oil had in farming out the prospect.

In March, 1999, Cyclone Vance hit Exmouth with winds gusting up to 280 km/hr, similar to the very destructive hurricane Katrina in the Gulf of Mexico. Fortunately nobody was killed but 70% of the houses and buildings in the town were damaged.

POINT TORMENT AREA CANNING BASIN

In 1996, Capital Energy Limited, a Sydney based company, launched a 4 for 1 script bid for Stirling Resources Limited, the operator of the EP-104 Joint Venture in the Canning Basin. The takeover was successful and the company was moved to Sydney. By November, 1999, Capital Energy decided to give up the operatorship of EP-104 and Gulliver was appointed the new operator. The timing was perfect as I had to go to Sydney for an OCM with Mosaic Oil on AC/P 26. I stayed an extra day and met up with Capital Energy and arranged for the technical data to be returned to Perth.

There had also been changes to the joint venture. Cultus Petroleum had been taken over by OMV Australia in September 1999 and since OMV had no interest in the Canning Basin, its subsidiary Basin Oil, put up its 30% interest in EP-104 for sale. Gulliver (18%) and Indigo Oil (12%) purchased this interest increasing their permit interests to 20% and 17.4% respectively. The sale price was one dollar plus taking on the liability of the plugging and abandoning of the Point Torment-1 well.

Capital Energy changed its name to E Com Multi Limited in July 2000 and in March, 2001, Chimelle Petroleum purchased E-Com Multi's 34% interest in EP-104. Voyager Energy then acquired all the shares in Chimelle Petroleum effective from December, 2000. While all this was going on, Austin Oil's interest was transferred to Kimberley Oil and Kiwi International's interest was transferred to Pancontinental Oil & Gas. The interests in EP-104 were then Voyager 34%, Gulliver 20%, Indigo Oil 17.4%, FAR, Pancontinental & Kimberley Oil 8% each and Pelsoil 4.6%.

EP-104 had just been renewed for the fourth time for a further 5 years in November, 1999. In addition, a three block Location had been taken out over the Point Torment gas discovery by the previous operator, Stirling Resources in July 1998. In July, 2000, the EP-104 Joint Venture applied for a two year extension of the Location to July 2002.

In February, 2000, Gulliver contracted Roy Hartley to undertake an analysis of the drill stem test of Point Torment-1. His evaluation of the test was that although the reservoir pressure had dropped, there was evidence that the tighter reservoirs had started to contribute to the flow. This was interesting and may have proved significant given the future advances in horizontal drilling and multistage stimulation techniques. I then prepared a technical review of the permit which was completed in April 2000.

My interpretation of the seismic indicated that the potential gas bearing sands extended along the Pinnacle Fault System to the northwest of Point Torment-1. Assuming that the Pinnacle Fault acted as a seal, there was the potential for the A, B and C sands to trap an estimated 80 Bcf gas and six million barrels oil and condensate. There was also the potential for a further 80 BCF gas-in-place trapped in the deeper, tight Lower Anderson sands. The seismic indicated that the highest point on the structure was 5 kilometres to the northwest of and 80 metres updip from Point Torment-1. My interpretation was similar to Stirling Petroleum's where they had a location called Pillambi-1. This would be the location of the future Stokes Bay-1 well.

I also evaluated the Valentine Prospect developed by Stirling Resources. This prospect was a large, high risk stratigraphic trap defined by a package of seismic reflectors within the Late Devonian on the downthrown side of the Pinnacle Fault. Similar aged sands were intersected in the nearby Puratte-1 well where they have log derived porosities ranging from 15 - 25%. The prospect covered a large area and had the potential to trap an estimated resource of 200 million barrels oil or 1.2 TCF gas.

Following an Operating Committee Meeting in March, 2001, my next task as operator was to farm out the drilling of the Stokes Bay-1 well with the Valentine Prospect as a high risk, high potential sweetner. This was not going to be easy as there was little interest in the Canning Basin by the general oil industry. Initially, Gulliver negotiated an Option Farm-in Agreement with Bounty Oil & Gas NL to carry out a seismic interpretation and geological evaluation of the area with the option to earn a 60% interest in EP-104 by contributing 100% to the cost of drilling one well in the permit. This Farm-in Agreement between the EP-104 Joint Venture and Bounty Oil was signed in May, 2001. When I received their report, their seismic interpretation and proposed location for a Stokes Bay-1 well was very similar to mine.

Now that I had a confident location to drill Stokes Bay-1, I needed to inspect the area for the logistics of drilling such a well. The location, as defined by Stirling's, Bounty's and my seismic interpretations, was about 200 metres south of the intersection of three seismic lines so there should be a permanent market at this intersection. The vertical drilling location was about 1.5 kilometres out into the mud flats from the hard ground. There was a shorter route across a tidal channel which also needed to be inspected. So I flew up to Broome, met up with George Morris and Brian Philp, affectionately nicknamed Stumpy. The next day we all headed off to Derby and north to the Point Torment-1 location. After six years, the causeway had only just started to deteriorate and we carefully drove out to Point Torment-1. The wellhead was still in good condition and the drill pad was littered with empty shells from crocodile shooters.

We then returned to the hard ground and drove northwest to the Stokes Bay area. Our first task was to inspect the shorter route across the tidal channel so we walked out across the mud flats, wound our way through the mangroves and reached the tidal channel. Our jaws dropped – instead of a meandering, shallow creek, we were staring at a wide channel of deep water. So scratch that idea and return to Plan A. As it was approaching "crocodile lunchtime", we returned to the hard ground to decide our next move. We needed a firm GPS location of the Stokes Bay-1 location and the permanent seismic marker – if we could find it. There was only one option – we had to walk it. So Stumpy and I headed off across the mud flats. George reckoned that we were both completely mad and stayed right where he was. It was a long walk but fortunately there was a mild breeze to keep us cool. We found the permanent marker with the numbers of the three intersecting seismic lines still visible on the tag. Stumpy took a GPS reading and then we walked 200 metres to the south and took a GPS reading of the Stokes Bay-1 location. Then we set off on the long march back. When we got back to the hard ground, George laughed at us both covered in mud and sweat and looking very weary. We were also an inch taller with all the mud stuck to the soles of our shoes. When we got back to Broome, the first beer tasted sooooo good!

When gas was discovered in Point Torment-1 in 1992, there was no significant market for the gas in the area. Point Torment-1 was deepened in 1994 to evaluation the oil potential in the Devonian limestones. However, in early 2000, the Office of Energy prepared a discussion paper on the "Policy for the Connection of Western Australia's Regional and Remote Communities to Western Power's Regional Electricity Supply System". At about this time, I was approached by WA Consolidated Power and The Smart Fuels Company about using Point Torment gas to supply a compressed natural gas project. These discussions were followed up but no progress was made. After some research, I concluded that this may not be the most efficient way of distributing the gas. I felt that mini-LNG plants might be a better option.

During late 2000, the tidal power project was promoted as a possible option for the West Kimberley. However, the electricity market in the area was simply not large enough to support the enormous capital costs of a tidal power station even assuming that it was going to actually work.

The 2002 APPEA Conference was held in April in Adelaide. At the PESA

Deal Day, I gave a farmout presentation on the Point Torment and Valentine prospects in EP-104. At the conference, the most interesting paper for me was on the John Brookes gas field which, as discussed earlier, was very dear to my heart from my Pancontinental Petroleum days.

On 30th April, 2002, Western Power formally tendered for the supply of new energy supplies for the towns of Broome, Derby, Halls Creek, Fitzroy Crossing and Camballin/Looma. At this point, I decided to contact Maurice Brand. During the 1990's, I carried out various consulting projects for his former company, Energy Equity. I knew that he was experienced in mini-LNG plant technology as Energy Equity has built a plant at Yulara. Following the takeover of Energy Equity by Energy World Corporation Limited in January 2002, Maurice had formed a new company, LNG International (LNGI). In May, I met with Maurice Brand and discovered that he was well advanced on tendering on the Western Power West Kimberley project. LNGI had estimated that the potential market was over 40 MW for Derby 7MW, Fitzroy Crossing 2.4MW, Halls Creek 2MW, the Pillara mine 12MW, the Sally Malay mine 8MW and the Platinum Australia mine 10MW). We developed the concept that if the EP-104 joint venture could prove gas reserves of 35 BCF in Point Torment, then LNGI would build a mini-LNG plant at Derby and take up to 6 MMcfg/d from the field. Gas would be transported via a pipeline from Point Torment to Derby, liquefied at the plant and then transported to gas fired power stations at the various town and mines. The 6 MMcfg/d was calculated to supply about 30 MW of power to supply the towns of Derby, Fitzroy Crossing, and Halls Creek plus the mines of Pillara, Sally Malay and Platinum Australia. At this time, power to these towns and mines was being supplied by burning diesel at an equivalent price of ~ $25 per Mcf of gas. At these prices, there was room for both our joint venture and LNGI to make a profit and still deliver considerable savings to our potential customers. A Memorandum of Understanding (MOU) was signed with LNGI on 30th September, 2002.

In September, 2003, the joint venture signed a farmin agreement with Bounty Oil & Gas and Avon Resources whereby each company would earn a 20% interest in the tenements on the same terms as the Bounty 2001 option. I was also negotiating with private investors but these would not commit unless LNGI was awarded the West Kimberley Power contract.

The MOU with LNGI terminated at the end of December, 2003. Western Power was supposed to have awarded the West Kimberley Power contract in July, 2003 but by this time, they had only reduced the tender list to two preferred bidders, Kimberley Power (LNGI) and Energy Developments. Western Power has still not made a final decision, and the EP-104 joint venture has missed the weather window to drill Stokes Bay-1 during the 2003 dry season.

In December 1998, the joint venture had also applied for a Production Licence over the West Kora oil discovery designated as Location L98-1. I had several meetings with the Department of Industry and Resources (DoIR) regarding this application and was informed that due to new HS&E conditions for onshore production licences, they considered that the West Kora oil field would not be commercial on its own. The DoIR agreed to put the application on hold until the results of the appraisal drilling on the Point Torment gas discovery were known. The gas in Point Torment-1 was rich in condensate which could be combined with the oil in West Kora to create a commercial operation. This proposal from the DoIR made a lot of sense and was readily agreed to by the joint venture.

An application for a three block Retention Lease over Point Torment was made on 10 July 2002. The DoIR notified the joint venture that the application was gazetted in February, 2003 for any Native Title claims. After four months, no objections were received and Retention Lease RL1 was granted on 15 August 2003 for an initial five year term.

In October, 2003, Gulliver negotiated and signed a very complex agreement with Voyager Energy to purchase its Canning Basin tenements. Voyager Energy had decided to concentrate its exploration in the USA and was selling its assets in Australia. As a result of this deal, Gulliver increased its interest in EP-104 and R1 to 54% and its interest in the West Kora application area to 49%. The purchase price was the payment of $250,000 to be paid out of future production.

APPLICATION AREAS AND NATIVE TITLE ONSHORE CANNING BASIN

In December, 1998, Gulliver, Maneroo Oil Company and Indigo Oil applied for an exploration permit in the Canning Basin designated 2/98-9. Maneroo was a company controlled by Craig Marshall, Mark Hadan and Mike Harrison, The area was mostly on the Broome Arch and included the

Hedonia-1 and Goldwyer-1 wells with excellent oil shows in the Ordovician limestones and dolomites of the Willara Formation. It was offered to the consortium by the DoIR in August, 1999 subject to negotiating a Native Title Agreement.

It was in 1992 that the High Court brought down the Mabo decision. This was to have enormous ramifications for the onshore oil industry in Australia. In my opinion, it was clear that the Mabo People had continuously occupied their land since 1776 and could claim title to their land. What I didn't understand was the concept of "terra nullius" being used as the basis for the British occupying Australia in 1776. The term "terra nullius" strictly means either "empty land" or "land belonging to no one" both of which were patently untrue, but this was not the basis of British occupation. This was based on Australia being "a land not under any sovereignty". In other words, Australia was occupied by numerous nomadic tribes with individual leaders but no single authority like a king or an emperor. Captain Cook had no one to negotiate with who could speak for all the tribes. If there had been such a person, then under the then British laws, he would have had to negotiate a treaty then and there. Actually if there really had have been a king or emperor of all of Australia, Captain Cook and his small party would probably have been annihilated then and there!

The Mabo decision led to the Commonwealth Native Title Act (1993), a hastily drafted piece of legislation that did not address the procedures for making it work. To make matters worse, the Federal government also conveniently forgot to fund and staff the organisations that were charged with administering them. The Act had to be considerably revised in 2000.

For our application 2/98-9, there were three claimant groups, the Karajarri People and the Rubibi People both under the jurisdiction of the Kimberley Land Council (KLC), and the Nyangumartu People administrated by the Pilbara Native Title Service (PNTS). The KLC send their model Native Title and Heritage Agreement in August, 2000. I sent my comments to them in October. I received nothing further from the KLC and contacted Peter Hwang, the Future Acts Legal Officer, as to when they might respond to my comments. He told me that the KLC were drastically understaffed and were also trying to sort through over 4,000 mining applications at the same time. As mentioned above, the Federal government conveniently forgot to fund the organisations that were charged with administering the Native Title Act. In any event, it didn't bother us as we were in no particular hurry to have the permit granted since as long as it was grinding its way through this tortuous bureaucracy, we had clear title and didn't have to pay any fees or spend any money on exploration. It wasn't until August, 2001 that I requested mediation by the National Native Title Tribunal ("NNTT").

During 2000, I had a meeting with John Karajas on the potential of Ordovician bioherms in south Canning Basin. He had been studying some of the core data collected by CRA from their appraisal drilling of the Admiral Bay base metal deposit along the Admiral Bay Fault Zone. He prepared a report on his findings and in February, 2001, Gulliver, Maneroo, Indigo Oil and Kjirt Exploration Services, applied for an exploration permit in the southern Canning Basin designated 14/00-1. Kjirt was a company owned by John Karajas. This area was located along the southern margins of the Willara and Kidson sub-basins with potential in the Ordovician limestones of the Willara Formation. It was again offered to the consortium by the DoIR subject to negotiating Native Title Agreements with the Ngangumartu and Martu people.

Following meetings with the NNTT in September and October, our consortium reached a verbal agreement with the PNTS regarding the Nyangumartu claims in November 2001 but they then changed the agreement and wanted a percentage of exploration costs. On 18 July, 2002, I travelled to South Hedland for a Native Title meeting on 2/98-9 and 14/00-1. The meeting was very interesting as the men and women were segregated. In terms of compensation, the women were very vocal in wanting education for their children and I was all in favour of structuring at least part of the compensation to be set aside for education. Unfortunately the PNTS didn't support this proposal. Back in Perth, there were further meetings with the NNTT in July.

The initial proposal by the PNTS was for the grantee parties to pay 6% of all exploration expenses. I countered by arguing that all exploration carried out in the Perth office (such as seismic reprocessing, seismic interpretations and geological studies) was not intruding on their land and therefore these costs should be excluded from the 6% calculation. I also argued that the 6% of ground disturbing exploration (such as seismic surveys and drilling) should be charged as a proportion of the time during the year that the operations were intruding on their land. For example, if a well took three months to drill, then the 6% of the costs should be divided by four. This was, after all, more in line with access compensation payments to private freehold landowners where native title had been extinguished. However, they did not agree to this and so our consortium requested a Section 35 Determination from the National Native Title Tribunal (NNTT). The PNTS then accused me of not negotiating in good faith. I was outraged – after all, they were the ones who wanted to change our original verbal agreement. However, I bit my tongue and submitted our Form 5 defence in October 2002.

My next meeting was with the Rubibi in Broome. This was a most remarkable meeting. After the normal introductions, we sat down to begin the negotiations. The tribal elder got up and began a tirade against the white man who had poisoned their water holes, stolen their women and killed their men. At first, I didn't know how to respond to this angry tirade and was inclined to walk out of the meeting. But this wouldn't have achieved anything and would have been unfair on the other members of the Rubibi. So I bit my tongue, got up and began my presentation as if he wasn't there and had never spoken. The other and younger Aboriginals in the group didn't say anything and the negotiations fortunately continued as if nothing had happened. Originally, the Rubibi claimants wanted 4% of the minimum annual expenditure for the permit. Again, I went through the same offer as that for the Ngangamartu but their response stunned me. They said that they didn't care what I had offered the Ngangumartu as they didn't like them. After this staggering comment, my initial reaction was that the Rubibi were running out of friends – first the white man, now the Ngangumartu. But I finally concluded that there's no such being as an Australian Aboriginal – there are 300+ tribes of Australian native peoples who have different languages, distinct cultures and don't necessary like each other! This after all is exactly what would be expected over such a large continent as demonstrated by the histories of all the other continents excluding Antarctica.

Meetings with myself, the DoIR, the lawyers representing the Claimants and mediated by the NNTT regarding the two application areas, 2/98-9 & 14/00-1, took place in November 2001 and March, June, July, August and September 2002. You can imagine how much all this was costing in legal bills – and all (except for my time) at the taxpayers expense! All these negotiations were unsuccessful so our consortium requested a Section 35 Determination from the NNTT. The Hearing was held in November 2002 and a Statement pursuant to directions from NTTT in respect of the Section 35 Determination was submitted to the NNTT in June 2003.

I also went through the same process with the Karajari and had my first meeting with them at the Bidyadanga Aboriginal Community some 180 kilometres south of Broome. This was very interesting as Bidyadanga was in complete contrast to the Jiggalong Aboriginal Community. The community was neat, tidy and very well administered and I soon found out why. Bidyadanga was run by a very strong elder who was respected and obeyed by all members of the community. No alcohol was allowed into the community. The negotiations were friendly but the demands were tough and I knew that it was going to be a slow process. However, soon after, I sold Gulliver to Empire Oil and they took over the negotiations. The permit was eventually granted as EP-438 on 23 December, 2004, a staggering six years after we had applied for the area.

Regarding Area 14/00-1, negotiations with the Ngaanyatjarra Land Council Aboriginal Corporation on behalf of the Martu eventually broke down in November 2005. They wanted a cash payment of $50,000 just to approve the grant of the permit and this would have been on top of all the other payments required under the Hertitage Agreement. They also wanted a meeting on country using the Telfer gold mine for accommodation. All this was not only unacceptable but also totally impracticable. As the Martu land was granted title and not a claim, a Section 35 Determination was not possible. So the consortium agreed to excise those application blocks covering Martu land. The reduced application area was eventually granted as EP-448 on 16 June, 2006, some five years after the initial application.

ONSHORE CARNARVON BASIN

Following our unsuccessful farmout efforts at the NAPE conferences, the EP-410 Joint Venture decided in August 2002 to combine our permit with exploration permits EP-369 and EP-405 in the onshore Carnarvon Basin. EP-410, operated by Gulliver, was in the Gascoyne Sub-basin and the permit interests were Gulliver, Indigo Oil and Falcore 33-1/3% each. EP-369, operated by Euro Pacific Energy (now Pancontinental Oil & Gas) was in the Merlinleigh Sub-basin and the permit holders were Euro 95% and Vigilant Oil 5%. EP-405, owned and operated by Empire Oil, was in the

Merlinleigh Sub-basin. All three joint ventures believed that the combined permits would create a critical mass of acreage which may be easier to farm out. The combined permits covered an area of almost 13,000 sq kms or 3.2 million acres. The operators of each permit remained the same but the equities in each permit became Empire Oil 33.33%, Euro Pacific Energy 31.67%, Gulliver, Indigo and Falcore 11.11% each and Vigilant Oil 1.67%. EP-410 was eventually farmed out to West Oil, Longreach Oil and FAR, and the Carlston-1 well was drilled in 2002. Carlston-1 was a dry hole and eventually all three permits were relinquished.

Prior to Carlston-1 being drilled, Gulliver, Indigo, Falcore, Empire Oil, Euro Pacific, Longreach Oil and Vigilant Oil applied for a large onshore block in the Gascoyne Sub-basin, onshore Carnarvon Basin in February 2001. Our consortium was the highest bidder and the application area was granted as 16/00-1 subject to negotiating a native title agreement. The only claimants over 16/00-1 were the Gnulli People. After the initial paperwork, I travelled to the town of Carnarvon in February 2003 to meet with and give a presentation to the Gnulli. I was accompanied by Marie Platt of the DoIR who was providing assistance on the tortuous path through the native title process. In the motel restaurant the evening before our meeting with the Gnulli, Marie recognised at least half a dozen other bureaucrats who were involved with native title negotiations with other mining companies. Again, all (except me) at the taxpayers' expense! At the meeting with the Gnulli, I gave my presentation and commenced the negotiations. I used the same formula I had used in the Canning Basin applications and the Gnulli seemed to be happy with that and the negotiations continued very smoothly. Later that year, I sold Gulliver to Empire Oil and they continued the native title process. Application area 16/00-1 was eventually awarded as EP-439 in February 2005.

Empire Oil Exmouth Area

In February, 2001, Gulliver was contracted by Empire Oil to prepare a risk analysis of four prospects in EP-41 in Exmouth area of the onshore Carnarvon Basin. This permit contained the old Rough Range oil discovery. Empire Oil had mapped four prospects, Elysia, Brooke, Tess and Jennifer and it was my task to rate their risk and put them in an order for drilling.

The Elysia Prospect was essentially updip Parrot Hill-1 which had encountered a 3 metre oil column in the Birdrong Sandstone. The Brooke Prospect was a mapped anticlinal closure updip from Rough Range-7 where oil shows had been encountered in the Birdrong Sandstone. The main risk for these prospects was the actual shape of the structure as the velocity variations within the overlying Tertiary limestones made depth conversion very difficult and uncertain.

The Tess Prospect was a mapped anticlinal closure updip from Lefroy Hill-1 where oil shows had been encountered in the Windalia Radiolarite. In this prospect, the main risk was timing of migration since the structure formed during the mid-Miocene compression and may have post-dated primary oil migration out of the Paterson Trough.

The Jennifer Prospect was a tilted fault on the east side of the Rough Range Fault. It had been mapped by Ampolex in the 1980's as the Parrot Hill East prospect. Ampolex had a theory that the Rough Range anticline, being a mid-Miocene compressional structure, was not present when oil was generated and migrated out of the the Jurassic source rocks in the Paterson Trough. Therefore, oil could migrate into the Jennifer fault block which, based on seismic interpretation was formed during the Cretaceous. It was an elegant theory and I agreed with their logic.

I completed my report in March and recommended that Empire Oil drill the Elysia and Jennifer prospects but not the Brooke and Tess prospects which were too high risk. Empire Oil didn't agree with my report as they drilled Brooke-1 and Tess-1 in 2001 and it wasn't until 2006-7 that the Jennifer (re-named Dune-1) and Elysia (re-named Parrot Hill-2) prospects were drilled.

HISTORIC NATIVE TITLE AGREEMENT FOR COOPER BASIN

In October, 2001, the South Australian government finalised a native title agreement with three native title/claim groups and seven petroleum exploration consortiums covering eleven exploration tenements in the Cooper Basin. This cemented the SA Department of Minerals and Energy as the premier mines department in Australia. Instead of wringing their hands in despair and hiding from the many serious issues created by the

Commonwealth Native Title Act, they rolled up their sleeves and took a proactive role in getting the exploration companies and Aboriginal groups together and solving the problems.

The best thing about this agreement was that they made it conjunctive, that is, it also provided for companies to obtain production licences in the event of a discovery. This contrasts with the situation in Western Australia where companies have to get a native title agreement for the exploration permit and then, in the event of a discovery, companies have to get yet another native title agreement to get a production licence. This not only doubles the cost but, more importantly, doubles the time to get into production.

The agreement took over two years to negotiate but once finalised, it provided a blueprint for future agreements. This again contrasts with WA where all native title agreements are confidential and each new application consortium has to "re-invent the wheel" and negotiate a new agreement.

In April, 2002, Beach and Stuart Petroleum announced an oil discovery in Acrasia-1 which intersected an 11 mertre oil column in the Hutton Sandstone. Outside of the Santos operated areas, this was the first petroleum discovery in the Cooper Basin for 25 years.

Seram

In January 2002, I met with Kalrez Energy who wanted me to prepare a report on their Production Sharing Contract ("PSC") on the island of Seram in Indonesia. They acquired this interest from Santos in 1999 by purchasing all the shares in Santos Petroleum (Seram) Limited. This PSC contained the Bula Oil Field which has an interesting history. The Bula Oil Field was discovered by the Dutch in 1896 and began production in 1919. During WW2, production operations were taken over by the Japanese who drilled 6 new wells and produced an estimated 210,000 barrels of additional oil. The field was heavily bombed by the Allies and was abandoned after the war. After independence, Indonesia issued the first PSC to Gulf & Western in 1969. In 1972, they sold their interest to Associated Australian Oilfields (AAR) who acquired the first modern seismic in 1974. In 1978, AAR was taken over by CSR who drilled 83 shallow development wells on the field between 1985 and 1989. In 1988, CSR's subsidiary CSR Petroleum was acquired by AGL Petroleum.

The interesting thing about the Bula Oil Field was that it had produced over 15 million barrels oil from Pleistocene sands in a low relief anticline covering an area of only 3 square kilometres. The amount of oil produced far exceeded the estimated original recoverable oil reserves. AGL Petroleum's interpretation of this situation was that the shallow reservoirs in the Bula Field were being replenished by leakage up faults from a deeper oil accumulation. The existing seismic was not of good quality and a deeper structure could not be mapped and defined. However, in November, 1985, a Kufpec operated consortium farmed into the PSC (excluding the Bula Oil Field), acquired new seismic and drilled a number of wells. In 1993, the consortium discovered the Oseil field with the oil trapped in the fractured Jurassic Manusela Limestones. A deeper Jurassic oil field under the Bula Field was therefore a possibility as the existence of this petroleum system had now been established by the Oseil discovery. Kalrez retained a 2.5% interest in the non-Bula Kufpec PSC area. The Oseil Oil Field commenced production in 2002 and was still producing at 3,100 bopd in 2015 after producing over 13 million barrels oil.

Interpretation of the existing seismic in the area excluded from the Kufpec Farmin Agreement, showed four, poorly defined structures, Solan, Dawang and Tingerang A & B. Additional seismic was required but Kalrez's dilemma was that the seismic would have cost a lot more than drilling the wells which had very shallow objectives. Kalrez also owned their own rig for drilling development wells and working over existing wells on the Bula Field making it even cheaper to drill any additional exploration wells. I completed my report in December, 2002, and recommended that they reprocess the seismic, remap the structures and try and farm out the PSC for the acquisition of detail seismic over these leads. I heard nothing more and soon after, there was a change in management and eventually, Kalrez Energy Limited became Lion Energy Limited in October 2003.

CARLSTON

During the 1990's, Western Geophysical had one seismic crew working onshore in Western Australia. It kept reasonably busy by working in the Perth Basin in the summer and the Canning Basin in the winter (the dry season). On one journey south from the Canning Basin, they had some time to kill before their next job in the Perth Basin. The manager of Western Geophysical, John Schulstad, decided to shoot some speculative seismic to sell later on a non-exclusive basis and he asked me where would be a good place to acquire this. I suggested the Gascoyne Sub-basin which extends from Kalbarri north to Minilya Roadhouse. This basin had very little existing seismic but had interesting geology with Silurian carbonates and salt, Devonian carbonates and reefs, and marine Carboniferous clastics. They acquired their speculative seismic over the north end of the basin where the prospective Devonian sediments were at estimated depths where any source rocks could be interpreted to be in the oil generation window.

On one of the lines, there was an anomaly that could be interpreted as a reef buildup. Our consortium of Gulliver Productions Pty Ltd, Indigo Oil Pty Ltd (Keith Thompson) and Falcore Pty Ltd (Charlie Morgan) bid on this area and Application Area 6/97-7 was granted as EP-410 in September, 1997.

The Western Australian Basin Symposium was held in Perth from 31 August to 2 September, 1998. The most interesting papers from my perspective were the papers on the Devonian and Silurian in the Gascoyne Sub-basin by Gorter, Mory and Nicoll. I had been studying this basin myself and had put together the consortium which applied for and was granted EP-410. These papers greatly assisted my evaluation of the petroleum potential of the permit.

At the end of 2001, the EP-410 Joint Venture farmed out the permit to West Oil, FAR and Longreach Oil by those companies contributing to the cost of drilling Carlston-1. The Carlston Prospect was interpreted by me as a possible Late Devonian reef similar to those intersected in Pendock-1D and Barrabiddy-1. In Pendock-1D, the reef was calcite cemented and tight but in Barrabiddy-1, the reef had excellent porosity due to the presence of platy *Stromatoporoids*.

The Carlston-1 well was located on a pastoral lease south of Minilya on the road to Exmouth. The pastoral lease was owned by the Heytesbury Pastoral Company and although we had no legal requirement to do so, I tried to arrange an access agreement with this company. I considered this to be a simple courtesy, particularly for the lease manager. However, after meetings with their lawyer in Perth in October and December, 2001, I received a ~ 40 page draft access agreement instead of the simple document

I originally gave them. I didn't have the time or the money to deal with this so we went ahead and drilled the well without an agreement.Carlston-1 was drilled during January-February 2002. I went up to the wellsite to supervise the drilling of the pre-Cretaceous section. The mudlogger was Ted McNally and the trainee mudlogger was Kane Marshall, Empire Oil Managing Director Craig Marshall's son who had recently graduated as a geologist. When I arrived at the wellsite, the temperature was 52°C but fortunately "cooled" down to the mid 40's next day. The section beneath the Cretaceous was interbedded sandstones and dolomitic siltstones which were interpreted as the Munabia Sandstone. The sandstones were porous and drilled fast but the dolomitic siltstones which were not predicted, were "as hard as the hobs of hell" and drilled very slowly at 1-2 metres per hour. I expected to be at the wellsite for a week but it ended up being almost three weeks. Fortunately I had two very long books to read "War and Peace" and "Lord of the Rings" which lasted me for the duration.

Now the area between Carnarvon and Exmouth is in between the temperate zone and the monsoon area. We drilled the well in the "summer" expecting dry temperate conditions but a tropical rain squall crossed the coast and dumped a heap of rain on the wellsite. I was amazed how quickly the water rose – I expected it to immediately soak into the sandy soil. The well site became a sea of mud **(Fig 41)** and the mudlogging shack was almost washed into the sump. I returned to the camp at the end of the day and a few of the crew were standing up to knees in the brown, murky water having a few beers. I joined them but eventually it dawned on me and I asked "Can King Browns swim?" Everyone slowly moved to higher ground.

The well eventually intersected a change from the hard silty dolomites and sandstones to the interbedded limestones and dolomites of the Gneudna Formation. I had been collecting large cuttings samples of the limestones for conodont analysis and when these were analysed, this lithology change was confirmed as the top of the Gneudna Formation. The conodonts changed from the Fammenian *Polygnathus communis collinsoni* zone to the Frasnian *Polygnathus asymmetricus* zone. My interpreted reef on the seismic was actually a buried hill and with no shale seal above the unconformity, and the well was a dry hole.

Figure 41
Anyone for a dip?

BALI

On 11th October 2002, Mo and our friend Louise Fitzgerald went on a holiday to Bali. It was their first time to Bali and they were staying at a very palacial resort at Legian away from the crowds in Kuta. On the morning of 13th October, I heard the news that a bomb had gone off in Bali. As bombs go off quite often in Indonesia, I wasn't immediately concerned. However, I phoned Mo immediately and asked her if she knew anything. She said that there was a loud noise during the night but they thought it was the air conditioning playing up. So after a brief chat on how the holiday was going, I hung up. On the next news, the commentator said that over 80 people had been killed. Now I was really concerned and rang Mo again. I told her that it was a "really big bomb" and to find out what was going on. She phoned me back and told me that the bomb had gone off in Kuta and a lot of people had been killed. She also said that a lot of tourists were leaving the hotel to go to the airport. I told her to stay where she was and see how things developed. If I was a terrorist, I would have planned to set off another bomb at the airport the following day to kill as many tourists as I could fleeing the island. In my opinion, Mo and Louise were much safer staying right where they were. Mo and Louise stayed in the resort until the 18th. The Balinese staff were horrified at what had happened and looked after the remaining guests with extra service. They couldn't understand why anyone would do such a thing. Neither could we! Naturally, all the tours were cancelled and all remaining guests were told to stay in their resorts. Mo and Louise did go into Kuta and were appalled at the devastation. They returned to Perth on 19th and I was very relieved to see them at the airport.

ATTEMPT TO LIST GULLIVER ON ASX

During 2002, I attempted to list the assets of Gulliver Productions on the ASX. These assets consisted of the interests in EP-104, RL1, the West Kora application, EP-438 and EP-448 in the Canning Basin; EP-410 and application 16/00-1 in the onshore Carnarvon Basin, and applications 3/02-3 and 4/02-3 in the onshore Perth Basin. Gulliver had applied 100% for the latter two areas and was the successful applicant subject to negotiating native title agreements.

Area 3/02-3 was located in the northeastern part of the basin where there was a reasonable amount of old Wapet seismic plus the Discovery Petroleum 1994 Mingenew Seismic Survey. My interpretation of these seismic surveys indicated an anticline similar in style to the Mount Horner Oil Field and with the same Jurassic objectives. I called it the Moriary Prospect and it was covered by sufficient seismic that it was ready to drill as soon as it could be farmed out. The area also contained part of the North Erregulla structure. This structure was drilled by Wapet in 1967 and recovered 38° API oil from two drill stem tests, one over the Early Triassic Arranoo Sandstone and one over the Late Permian Wagina Sandstone.

Area 4/02-3 was located in the central western part of the basin and mostly covered the old Barrack Energy EP-323. Barrack acquired two seismic surveys over this permit, the Duggan survey in 1989 and the Woolka survey in 1990. In 1992, following my departure from Barrack Energy, the company, now called Arrow Petroleum, drilled the Mullering-1 well on the large anticline defined by these seismic surveys. During my time at Barrack, I wasn't keen on this prospect as the structure was heavily faulted on the crest so cross fault seal was a major risk. However, Mullering-1 encountered over 200 metres of lacustrine shales at the base of the Jurassic Cattamarra Coal Measures before getting stuck in the hole and abandoning the well. The thickness of these shales was unexpected and considerably increased the chance of cross fault seal. In 1994, Discovery Petroleum who now owned Barrack Energy, drilled Cataby-1 on the Mullering anticline. This well encountered a 2.5 metre sand in the Cattamarra Coal Measures and this sand tested oil at 25-30 bopd on drill stem test. After EP-323 was relinquished, I decided to apply for the area. What the Mullering anticline needed was a 3D seismic survey to better define the location and the amount of throw and the orientation of the many faults, and delineate more structures for drilling.

In order to attract sufficient risk capital for an IPO, Gulliver needed to buy some production. I initially tried to purchase the Mt Horner Oil Field from Petro Energy Pty Ltd owned by Clive Willis but this was unsuccessful. So I began negotiations with John Begg the CEO of Voyager Energy. Voyager Energy had decided to concentrate its exploration in the USA and was selling its assets in Australia. I was interested in purchasing their 40.936% interest in the Nockatunga Project in the Eromanga Basin. These fields were dear to my heart as they had been discovered by Pancontinental Petroleum and production was still pottering along at 350 bopd. Since my days at PPL, Production Lease 33 had been granted in 1986, an agreement had been signed in 1986 settling the disputes over the payment of the Banner carry, Banner itself had been bought by Chimelle Petroleum in 1987 and a new JOA had been signed in 1988. The surrounding exploration permit ATP-267-P was renewed in 1999 and Voyager Energy had bought Chimelle in 2000.

The fields were operated by Santos Limited and Voyager supplied the latest reserves report dated January 2001. The remaining oil reserves net to Voyager ranged from P1 of 0.8 million barrels, P2 of 2 million barrels and P3 of 10 million barrels. The major upside was in the Maxwell Field where remaining oil in place was high but recovery factors were low. Voyager had proposed to the operator to try a horizontal well to see if production rates and oil recoveries could be improved. At this time, there was no budget but they were interested in the concept.

With the assistance of Maurice Brand, we gave presentations to brokers but due to the low oil price, there was not a lot of enthusiasm. We then tried back door listing the assets into listed shells; firstly negotiating with Asia Multi-

Foods & Oils (Australia) Limited and then with Tasmania Mines Limited. But in the end, no agreement could be reached with these companies. In July, 2003, Voyager Energy sold its interest in the Nockatunga Project to Magellan Petroleum for $2 million.Voyager Energy itself was taken over by Arc Energy Limited in August, 2005

INVASION IRAQ

In March 2003, a select coalition lead by the USA invaded Iraq. They certainly gave Iraq plenty of warning as they began lobbying the UN Security Council to invade Iraq in September 2002. The pretense for invading Iraq was that Iraq was supporting al-Qaeda and that they had also renewed their work on producing weapons of mass destruction. The first of these was extremely unlikely as Saddam Hussein and Osama bin Laden despised each other. The second could not be confirmed by UN weapons inspectors when they visited Iraq in February 2003.

Prior to the invasion, the oil price rose to $US40 per barrel but with the swift military success, the oil price fell to $US25 per barrel in the expectation of the lifting of sanctions and increased oil production from Iraq. Alas, the Americans won the war but lost the peace and there were ongoing insurgencies by disaffected Sunnis and Al Qaeda groups until American troops withdrew in 2011. While all this was going on, Iraq's oil production dropped from its pre-war peak of about 6 million bopd to 1.5 million bopd. Since the defeat of ISIS, their oil production had increased to almost five million bopd by 2018.

KIMBERLEY ENERGY GROUP PTY LTD

In November, 2003, I was commissioned by Kimberley Energy Group to evaluate the Weaber Gas Field in Retention Lease R1 in the Northern Territory and Exploration Permit EP-386 in Western Australia in the onshore Bonaparte Basin. These tenements were owned by Amity Oil and were up for sale. The Weaber Gas Field had an interesting but frustrating history. Weaber-1 was drilled by Australian Aquitaine Petroleum in 1982. There were good gas shows over the Early Carboniferous Enga Sandstone but Aquitaine wasn't interested in gas and the zone was never tested. In 1985, Santos re-entered Weaber-1 and tested the Enga Sandstone which flowed at a stabilised rate of 4.5 MMcfg/d through a 1/2"' choke. Due to poor hole conditions, the well had to be plugged and abandoned. Santos also drilled Weaber-2 in 1988 but it was outside structural closure, the Enga Sandstone was water-wet and the well was plugged and abandoned. Potential gas reserves were reduced by this well result and Santos relinquished the area.

Weaber-3 was drilled by Capital Energy in 1994 but severe hole conditions prevented the well from being logged, tested or completed. Gas shows indicated that the Enga Sandstone was gas bearing.

The area was then picked up by Amity Oil who drilled Weaber-4 in 1997. The Enga Sandstone was gas bearing and tested at a stabilised rate of 2.1 MMcfg/d. They then drilled Weaber-5 in 1997 but it was outside structural closure, the Enga Sandstone was water-wet and the well was plugged and abandoned. Amity then conducted a production test of Weaber-4 but a pressure drop of 98 psi indicated recoverable reserves of only 0.15 BCF gas based on material balance. This did not compute with my estimate of 2.35 BCF for the field based on Amity Oil's seismic interpretation. The seismic quality was poor so small faults may have compartmentalised the Weaber-4 well.

Exploration Permit EP-386 had three gas discoveries with uncertain potential reserves. Vienta-1 encountered good gas shows in the Enga Sandstone and these were drill stem tested at 0.258 MMcfg/d declining to 0.169 MMcfg/d after two hours. Analysis of the test results indicated that the Enga Sandstone had low permeablity but there was no obvious pressure depletion. I estimated that there was a potential resource of 4.8 BCF gas in the structure but wells would have to be stimulated in order to flow at commercial rates. There were also gas shows in a sand in the deeper Septumus Limestone but due to the onset of the wet season, the well was suspended without testing this sand.

Waggon Creek-1 was drilled on a stratigraphic trap and tested thin sands in the Carboniferous Milligans Formation at 1.34 MMcfg/d over the interval 378-395 metres and 0.95 MMcfg/d over the interval 588-601 metres. The latter test also recovered 3.4 barrels of oil. Analysis of the test results indicated that these sands had low to fair permeablity with considerable formation damage but no obvious pressure depletion. These shallow sands cannot be seismically mapped but could cover an area of 600 hectares. In addition, they are also discontinuous with variable reservoir quality and some may be oil bearing. Potential resource estimations are very uncertain but could be up to 12 BCF gas.

Bonaparte-2, drilled by Alliance in 1963, tested a package of sands in the lower Milligans Formation at rates of 1.54 and 1.15 MMcfg/d. I estimated about 3.5 metres of net gas bearing sand and based on Amity Oil's mapping, an estimated potential resource of 4 BCF gas was calculated.

Due to all this uncertainty, Kimberley Energy purchased the tenements from Amity Oil in December for only $6,000 and took over the $50,000 bond. R1 had been renewed in June 2001 for five years but EP-386 had to be renewed in July 2004. There was one exploration well remaining to be drilled prior to renewal but by agreement with the government, this commitment was waived and the permit was renewed in October, 2004 with a firm commitment in Year One to conduct long term production testing of Vienta-1 and Waggon Creek-1.

I had no further involvement with the project after this but the tenements were transferred to Onshore Energy Pty Ltd which is now a wholly owned subsidiary of Advent Energy Limited. In October, 2012, Advent Energy conducted a production testing program on the Waggon Creek gas field. The interval tested was 378 to 595 metres and flowed at 0.5 - 0.9 MMcfg/d.

LIQUEFIED NATURAL GAS CORPORATION — TANZANIA

In June 2004, I was contacted by Maurice Brand to prepare a report on the Mnazi Bay Gas Field in Tanzania. He had formed a new company, LNG The Energy Link, to search for stranded gas fields which could be commercialised using gas fired power plants to generate electricity. The Mnazi Bay Gas Field was located in the Rovuma Basin and was situated in shallow water. The discovery well was drilled by AGIP in 1982 to a total depth of 3,489m and tested at rates of 7.5-14 MMcfg/d from two Oligocene sands between 1,875 to 1,909m.

A Canadian company called the Artumas Group had acquired the leases over the gas field and planned to conduct 125 kms 2D transition seismic over the Mnazi Bay structure at an estimated cost of $US 2.5 million. Interpretation of the existing seismic indicated a large anticlinal structure covering about 150 square kilometres with two culminations. Mnazi Bay-

1 was drilled on the highest culmination. Apart from the two sands that were tested, there were another five sands between 1,608 to 1,772m which, based on log analysis, may have also been gas bearing. The Artumas Group also planned to drill three appraisal wells from Msimbati Island in 2005 for an estimated cost of $US10.5 million. The standard Tanzanian PSA Model Agreement had a royalty rate of 12.5% and a production sharing split of 65% for the Government and a 35% for the contractor split after cost recovery of 60%. Maurice had reached agreement with the Artumas Group to purchase 40 MMcfg/d for 15 years indicating that the Mnazi Bay gas field needed gas reserves in excess of 220 BCF.

I estimated that proved gas-in-place reserves around the well were 27.6 BCF with a probable gas-in-place estimate of 450 BCF. Potential gas-in-place could have been over one TCF if the whole structure was full to spillpoint. A proved plus probable recoverable resource of 340 BCF would have been sufficient for the gas sales contract with the Artumas Group. Wells were expected to each produce at initial rates of 20 MMcfg/d. Nick Papalia's Global Capital Resources Pty Ltd was commissioned to prepare an Independent Geologists Report for LNG Energy Link's due diligence.

An acceptable gas price could not be negotiated with the Tanzanian Government so the project never went ahead and the Artumas Group eventually relinquished the area. The discovery lay dormant until 2005 when Wentworth Resources Limited, a UK listed company, drilled two appraisal wells and one exploration well in the area, all of which were successful. It went on to build a 27 km 8" pipeline to Mtwara and an 18 MW gas-fired power plant and commercialised the field in 2006.

Matrix Oil

In June, 2004, I prepared a valuation of the assets of a delisted company called Matrix Oil NL ("Matrix"). Matrix had a long history. It was incorporated in 1969 as Eastern Copper Mines NL and changed its name to Maple Oil NL in 1993. Maple Oil was listed on the ASX in 1994 but after drilling a number of exploration wells, eventually ran out of money and was suspended in 1998. The company was re-listed as Matrix Oil in 2000 with a 90% interest in the Langsa Offshore Technical Assistance Contract. The Langsa Field was developed but the two production wells experienced opcrational problems and production ceased in October 2002. The company immediately requested suspension of its shares and appointed an administrator. The administrator sold off most of the assets leaving only a 5.0% interest in NT/RL 3, containing the Barnett Oil Field, and a 5.89% interest in WA-13R, containing the Turtle Oil Field. The remaining interests were held by OMV (the operator) and Basin Oil. My report valued the assets at up to $1.5 million with most likely value of $350,000.

Matrix entered into a Deed of Company Arrangement with the Administrator whereby the company would be re-capitalised by exchanging every 20 existing shares for one new share and an attempt was made to re-list Matrix by issuing 200 million shares at 1.4 cents to raise $2.8 million. This proposal was rejected by the shareholders at a meeting in October, 2004. Matrix Oil eventually became Emerald Oil & Gas NL in 2006.

GRYPHON MINERALS LIMITED

In November, 2004, I prepared a short report on Frontier Oil & Gas' permits and applications in the Amadeus Basin for Gryphon Mining Limited. Mining exploration was very slow and they were looking into getting into the oil industry. Based on my knowledge of the basin, I told them that this area was not the best choice to start your first venture into the oil industry. The basin was very remote, expensive and had a poor exploration success rate. I followed this up in early 2005 with an overview of the Cooper Basin where, although the deals were very tough, it was a proven oil and gas province and the success rates were good. In the end, they didn't get into the oil industry but went off to Africa where they discovered and are developing the Banfora gold deposit in Burkina Faso.

VARIOUS US PROJECTS
LIVINGSTON PETROLEUM LIMITED

This company was formed in December, 2004 and I was contracted to evaluate a number of opportunities in the USA with the objective of listing on the ASX.

I reviewed the Port Acre and Deep Lake prospects in the Gulf Coast Basin in

Texas and Louisiana. The Port Acre Prospect was a Frio sand test of a tilted fault block. Although the prospect was technically sound, I calculated that the company under the terms of the farm-in, would be risking $500,000 to make only $1.0 million over 9 years which was certainly not commercially attractive. The Deep Lake Prospect was a Miocene test in the centre of Deep Lake Gas Field which had produced in excess of 1 TCF gas. The well was located to drain a small fault block updip from an existing and depleted gas well. Like Port Acre, this prospect was technically sound but commercially unattractive. Both these projects were rejected.

I reviewed the Forbes Project in the Sacremento Basin in California. This project consisted of 30,000 leased acres covered by 3D seismic surveys. The main objectives were the Late Cretaceous deep water sands of the Forbes Formation and potential accumulations are complex structural –stratigraphic traps. Gas accumulations also show up very well as amplitude anomalies on the 3D seismic. The deal was for Livingstone to pay $US840,000 for a 50% working interest and 37.5% net revenue interest in the whole project and then paying 66% of the cost of drilling the next three wells. Although the deal was tough, the project had the potential to pay out in three years and would ultimately give Livingston an interest in 30,000 acres in this highly prospective basin. A prospectus was lodged in August and raised $4.8 million by the issue of 12 million shares at 40 cents per share. Livingstone Petroleum listed on the ASX in September, 2005. The exploration program got off to a good start with the Larner 2-12 well encountering 39 feet (12 metres) of net gas pay. The Edge 1-10 well encountered 25 feet of gross gas pay but the sands were very shaley and net pay was difficult to estimate.

In April, 2006, I was commissioned to prepare an update of the Independent Geologists Report in the August Prospectus. This was my final report for Livingstone Petroleum. In May 2007, they merged with a private US company with a large portfolio of acreage in the San Joachin and Ventura basins and changed their name to Solimar Energy Limited.

Transerv Energy

I reviewed the Cymric Monterey Project in San Joachin Basin, California for Transerv Energy Limited in February 2006. The project consisted of

160 acres on the flanks of the Cymric Field. Oil is trapped in Late Miocene diatomaceous laminated sands with high porosities of 30% but low permeabilities of 0.1 to 2 millidarcies. The average well recovered only 56,000 barrels of 8-18° API oil. I recommended against participating in the project due to small acreage, small potential reserves and low flow rates. It simply wasn't meaningful to a company listed on the ASX.

In August, 2006, I reviewed the Wind Gap Project in the San Joachin Basin, California. The proposal was quite complicated but essentially Transerv would acquire a 12.75% working interest in the first lease by contributing 15% to the cost of drilling and completing the Wind Gap 42-36 well. Transerv would also have the option to drill a second well to earn 12.75% working interest in the second lease.

The main objectives in the Wind Gap 42-36 well were Oligocene deep water turbidite sands with secondary objectives sands in the Eocene. The well was drilled during September-October, 2006 and was cased at 12,015'. Sands with gas shows were tested but did not flow to surface. In my opinion, the better sands may have been damaged by the water based mud system but were not fracced prior to testing. The joint venture decided not to drill to the deeper Eocene objectives and to P&A the well. Transerv wanted to exercise its option to drill a second well to earn its interest in the second lease but the other joint venturers did not. The original lease holder told Transerv that as the other parties did not exercise the option, the option was therefore terminated. He then immediately farmed out the second lease to another company. Transerv was appalled by this behaviour but wasn't prepared to take on a law suit in California. But it reinforced what a lot of companies say about the US oil industry – you have to be very careful who you deal with there!

BARAKA PETROLEUM PTY LTD

In early 2005, I was asked by Tony Gates' Pacific Exploration Consultants to assist him in preparing an independent geologists report on the assets of Brimax Petroleum International Limited. Brimax had exploration leases in Mauritania and Mali and it was intended that these assets would be purchased by Baraka Petroleum Pty Ltd for a potential listing on the ASX.

Tony was going to prepare the report on the Palaeozoic basins in Mali and I was to prepare the report on Block 20 in Mauritania.

Block 20 was located in the Northwest African Coastal Basin, onshore but bordering the Atlantic coastline. The tenement was originally granted as an exploration study agreement in 2000 and following a 50% relinquishment, a Production Sharing Contract (PSC) over the southern half was submitted to the Minister of Mines, Energy and Industry in February 2005. The terms of this PSA were a production sharing split of 30% for the Government:and a 70% for the contractor split after cost recovery of 60%. There was no royalty but a 27% tax rate.

The main lead in Block 20 was the Heron Lead, a north-south trending faulted anticline mapped by Texaco in 1992. The nearest control was Autruche-1, an offshore well drilled by Texaco in 1989. This well encountered good oil shows in Cenomanian and Aptian sands which are the main objectives in the Heron Lead. Aptian black shales are the main source rocks in the area and also act as the seal for the objective Aptian sands. The lead was conceptually very attractive but needed additional seismic to mature it for drilling. A 200 kilometre onshore seismic survey was planned for the first year of the PSC.

My report was completed in February and Baraka Petroleum was listed in May 2005.

Coogee Resources

In February 2005, I was commissioned by Coogee Resources ("Coogee") to prepare an independent review of the Magnolia Prospect in AC/P 32 in the Timor Sea. Coogee had entered into a Farm-in Agreement, subject to due diligence, with the existing joint venture of Norwest Energy 49%, Bounty Oil & Gas 26% and the Chinese Petroleum Corporation 25%. The Magnolia Prospect was a faulted anticline defined by 3D seismic and with its main objectives being the Jurassic Montara Sands sealed and sourced by Jurassic marine shales. The main risk was cross fault leakage but all other risks were minimal. Given an estimated potential resource of 67 million barrels oil, I recommended that they proceed with the farm-in. Alas, Magnolia-1 was a dry hole.

HUNGARY

During 2005, I undertook a consulting project on the Pannonian Basin in Hungary. A private company Inke Petroleum Pty Ltd had entered into an agreement with Gemstone Properties Limited, Blue Star 95 Kft to acquire the Inke Concession in the Western Pannonian Basin in November 2005. The concession covered 2,300 sq kms and was located 120 kms southwest of Budapest. It also contained a number of non-commercial oil and gas discoveries and was surrounded by producing oil and gas fields.

I was not aware that Hungary had so many producing oil and gas fields. Exxon, among other companies, had explored the country before WW2 and had found some significant oil and gas fields. Of course, these were all nationalised after the Soviet conquest of the country in 1945. The other interesting thing about Hungary is that although it is small in area, it borders seven other countries: Austria, Slovakia, Ukraine, Romania, Serbia, Croatia and Slovenia.

I prepared reports on the Blue Topaz and Nagyatad prospects in the Inke Concession and also evaluated the Nagylengyel Block in the West Danube Basin. The Blue Topaz 1-8 prospects targeted Miocene marine sands in small structural closures, one of which was updip from the Somogysamson-3 gas discovery made in 1983. The Blue Topaz 9 Prospect was a stratigraphic trap of Miocene marine sands in a distinctive mounded package which were interpreted to pinch out updip. This prospect also had secondary objectives in Triassic carbonates trapped in an eroded fault block. The Nagyatad Prospect was a stratigraphic trap of Early Miocene marine sands pinching out updip over an anticlinal nose. Oil had been already found in three wells but flow rates were low, less than 40 bopd, possibly due to formation damage. An independent evaluation of the Inke Concession was made by RPS Energy in December 2005.

The Nagylengyel West Block had only one drillable prospect. The Csdor Prospect was a simple anticline draped over an eroded high of Mesozoic carbonates. The main objectives were Miocene marine sands with secondary objectives in the underlying Mesozoic carbonates. It was decided not to farm into this block.

In February, 2006, there was a placement of 20 million shares which raised $1.4 million to fund the acquisition of the Inke Concession. Upon completion of this purchase, Inke Petroleum was acquired by Matra Petroleum Plc by purchasing each Inke share for one Matra share. Following the acquisition of Inke Petroleum, Matra Petroleum listed on AIM in London in April, 2006 following the issue of an additional 160 million shares and 80 million warrants (options). The company was then run from London and my involvement with the project ended. Blue Topaz-9 was eventually drilled in May 2006 and encountered thick water bearing sands in the Miocene section and oil shows in Triassic dolomites. These oil shows were tested and although exhibiting high permeability, they flowed only salt water.

In June, 2007, Matra farmed out a 60% interest in the Inke Concession to Aspect International LLC for Aspect to pay the next $5.5 million of costs. In July, 2007, Matra drilled the Horvatkut-1 well approximately 400m northeast of the Somogysamson-3 gas discovery. This well encountered good gas shows at 1640m and an open hole DST was conducted. The well tested 3.9 MMcfg/d on 12mm choke (1/2") with WHP of 1015 psi.

In April, 2007, Matra acquired rights to the Arkangelovskoe Licence in Orenburg Russia. As a result, the Delek Group became a significant shareholder of Matra. In late 2007, Matra discovered the Sokolovskoe Oil Field in Devonian carbonates. Subsequent wells proved up 2P reserves of 13.5 million barrels oil and a production licence was granted in December 2010. However, in 2012, control of the company changed hands and in July 2013, the Sokolovskoe Production Licence was sold and Matra became an investment company.

AAPG FIELD TRIP

I had been a member of the American Association of Petroleum Geologists ("AAPG") since 1976 but had never been to one of their annual conferences. I was unlikely to do so as they nearly always clashed with the APPEA conferences in Australia. However, each year they organised various courses and field trips around the world – and this year, they were running a structural geology course in June 2006 on the Spanish side of the Pyrenees Mountains. So I decided to treat myself and booked to attend this field trip. The trip started in Barcelona but because of poor connections, I flew to London via Dubai and the following day caught the short flight back to

Barcelona. I got there a day early and took a tour of the city. Barcelona is a very historic city with bits of the old Roman walls still preserved. The old cathedral was being repaired and was covered in scaffolding but inside, the architecture, carvings and decorations were just magnificent. The choir was singing and the music reverberated around the cathedral. Now I'm not remotely religious but the whole effect was just awesome compared to what I had ever experienced inside modern churches.

The next day, I caught up with the rest of the participants and Antonio, the leader of the course. All the participants were younger than me except for a retired American who turned out to be fitter than me! We all piled into a bus and drove to the town of Jaca at the foot of the Pyrenees. On the drive there, the country side was very dry and brown, just like Australia until we came over a hill and there was a large medieval castle sitting on top of the next hill. Not quite like Australia! After arriving in Jaca, Antonio checked us into our hotel and we cleaned up and went out to dinner. As we finished and headed back to our hotel at about 10pm, the locals were just going out to dinner, kids and all.

The first day of the course we drove to the Belagua Valley and inspected the Lakora imbricate thrust sheets repeating the Cenomanian and Santonian limestone sections. We ended up wandering into France but there was nobody there to inspect our passports. On the second day, we drove past the little town of Hecho north to the foot of the high valley of the Aragues and Antonio told us to start walking up the valley. It took all day but fortunately he brought some sandwiches otherwise we would never have made the peak. We walked up the full section starting at the bottom with outcropping Santonian limestones, passed an Alpine hut before reaching the light orange brown Maastrichtian sandstones. These were in turn overlain by white Eocene limestones and eventually an excellect outcrop of brown Miocene limestones overlying the Eocene. When we reached the top of the mountain, there were patches of snow on the ground.

On our return down the valley, Antonio showed us an amazing outcrop of brown Maastrichtian sandstones thrust over white Eocene limestones. The most remarkable part of this outcrop was that there was virtually no breccia along the thrust fault – trillions of tonnes of rock simply slid over each other as if they were plastic **(Fig 42)**.

Figure 42
Thrust fault with Maastrichtian overlying Eocene

On the third day, we ended up south of Hecho to view the Eocene carbonate megabreccias caused by collapse of the carbonate platform and turbiditic transport into the basin. On the last day, we drove south to the town of Riglos where spectacular cliffs of Oligocene conglomerates form a backdrop to the town itself **(Fig 43)**.

We also visited a monastery which was carved out of and imbedded into these conglomerates. The worst part of this day was that bloody Antonio made us climb to the top of those cliffs. The best part of this day was that view from the top was breathtaking and Antonio redeamed himself by spending an hour describing the complex structural geology spread out below us.

The next day, we drove back to Barcelona airport, said our goodbyes and I flew back to London. When I arrived at Heathrow, I initially confused immigration because I had already arrived in the UK but officially hadn't

Figure 43
The town of Riglos with spectacular cliffs in background

left since I had travelled within the EU. It was quickly sorted out and I passed through customs and met Brian and Eve Armitage, our old neighbours in Broxbourne, who took me back to their place to stay overnight. We had a great meal that evening with our old neighbours and caught up on old times. After that, I headed off for Cork and spent a week with Bob on his farm just northwest of Glengarriff before returning to Perth.

Bob was progressively converting most of his farm back to its natural state. He still kept various orchards but large areas of his property were infested with bracken and brambles so his task wasn't going to be easy. Driving around this part of Ireland is very enjoyable (when it's not raining) and the area also has the odd archaeological sites of standing stones.

BLACK ROCK PETROLEUM NL

At the invitation of Ivan Burgess, I became a director of Black Rock Petroleum NL in August, 2006. In October, 2005, Black Rock had sold all of its shares in Black Rock Resources (UK) Limited to Black Rock Oil & Gas Plc, a company listed in the UK. Black Rock thereafter had no liability for exploration permits in the UK held in the name of Black Rock Resources.

After the sale, Black Rock's only remaining asset was Wildlook Enterprises Pty Ltd which held a 15% interest in Retention Lease R3 containing the Cyrano oil discovery in the offshore Carnarvon Basin. The Cyrano structure is an elongate three way dip closed structure trapped against the downthrown side of the Flinders Fault. Two wells have been drilled on the structure; Cyrano-1 intersected 31.5 metres of gross hydrocarbon column but Cyrano-2 only intersected a 4 metre oil column. The oil and gas pay was mostly in the Mardie Greensand, a high porosity but poor permeable reservoir. The operator had estimated the recoverable reserves to be about one million barrels. I estimated that the oil-in-place resource could be as high as 17 million barrels if the structure extended to the northeast and southwest. However, given the expected low recovery factor expected for the Mardie Greensand, this would still not be economic.

The company was wound up after transferring its interest in R3 to Oil Basins Limited who now hold this interest 100% after buying Tap Oil's interest in October 2010.

BLOCK M, BRUNEI

In early 2007, I was commissioned by Verona Capital to evaluate an onshore concession in Brunei. Block M was awarded to Valiant International Petroleum 25%; China Oil USA 65% and Jana Corporation 10% in February 2006. Block M contained the Belait Gas Field which had never been fully developed. The field was a complexly faulted compressional anticline with the hydrocarbons trapped in Miocene sands. The seismic was very poor and the exact nature of the structure was poorly understood. There were 19 wells drilled on the field and these were used to complement the poor seismic to give a clearer picture of the structure.

I submitted my report in March 2007. It concluded that estimated reserves in the field were of the order of 15 BCF gas with considerable upside potential in the subthrust structures. Verona Capital did not take up the opportunity but in February 2008, Tap Oil bought a 39% interest in Block

M. They in turn sold their 39% interest to Polyard Petroleum International Exploration and Production Pte Ltd in September 2011. The term of the Block M contract expired in August 2012.

ELIXIR PETROLEUM LIMITED

In April, 2007, I was commissioned by Elixir Petroleum Limited to provide an Independent Geologist's Report for the acquisition by Elixir of Gawler Resources Limited by way of a Scheme of Arrangement. The purpose of the report was to evaluate the prospectivity of Elixir's leases in the North Sea and provide a valuation of these leases. There was an incredible amount of technical data (some 1.15 GB to be precise) but this I retrieved from their ftp site.

My report was completed in May and led to a further commissioning by BDO Consultants (WA) Pty Ltd for my report to be incorporated in their Independent Expert's Reports for the merger of Elixir Petroleum Limited and Gawler Resources Limited. As these reports would be included in a Prospectus to be put the Elixir shareholders, they could only be provided by a company with an AFS Licence.

My initial report was submitted at the end of May with a valuation of minimum $2.1 million, most likely $13.34 million and maximum $80.5 million. However, there were issues with ASIC on the use of the EMV method for valuations. In obtaining the NPV for the development of similar potential resource sizes to Panther and Leopard prospects in the North Sea, I had referred to an independent engineering consultant commissioned by Elixir.The ASIC wanted a full explanation of how these numbers were calculated, but after providing the technical credentials of the engineering consultant, they finally accepted the report.

CAMBODIA

In July-August, 2007, I got involved in a study of onshore Cambodia. A private company Focus Energy Pty Ltd was negotiating an exclusive 12 month scoping study of the onshore oil potential of Cambodia from the National Petroleum Authority with the option to take up two tenements.

Apart from mapping by the Russians in the 1990's, the only other available surveying was by the Japanese National Oil Company ("JNOC"). In the late 1990's, JNOC acquired 21,675 kms of airborne gravity and magnetic data coving some 48,000 sq kms of onshore Cambodia. Interpretation of these surveys indicated two deep sedimentary sections, one to the southwest of the Tonle Sap Lake and one some 100 kms to the northeast of the lake. If these basin-like features reflected deeper Tertiary grabens, then their oil potential could be significant as similar grabens in Thailand had proved to be oil bearing. Since the JNOC survey was acquired at only a three by ten kilometre spacing, Focus recommended that the initial exploration program for any PSC's would be to acquire a closer spaced aerogravity survey to confirm the deeper sedimentary sections. If positive, these surveys would be followed up by more expensive seismic surveys.

Being a private company, Focus needed funding from a reputable company in order to make a formal application to the Cambodian National Petroleum Authority for a PSC. They contacted Chalice Gold Limited, a company listed on the ASX and with funds available for new ventures. I was contacted by them to provide an independent evaluation of the Focus proposal. I travelled to Adelaide to discuss the geology with Jim Allender who was involved with Focus, and collect copies of what data was available. My evaluation was that although it was a frontier area, there was definitely oil potential in the area. In addition (and most importantly for a small company) it would not cost a lot to upgrade (or downgrade) its potential. Initially it would cost ~$100,000 to visit Phnom Penh and apply for the PSC's. When (and only if) a PSC was granted, would a further $500,000 be committed to acquiring the more detailed aerogravity survey. If the interpretation of the gravity survey confirmed the thick sedimentary section and the presence of basement highs, then the joint venture could decide to proceed to acquire, or farmout, the seismic survey. At any time during this process, if the oil potential was not upgraded, then Chalace Gold could cut its losses and withdraw from or relinquish the PSC.

There was however, one difficult commercial problem. This was the requirement to have a local company as a 20% partner. Chalice would have to initially do a deal with Focus on its 80% interest and then, if the gravity results were positive, then farm down again to fund the seismic. Given the high risk and frontier nature of the oil prospectivity, this was a big ask and in the end, Chalice declined to participate. This was a real shame as I still believe that Tonle Sab Lake has the potential to be a major oil producing area for Cambodia.

As at 2014, there were only three companies, Medco, PetroVietnam and JOGMEC operating onshore Cambodia.

MURPHY OIL

In 2007, I had a meeting with David Wood, my old boss at Ashland Oil. He was now Vice President International Exploration for Murphy Oil Corporation, an independent US petroleum exploration and production company. Murphy Oil was looking to invest in Australia and he asked me to assist the company in setting up an exploration portfolio. They set up a subsidiary company, Murphy Australia Oil Pty Ltd, and as they needed one domestic director, I consented to become a director and was appointed in October 2007.

At this time, Finder Exploration Pty Ltd had a 100% interest in AC/P 36 in the northern Browse Basin and they were trying to farm out this permit. Permit AC/P 36 was very dear to my heart as it contained the Bassett Deep Prospect which I had worked on in my days at Burmah Woodside. This prospect covered an area of 700 sq kms with 400 metres of vertical closure. It was so large, it extended from AC/P 36 south into WA-343-P operated by Inpex. It also extended into a vacant area which had been gazetted by the Federal Government as W07-6. Of course, the interpretation of the structure was based on poor seismic and it was very unlikely that it was one simple structure. It would be necessary to cover the entire structure with 3D seismic to determine the number and size of individual prospects expected to be present in the Bassett Deep complex.

I had a presentation by Finder Exploration in August 2007. They had purchased 7,000 sq km 3D seismic from Western Geco and had PSDM reprocessed selected portions of it for interpretation. This work had resulted in the definitions of a number of prospects including the Abalone Deep Prospect which was solely within AC/P 36. This prospect was an anticlinal structure with the *M. australis* Sand as the main objective. Abalone-1 was drilled by Japex in the late 1990's. The well encountered very high pressure gas sands at 4,768m just above the *M. australis* Sand and had to be prematurely abandoned. There were also high seismic amplitudes beneath the total depth of this well. The structure covered an area of 80 sq kms with vertical closure of 100 metres.

There was a further presentation by Finder to Dan Hanchera from Murphy Oil's Houston office and Hoh Swee Chee from Murphy Oil's Singapore office. Following this meeting, we all agreed that although the Abalone Deep Prospect was very large, it was also very high risk. However, the main prize in AC/P 36 was the Bassett Deep Prospect but this could not be drilled until it had been detailed by 3D seismic. Of course the 3D seismic couldn't be acquired until there was a unitisation agreement with the WA-343-P joint venture to the south. In addition, an application on behalf of the new joint venture for W07-6 needed to be prepared and submitted to the government. All these comments and recommendations were passed on to Dave Wood.

In addition to all this, the Browse Basin was hotting up. In April 2006, Inpex had announced the development of the $US8 billion Ichthys LNG project. The Ichthys Gas Field was originally found by Brewster-1 drilled in the Browse Basin in 1980 by Woodside. It seems incredible today but the original permit was relinquished as there was at that time no market for natural gas. The area was granted 100% to Inpex as WA-285-P in 1998. Inpex drilled three wells, Dinichthys-1, Gorgonichthys-1 and Titanichthys-1, during 2000 and early 2001 and encountered condensate rich gas in the Jurassic Brewster Sandstone and Plover Formation. Titanichthys-1 was tested at a rate of 40.6 MMcfg/d plus 2305 b/d condensate. They reported the reserves of the field to be an estimated 12.8 TCF gas plus 527 million barrels condensate. Production from the field was forecast to be 8 Mta LNG, 1.6 Mta LPG and 100,000 b/d condensate.

The Ichthys LNG project, being in Western Australia, should have been developed via an LNG plant on Maret Island off the Kimberley coast but due to the appallingly slow approval process in Western Australia and objections from the Green movement and some Aboriginal groups, Inpex concluded that it wasn't welcome in WA and took the project to the pro development Northern Territory. The gas from Ichthys will now be taken via an 885 kilometre pipeline to Darwin where there was plently of land available for a new LNG plant, a supportive government and no objections from the local communities. This new development plan would open up the northern Browse Basin, including the Bassett Prospect, to further LNG projects given exploration success. You have to feel sorry for the Kimberley aboriginal groups and their frustration at this fiasco. Their potential benefits would have been $1.5 billion over 30 years. Rita Augustine, on behalf of the Jabirr Jabirr traditional owners, wrote a published letter that the deal could have saved a community in despair as it would have provided badly need funds for education, jobs, health and housing for her people. But I'm sure the Green movement is very proud of what they achieved.

Murphy Oil also needed another major company to join the farmin and share the costs of drilling Abalone Deep-1. Murphy Oil had a relationship with PTTEP, the Thailand national petroleum company. In September, a PTTEP delegation visited Perth for meetings on AC/P 36. Following their due diligence, they agreed to join Murphy Oil in farming into AC/P 36. This was followed up by further meetings to finalise the Joint Participation Agreement (JPA) with Finders together with the Joint Operating Agreement (JOA). Regarding the JPA, the main issue was the agreement with Western Geco for the purchase of the 3D seismic in AC/P 36. This agreement required additional payments to be made if new companies came into the joint venture. Upon the execution of the JPA, these payments would have to be made to WesternGeco.

Although the JOA was to be based on the AIPN standard, there were still the usual arguments over Pre-emptive rights versus First Right of Refusal. But the main sticking point was that PTTEP didn't want the operator to have the final vote on obligatory programs for renewals. But at the end of the day, the operator must have the final say otherwise the permit would not be renewed and all the other joint venture companies would be unfairly penalised by loosing their interest in the permit. If a company is that opposed to the renewal program, it still has the right to withdraw from the permit.

In November, a meeting was held to finalise and execute the JPA and JOA. The Deeds of Transfer for Murphy Oil and PTTEP to become permittees of AC/P 36 were signed after this meeting. Since the JPA involved the payment of past costs, there was a registration fee of 1.5% payable on these costs. At the request of Murphy Oil, the transfers for AC/P 36 were promptly approved by the government but registration was held up due to an administrative problem in paying the registration fees. To promptly expedite the registration of the transfers, I wrote out a personal cheque for $28,800 for the registration fees. The following month, my expense account was the largest that I have ever submitted to a client! The transfers were registered by the Department of Primary Industries, Fisheries and Mining in April 2008.

The application for W07-6 was prepared based on the Finder studies and there were many arguments over the primary work program for the first three years of the permit term. These commitments were obligatory so it was a delicate balance between the costs of the program versus it being sufficiently high enough to win the bid. Once this was finally agreed and the application prepared, I submitted it to the DoIR on the closing date of 18 October 2007. I waited until 15 minutes before the closing time of 4pm so we could make an estimate of the number of applicants. Each bid is given a number and at 3-45pm, our bid was number 8 meaning that there had been seven other applications. These may of course not been all for W07-6 but it gave us an indication. The change in Federal Government also caused delays in the processing and award of all the October gazettals. When I enquired from the DoIR, I was told that W07-6 was very competitively bid and I told Murphy that their bid was unlikely to be the winner. This turned out to be the case with Total winning the bid with a primary exploration program of 1300 sq km 3D seismic plus three wells. The permit was granted as WA-408-P and soon after, Total farmed into WA-341-P, WA-343-P and WA-344-P to earn a 40% interest in each permit from Inpex.

The most significant industry activity in Australia during 2007 was the granting of production licence WA-34-L to Woodside over its Pluto gas field and the go-ahead for the $12 billion Pluto development. This was a monumental decision for Woodside as they had to fund the entire project themselves and it reminded me of BP's 100% development of the Forties Oil Field in the North Sea during the 1970's. By June 2011, the cost of the project had increased to $14.9 billion. Despite this, they eventually reaped the rewards of massive cash flow when the LNG project finally came on stream in May 2012. It was not all good news as, in August 2012, Woodside reported that its 25 well exploration program had failed to find sufficient gas reserves for Pluto Phase 2.

In January 2008, AED Development was looking to sell down part of its interest in AC/P 22 and production licence AC/L6. I was asked my opinion of the value of these assets given the fact that I had worked on these fields during my time at West Oil. I recommended against Murphy Oil purchasing an interest in the Puffin Project on the basis that, in my opinion, the independent reserves estimate of 40 million barrels was very optimistic given the velocity uncertainties in depth conversion. As discussed earlier, a figure of 20 million barrels might be possible but should be considered as a 3P upper limit. Fortunately, Murphy Oil accepted my recommendation as the field ceased production in May 2009 after recovering a total of only 2.2 million barrels oil.

The drilling manager for Murphy Oil also visited Perth to start preparations for the drilling of Abalone Deep. The permit had a well commitment in Year Four so Abalone Deep-1 had to be drilled by October 2009. I set up meetings for him with various drilling management consultants in Perth. I had previously contacted Tom Brand of Labrador-Petro-Management who I knew and discovered that, by an amazing coincidence, he had been on the Abalone-1 well.

After the disappointment of missing out on W07-6, Murphy Oil and PTTEP agreed to form a study group on the latest Federal offshore gazettals W07-8 to 11 in the Browse Basin. Murphy and PTTEP prepared a joint application for W07-11 which I submitted to the Designated Authority in April. This time Murphy Oil and PTTEP were successful and WA-243-P was awarded in December 2008.

In May, 2008, I assisted Murphy Oil in evaluating various to farmin opportunities in the Gippsland, Bonaparte, Carnarvon and offshore Perth basins and the offshore Cantebury Basin in New Zealand. Of these, the only one I liked was the Yngling Prospect in the offshore Perth Basin. The prospect was an anticline, as opposed to a tilted fault block, and fault seal risk was therefore reduced. The Early Cretaceous Gage Sandstone is an excellent reservoir with the marine shales of the South Perth Shale acting as great seal. The thick Parmelia Formation contains abundant lacustrine source rocks which are buried sufficiently to be in the oil generation window. Unfortunately the seismic quality was atrocious and the crest of the structure was difficult to map with any certainty. Today, with modern acquisition and processing techniques, I believe this prospect could be better defined and a confident drilling location selected. Murphy

Oil didn't farm into any of these projects and the Yngling permit was eventually cancelled by the government in 2011. The Yngling Prospect remains undrilled to this day and awaits not only an intrepid explorer but also someone patient enough to battle the approval process to drill an oil exploration well off the northern beaches of Perth!

VARANUS ISLAND

In July 2008, Apache's Varanus Island gas plant exploded and removed 330 MMcfg/d from the domestic gas market and also 8,000 bopd from Apache's oil production. In another classic example of Murphy's Law, this accident happened at the same time as WA's energy supply was reduced by damage to the Collie coal fired power station and winter maintenance on the Kwinana power station. WA's power supply was reduced by 30% and cost the state's industries an estimated $2.5 billion. The WA inquiry into the incident was eventually released in April, 2012. The report claimed the incident was caused by low staffing and inadequate internal systems. The explosion itself was caused by external corrosion of the high pressure 12" gas pipeline in the tidal zone that was exposed to wet and dry conditions. It exploded and triggered the explosion of five other nearby pipelines. The WA government had to drop its criminal case against Apache on a technicality – the use of the term "pipe works" rather than "pipelines" was used in the legislation. In any event, the fines were only $30,000.

Apache Energy was sued for $200 million by a group of its customers affected by the loss of their contracted gas supplies. These included Alcoa, Wesfarmers, Newmont Mining, Iluka Resources and Harvey Industries Group. This action was finally settled in April 2015 in order to facilitate the sale of Apache's gas production assets to the Brookfield Asset Management-Macquarie Capital consortium (now Quadrant Energy). In December 2015, Alcoa entered into a 12 year sales gas agreement with Quadrant Energy to supply 120 TJ/d.

In August 2008, BHP Petroleum decided to put up their Griffin oil and gas field assets for sale. The Griffin oil field was discovered in 1989 and commenced production in 1994 at ~ 80,000 bopd. As the sale would have to be handled through their Houston office, Murphy Oil asked me if they should invest the time and cost to review these assets. I did a quick review and estimated that the remaining reserves were of the order of 30 million barrels oil-in-place so it was definitely worth the effort to review these assets. Producing these remaining reserves would not be easy and would require workovers and infill drilling. But the average production during 2007 was a respectable 5,500 bopd plus the associated 10 MMcfg/d of gas being taken via a pipeline to the onshore Tubridgi Field and from there into the Dampier to Bunbury pipeline. I don't know whether Murphy Oil actually made a bid but BHP Petroleum did not get any acceptable offers and at the end of 2009, the fields were shut-in by at ~4,000 bopd after producing 178 million barrels oil.

In October, 2008, I assisted Murphy Oil to lodge applications for AC08-7 and AC08-8 in the Timor Sea. The bidding consortium consisted of Murphy Oil, Mitsubishi and PTTEP. Application area AC08-7 contained three leads of which the SW Lead was the best defined. The lead was located in water depths of 900 metres with the main objectives being the Jurassic Plover Formation sandstones below 5,000 metres depth. Application area AC08-8 contained the Phrixus Deep Lead located in water depths of 500 metres. The main objectives were the Jurassic Plover Formation sandstones at about 5,000 metres depth.

The seismic reprocessing had been completed in AC/P36 and the depth conversion finalised. An Operating Committee Meeting was held and the Abalone Deep prospect was selected to be drilled using the Stena Clyde. Abalone Deep-1 was drilled from November, 2008 to February 2009 but the *M. australis* Sand, although gas bearing, was tight and the well was plugged and abandoned.

During late 2008 Chevron decided to develop the Wheatstone gas field independent of Woodside's Pluto gas field by taking the gas to an LNG plant to be built near Onslow. To me, it would have made more sense to develop Wheatstone via Woodside's Pluto LNG facilities since the fields are right next to each other and Woodside had failed to find enough gas for a second train at Pluto. Chevron also decided to build the Gorgon LNG plant on Barrow Island and sequestrate the carbon dioxide in the gas by injecting it into deeper water bearing Jurassic sands beneath the island. For this scenario, it made more sense for Chevron to build one large LNG plant onshore for both Wheatstone and Gorgon and send the carbon dioxide to Barrow Island by pipeline. A pipeline from Barrow Island to shore had to be built anyway to fulfil their domestic gas obligations. Building the

Gorgon LNG plant on the environmentally sensitive Barrow Island no doubt significantly contributed to the cost overrun of $15 billion for the Gorgon project. Because these costs are tax deductable from company and the resource rent taxes, the Australian tax payer will also be contributing a significant proportion of these cost over-runs. Gorgon was eventually completed at a final cost of $US54 billion. The first cargo of LNG was shipped in March 2016.

MONTARA

In August, 2009, a development well being drilled on the Montara Oil Field in the Timor Sea blew out and caught fire. The attached photo **(Fig 44)** is a brutal reminder of what Mother Nature will do to you unless you are VERY careful! The Western Atlas drilling rig was severely damaged but fortunately no one was killed. This was the first offshore blowout in Australia for 40 years since the Petrel blowout in 1969. Another rig, the West Triton, was mobilised some 1700 kms to the site arriving in September 2010. After several attempts, the blowout was finally killed in November, 2010 at cost of $180 million. Ouch!

Figure 44
How not to drill an oil well

The Montara Commission of Inquiry was completed in November 2010 and noted that over the past 40 years, 3,000 wells had been drilled safely offshore Australia. It found that the cementing of the production casing was botched, a pressure test of the casing had failed and a 9-5/8" pressure containing corrosion cap, rather than a cement plug, was installed. The main conclusions of the Inquiry were that PTTEP (the operator of the production licence) "did not observe sensible oilfield practices at the Montara oilfield", the Northern Territory Department of Resources "was not a diligent regulator ..." and "the proposal to establish a single national offshore regulator should be pursued". In my opinion, the unfairest criticism was that of the NT Department of Resources. You can't have "a diligent regulator" unless you fund it properly so that suffiently experienced technical staff can be hired to handle the petroleum operations taking place in the vast and remote expanse of the Timor Sea.

I assisted Murphy Oil in evaluating Application Areas W09-17, 18 & 19 along the inner Dampier Sub-basin for Murphy Oil and Petronas. These areas mostly covered by the old WA-243-P permit which I had worked on with Ashland Oil in the 1990's.

MACONDO

In April, 2010, the Deepwater Horizon rig drilling development wells on the BP operated Macondo Oil Field in the deep water Gulf of Mexico sank after an explosion killed 11 workers. The three affected wells were estimated to have been flowing oil into the ocean at rates of 12,000 to 20,000 bopd. BP set up a $US20 billion fund to pay for the estimated cost of the environmental damage expected to be caused by the blowout.

After capping one of the wells, they tried using new technology to control the oil flow. This was a giant dome which was lowered over the flowing wells. The oil, being lighter than the sea water inside the dome rises to the top of the dome where it is then piped to the surface into a tanker to collect the oil. In an excellent example of the universality of Murphy's Law, this technology didn't work because the extremely low temperatures at the sea floor (1,500 metres deep) froze the oil and it was unable to flow out of the top of the dome and to the surface. Good grief!

By drilling a relief well, the blowout was eventually contained by August 2010 after spilling an estimated 5 million barrels oil. This makes it the second largest oil spill in history after the 1991 Gulf War (8 million barrels) and just ahead of the Ixtoc-1 blowout offshore Mexico in 1979 (3.4 million barrels).

Based on my experience working in the Canning Basin, I assisted Murphy Oil on in evaluating the main prospect in WA-243-P. Interpretation of existing seismic and the seismic acquired by the joint venture in November 2009, showed the regional base Grant Group unconformity with Anderson Formation sediments below. I interpreted the next strong reflector to possibly be the top of the Devonian limestones. The critical issues are for the timing of structures to predate primary hydrocarbon migration. The location for the well, called Eupheme-1, was approved by the joint venture in July, 2011 but the well wasn't drilled until December 2012. It was plugged and abandonded as a dry hole and the permit was relinquished in May 2013.

In November, 2010, Murphy Oil held a strategic meeting on future potential in Australia. I was asked to attend this meeting but having just come out of hospital after my bowel operation, I was not my usual sparkling self. I did point out that Geoscience Australia had recently published in PESA News some information on samples dredged from the edge of the continental shelf in the Great Australian Bight. These samples were dated as Santonian to Turonian and had TOC's of 3-7% and hydrogen indices of up to 500. These source rock characteristics indicated restricted marine environments which may be related to the world-wide mid-Cretaceous anoxic event. I suggested that the Ceduna Sub-basin may have the potential to be the next major petroleum province in Australia. The downside was that it was located in one of the world's worst marine envirnonments with terrible storms, strong winds and constant giant swells. They may have taken some notice of what I said because the next thing I knew was that in October 2013, Murphy Oil and Santos were awarded EPP-43 in the Ceduna Sub-basin. They joined BP who, in early 2011 was awarded four permits covering 24,000 square kilometres with a staggering $600 million exploration program bid.

Post well studies of Abalone Deep-1 indicated that the Jurassic Plover-Brewster sandstones were the only remaining play in the permit. A well could not technically be drilled in AC/P 36 until 3D seismic surveys were acquired over the Bassett Deep prospect, in conjunction with the other joint ventures in WA -343-P and WA-408-P. The joint venture applied for and was granted a suspension and extension of AC/P36. Murphy, Inpex and Total eventually agreed to the unitisation of the three permits and at the end of 2010, Total acquired the 1150 sq km Bassett Deep Joint Venture 3D seismic survey using the Ramform Explorer.

The main industry news during 2010 was Chevron's Clio gas discovery believed to be up to 8 TCF; BHP & Apache approving the $1.6 billion development of Macedon Field; and Shell selling 78.3 million Woodside shares for $3.3 billion and reducing their equity in company from 34.3% to 24.3%. BHP certainly got their timing right by reporting that production from their Pyrenees Oil Field had commenced at up to 96,000 bopd. The oil price reached a high of $US115 per barrel in February 2011. You do the maths!

Despite the fiasco of Inpex's failed development proposal for Ichthys in 2006, Woodside continued to negotiate with NT groups for its Browse Basin gas fields to be developed via an LNG project at James Price Point in the Kimberleys. Although Woodside and the State government reached a $1 billion agreement with traditional owners for the James Price Point Browse gas development, this option was not approved by the joint venture and in September 2011, Woodside announced that Floating Liquified Natural Gas ("FLNG") was the preferred option for Browse Gas.

During 2011, PTTEP withdrew from AC/P 36 and Murphy Oil bought Finder's interest in the permit. The permit was formally renewed by the government in February, 2012 and Inpex farmed in to earn 50% and become the operator in July 2012. Total drilled Bassett West-1, the first well on the greater Bassett structure, in WA-408-P where Murphy Oil had a 20% interest. After encountering a 7.5 metre gas column in tight Jurassic Plover sands, the well was plugged and abandoned in March 2013. After waiting the 35 years since Bassett-1 was drilled, I was naturally very disappointed but given the size of the structure, there may be other fault blocks which might have better reservoir quality.

By mid 2013, the Perth office of Murphy Oil had reached a size where my services were no longer required. It was a great contract and I very much enjoyed working with their team. It also kept me involved in the offshore activity in Western Australia.

Empire Oil

Sale of Gulliver Productions Pty Ltd

With the low oil price of $US30 per barrel, the market for new oil exploration floats was very poor. I was approached by Craig Marshall of Empire Oil to sell Gulliver to Empire and join the board. In April-May 2003, the EP-389 Joint Venture had drilled Eclipse-1 on a fault controlled anticline in the southern part of the permit. Although there were some thin oil bearing sands at the base of the well based on electric log analysis, the well did not encounter commercial hydrocarbons and was plugged and abandoned. This was a devastating result for Empire Oil and two of the directors resigned. A third director wanted to retire as he now lived in France but this would have left only one director thereby breaching Empire Oil's constitution.

I looked at Empire Oil's assets and there was definitely some synergy with Gulliver – both companies had permits in the onshore Perth, Carnarvon and Canning basins in Western Australia. I was particularly interested in the onshore Perth Basin where Barrack Energy had had some exploration success at Mount Horner and Beharra Springs. In addition, the Arc Energy/ Origin Energy Joint Venture had discovered the Hovea oilfield in L1 in October 2001. This field had reserves of ~ 9 million barrels in the Late Permian Dongara Sandstone and kick-started another bout of exploration in the basin. However, Empire Oil did not have a lot of funds left after the drilling of Eclipse-1 but then neither did Gulliver! Of course, being listed on the ASX, it was much easier for Empire Oil to raise exploration funds than for me to try to do a listing of Gulliver on the ASX.

So I decided to go ahead with the sale and prepared a valuation of Gulliver. The value was $350,000 and was based mostly on the farmout of the Point Torment and Valentine prospects in the Canning Basin. Empire Oil did its own independent valuation which concluded that the price of $350,000 was fair and reasonable. At the next Empire Oil board meeting on 8th October, 2003, the proposal to purchase Gulliver was discussed taking into account the independent valuation. The Resolution proposed to purchase Gulliver was approved unanimously by the board. David Williams agreed to retire as a director and I agreed to be appointed as a director of Empire Oil.

The new board went on to discuss all the company's (and now Gulliver's) applications for exploration permits going through the native title approval process. These included four (5/98-9, 3/02-3, 4/02-3 and 12/00-1) in the Perth Basin, two (13/00-1 and 16/00-1) in the Carnarvon Basin and two (2/98-9 and 14/00-1) in the Canning Basin. Native title agreements for applications 5/98-9, 12/00-1, 16/00-1 and 4/02-3 had been settled and these areas were awaiting grant as exploration permits by the government. Areas 3/02-3 and 13/00-1 were still going through the Native Title application process.

At the Annual General Meeting held on 28th November 2003, my appointment as a director of the company and the issue of 44,871,795 Empire Oil shares to the Warris Family Trust was ratified by the shareholders. These shares would be held in escrow until 17 October, 2004 as per the ASX Listing Rules. The November placement of 60 million shares and 30 million options to sophisticated investors was also ratified by the shareholders.

At the next board meeting in January 2004, the scope of the combined assets and Empire and Gulliver became clear. Empire Oil and its subsidiaries had seven permits and applications in the onshore Perth Basin (covering approximately half the onshore basin), five permits and applications in the onshore Carnarvon Basin and five tenements and applications in the Canning Basin. All these tenements covered an area of about 15 million acres. However, the company had only $670,000 in cash so a major farmout effort would be required to fund the exploration programs to keep these permits in good standing. Memories of SOCDET flooded back into my mind. Empire Oil raised additional exploration funds via a placement of 240 million shares at 0.8 cents per share. This capital raising increased the issued capital of Empire Oil to over a billion shares, an unenviable milestone for the company.

Regarding Empire's onshore Perth Basin acreage, during the late 1990's and early 2000's, this basin had been having a bout of exploration success. In late 1997, Arc Energy had acquired the Dongara Gas Field including the two production licences, L1 and L2, from CMS Gas Transmission for $35 million. CMS had only just recently bought the field and the WANG natural gas pipeline from WAPET in July for $120 million. Arc Energy then raised $34 million from the equity market in March 1998 and also negotiated a $20 million debt facility from Rothschilds.

Arc Energy had then farmed out L1 and L2 to Origin Energy and this joint venture had discovered the Hovea oilfield in October, 2001. This field started production in September, 2002 and by the end of June 2004, cumulative production had reached 2.8 million barrels. Over this time, the oil price had risen from $US30 to $US40 per barrel, yielding Arc Energy a very nice cash flow as well as confirming the Perth Basin as an important petroleum province. In addition, the Jingemia oilfield was discovered by Origin Energy in December 2002 begining production in May, 2004. In the offshore Perth Basin, the Cliff Head oilfield was discovered by Roc Oil in January 2002 begining production May 2006.

During early 2004, two significant wells were drilled in the onshore Perth Basin. Arc Energy drilled Apium-1 which tested 4 MMcfg/d from the Dongara Sandstone. Origin Energy also drilled, a well, Tarantula-1, which blew out at an estimated rate of 20 MMcfg/d. My son, Alex, had graduated as a geologist in 1999 and it was not a good time to be looking for work in the oil industry. He had worked initially in the offices of OMV and Troy-Ikoda but his first operational job was with Baker Hughes as a trainee mudlogger. Tarantula-1 was his first well as the trainee mudlogger. The well had reached total depth and had encountered excellent gas shows in the objective Dongara Sandstone. The operator had decided to run a Totco survey to determine the angle of deviation of the hole from the vertical. Alex noticed that mudpit level began to rise and immediately reported this to his supervisor. His supervisor immediately notified the driller and then returned to the mudlogging shack to pick up all the data files and the two of them then headed off into the bush to get as far from the rig as possible. The well blew out with the Totco tool being the first thing out of the hole, fortunately landing well away from the rig and any people. Alex told me that the noise of the blowout was like standing next to a 747 taking off. I told him that he had the most extraordinary start in the oil industry. Not only had he seen a discovery in his first well but he had also witnessed a blowout. These are, fortunately, extremely rare events and I have never been involved in one during my 50 years in the industry.

With the recent exploration successes by Arc Energy, we were hopeful of being able to at least farm out some of the Perth Basin permits. The most urgent programs were the drilling of Moriary-1 in EP-426 and the acquisition of the Gingin West 3D seismic survey in EP-389 and the Mullering 3D seismic survey in EP-432. We gave farmout presentations at the 2004 NAPE Conference in Houston in February and at the PESA Deal Day at the 2004 APEA Conference in Canberra in March. Due to the urgency of attracting other companies to assist with the exploration program, we offered a soft deal of farminees paying a third of the cost to earn a 25% interest in the permit.

There was considerable general interested at the NAPE Conference with Finley Resources being the most promising. Empire Oil also entered into a Brokerage Agreement with Dr Milam Jr who claimed that he had contacts with Asian companies looking for petroleum assets. There was no initial cost to Empire Oil, only a success fee to Dr Milam of 1% of the exploration costs and a 1% royalty. These farmout efforts were ultimately unsuccessful so things were not looking too flash for Empire Oil. The company had to then raise additional funds on the equity market, using 708 placements to sophisticated investors.

Application Areas 12/00-1 and 4/02-3 in the Perth Basin were finally granted to Empire Oil by the DoIR as Exploration Permit EP-430 on 20 September, 2004 quickly followed by Exploration Permit EP-432 on 12 October 2004. Application Area 12/00-1 had been applied for by Empire Oil before I joined the company. It consisted of 64 blocks covering a narrow strip which ran north-south along the western side of the Darling Fault. My view was that the most prospective part of the area was in the north around Carnamah where limited seismic indicated a terrace on the western side of the Darling Fault. On this terrace, prospective Jurassic and Permian sediments were interpreted to be at commercially drillable depths.

August, 2004 was a busy month for Empire Oil. A mediation meeting was held with the Thalanji People, the claimants for the northeastern part of 13/00-1 in the onshore Carnarvon Basin. We had already reached a Native Title Agreement with the Gnulli People who were the claimants for the southwestern part of the application. The Thalanji lawyer was making outrageous demands which were totally unacceptable for the grant of a simple exploration permit. It was decided to apply for a Section 35 Determination. Application Area 13/00-1 was eventually granted as EP-444 on 20 February, 2006.

Also in August, 2004, Rough Range Oil conducted geochemical surveys in its Carnarvon Basin permits EP-359, EP-435, EP-439 and EP-412. Petrofocus was contracted to conduct the surveys and myself and John

Conolly travelled to Carnarvon and Exmouth to operate the surveys. On the drive up to Carnarvon, John suddenly asked me to stop the car. He jumped out and took a picture of a road sign that stated that the next service station was 230 kilometres away. Most people have absolutely no idea just how big Western Australia really is! But that was the only excitement as the road from Northampton to Minilya Roadhouse is dead flat and undoubtedly the most boring countryside I have ever seen in all my travels. It occurred to me that there just had to oil here as there didn't seem to be any other use for this godforsaken landscape.

In Carnarvon, we stayed at the Best Western Motel and John set up the gas chromatograph in his room. He asked that the manager made sure that the cleaners did not spray the rooms with anything as this would cause the chromatograph to give false readings. The next day, we drove out to the blowholes and passed the great sign "King Waves Kill" **(Fig 45)** to warn any unwary rock fishermen. John was amazed at the amount of diagenesis of the surface limestones and how hard and sharp they had become in such a short geological time. We drove north to Quobba Station, chatted to the lease holder Mr Meecham and got permission and directions to get to the

Figure 45
Fishermen beware!

Quobba-1 well site. The well bore had been converted to a water well drawing from the Cretaceous Birdrong Sandstone. Quobba-1 was drilled by Canada North West in 1983 and encountered a section of Devonian limestones, dolomites and shales called the Gneudna Formation, overlying the Nannyarra Sandstone. There were good oil shows in the dolomites and a drill stem test recovered fresh water and indicated good permeabilies.

In EP-439, there was a large anticline, the Lake Maclead Prospect, defined by a seismic line acquired in 1982 along the main road which runs just south of Lake Macleod from the blowholes back to the North West Highway. We conducted the first geochemical survey along this road and other public roads before returning to Carnarvon to analyse the samples.

John's geochemical sampling technique was very basic but quite effective. It consisted of a metal probe which was driven into the ground and any gases drawn up into the probe. A syringe was inserted into a rubber seal and any soil gases in the probe were sucked into the syringe. The syringe was sealed, labelled, and stored for analysis. I would keep a log of each sample number and where it was located. Surveying was done in the morning and analyses were done in the afternoon and evening to prevent degradation of the sample or loss of gases by leakage.

Our next stop was the town of Exmouth. On the way up, we detoured into Coral Bay, a popular holiday resort which neither of us had visited. There was very little development there – a pub, a caravan park and a few stores but the bay was breathtakingly beautiful. The Ningaloo reef itself was only a short distance offshore and the water was crystal clear and unpolluted. We had a quick swim and were intrigued by the number of sea slugs on the narrow shelf before we reached the reef itself.

The road from Minilya Roadhouse to Exmouth was slightly more interesting as it was dotted with hundreds of termite mounds, some of considerable size. The drive itself was certainly more exciting as you had to continuously dodge kangaroos and sheep wandering across the road. It was still very boring to the extent that there was actually a sign for tourists to detour off the road to view the termite mounds!

Our next survey was in the Rough Range permit. We dropped into the Lefroy's at Exmouth Gulf Station to introduce ourselves and proceed onto the Rough Range-1B well. We tried to survey over the structure but it was impossible to get a useful sample as the limestone was very hard and the natural fractures were filled with loose sand. We then drove southwest to the Parrot Hill-1 well drilled by Ampolex in 1984. Again we couldn't get samples on the surface limestone but on the pindan plain to the southeast, we got excellent samples over Ampolex's Parrot Hill East Prospect. I had remapped this prospect as an elongate, narrow fault block which extended from EP-435 into EP-359. Although the existing seismic over the Rough Range Anticline was very poor, the seismic on the pindan plain was excellent. It could be clearly seen that this prospect was not a Late Miocence structure like Rough Range but a much older structure formed during the Early Cretaceous. As originally proposed by Ampolex, oil generation and migration from the Jurassic source rocks to the west could have occurred prior to the formation of the Rough Range Anticline and migrated across the fault and into the Parrot Hill East structure. On returning to Exmouth, John got a very significant propane anomaly over this prospect.

The next day we ventured south into EP-412 where there was a strong amplitude anomaly on one seismic line over the Thornbill Prospect. There were only two seismic lines over the structure, one with the anomaly and the other through the Whitlock Dam-1 well. This well intersected the Birdrong Sandstone but did not reach the deeper Permian objectives. The main objective for the Thornbill Prospect was the Early Permian Lyndon Sandstone sealed by the marine shales of the Callytharra Formation. The Lyndon Sandstone had good porosities of 20% in Remarkable Hill-1 to the south. But getting to the prospect turned out to be an adventure. The prospect was in the Marrilla Pastoral Lease but access was easiest through the Giralia Pastoral Lease then owned by Conservation and Land Management. I had earlier contacted the manager of Marrilla and told him that we would be surveying on his lease but would come in via Giralia. We drove to Giralia Homestead accompanied by Doug Swift as, due to the remotness of the location, it was prudent to have two vehicles. Doug worked part time for Empire looking after the Rough Range-1B site and generally assisting with operations in the area. We contacted the Giralia manager to let him know what we were doing and where we were going to be on his lease. He told us that the country was very bad and to contact him again on our way out (which we did). So we headed south to the Whitlock Dam bore but further south the old 1980's seismic lines had been heavily eroded and it was a tortuous path to the Whitlock Dam-1 well site. After numerous flat tyres, we eventually reached the well, found the relevant seismic lines and carried out the survey. But it took most of the day and we were exhausted by the time we got back to Exmouth.

You can't visit Exmouth without going fishing. Doug had a boat and took us into Exmouth Gulf for a day of fishing. Now I'm not much of a fisherman but Exmouth Gulf was different. I dropped my line over the side and within five seconds I had a fish. This continued for the next hour but most of the fish were either too small or not good eating. Eventually we collectively ended up with a couple of perch, two snapper and various other fish. The tide eventually turned and we headed back to the house for a feast of barbequed fish washed down with beer and wine. The following day we headed off to Newman to conduct geochemical surveys for Amadeus Energy as discussed earlier.

In October, 2004, Empire Oil announced that it had entered into an agreement with First Diversified Financial Services ("FDFS") for a capital raising in the USA. The proposal was to form a private placement company to raise US$11.5 million to contribute 100% to cost of Empire Oil's 2004 seismic programs and 2005 drilling programs in the Perth Basin to earn 80% of net revenue from any hydrocarbons discovered until payout and then 40% thereafter. Craig had also arranged for John Todd to check out FDFS to ensure that they were a bona vide financial institution. John had worked with Empire before I joined and was the son of the famous Don Todd whose company IIAPCO negotiated the first production sharing contract in Indonesia in 1966 and went on to discover many oilfields off the north coast of Java. A representative from FDFS arrived in Perth and Craig and I took him on a field trip to visit our various permits in the Perth Basin. We stayed overnight in Dongara and on our return to Perth, stopped in to view the newly constructed Hovea oil field facilities. FDFS prepared the offer documents and sent them to institutions in February, 2005. By June, 2005, FDFS had failed to raise any funds for exploration in the Perth Basin. We concluded that this may have been some sort of scam but in the end, all they got out of Empire Oil was a few expenses.

At the November 2004 Board Meeting, it was agreed to transfer Gulliver's interest in EP-439 to Rough Range Oil Pty Ltd and its interests in EP-426 and EP-432 to Empire Oil Company (WA) Limited; and Empire Oil & Gas NL would also transfer its interest in EP-439 to Rough Range Oil Pty

Ltd. As these three private companies were all wholly owned subsidiaries of Empire Oil, this did not require shareholder approval. The result was that Gulliver would hold all Empire Oil's interests in the Canning Basin, Rough Range Oil would hold all Empire Oil's interests in the Carnarvon Basin and Empire Oil Company (WA) Limited would hold all Empire Oil's interests in the Perth Basin. If it was decided to sell any these assets in the future, it would be a simple matter of selling the company rather than having to go through the tedious process of transferring each permit interest and registering them with the Mines Department.

In February, 2005, Application Area 16/00-1 in the onshore Carnarvon Basin was finally granted as EP-439. The transfer of Empire's and Gulliver's interest in the permit to Rough Range Oil could now proceed.

On the personal front, we sold our beach house in Falcon in February 2005. We realised that it had become somewhat of a chore. Every fortnight, we would drive down and Mo would spend all weekend cleaning the house and I would spend all my time cleaning up the garden. And without the kids, it didn't seem to be fun anymore.

The same month, Mo and I went on a cruise out of Sydney during March around NZ on the Pacific Princess. The Pacific Princess was an elegant ship, very cosy with only 750 passengers, unlike the 3-4 thousand passenger monsters they build these days. The cruise left Sydney and sailed southeast across the Tasman Sea. The next morning, I sat on the balcony and looked to the west and you could see the huge swell coming toward the ship driven by the roaring 40's. The waves hit the ship at 45 degrees causing it to rock and roll and caused many of the passengers to be sick – not Mo and I of course. But it did occur to me that I wouldn't want to be cruising this route in the middle of winter! Despite the swell, the weather was fine and when we got to Milford Sound, the captain took the ship right into the fjord. It was a magnificent sight with numerous waterfalls spurting out of the sides of the cliffs on either side. After Milford Sound, we sailed around the southern part of the South Island and as soon as we got to the eastern side, the swell stopped and the ship settled down to the relief of the sea-sick passengers. We docked in Dunedin and spent a day touring the town. It was like stepping back into the 1950's but the city has a certain charm. The railway station was certainly a very impressive building. From Dundedin, we sailed north and docked in Picton. The following day, we did a tour of the Wairau Valley and visited the huge Montana Winery. On the drive down to the vineyard, we could see new vines being planted up the sides of the valley. The guide at Montara told the tour that they were planting a million new vines. From Picton, we sailed around to Aukland and spent a day touring the city. From there we sailed to the Bay of Islands and visited the site of the signing of the Treaty of Waitanga in 1840 between the British and the Maories.

On our return to Sydney, we hired a car and drove down to Kangaroo Valley, our old favourite camping spot. Chris was working at the Academy of Science in Canberra so he drove up to meet us there. Nothing much had changed at the site but a dramatic drop in the water level of the Tumut River had completely spoiled the area as a camping spot. No more kids swinging off the rope into the river.

In November, 1998, Empire Oil had applied for an area west of EP-389 in the Perth Basin. This area also had limited seismic with the only modern seismic being the 1990 Badamina Seismic Survey by Barrack Energy over the Badamina High. This high was drilled by Wapet in 1967 and encountered only poor shows through the Jurassic section. Empire Oil was successful and the area was offered to them as 5/98-9 subject to negotiating a native title agreement. This agreement had been completed but the government was still reticent to grant the permit. It had been six years since the area was offered for application and during this time, the suburbs of Perth had encroached into the southern two blocks of the permit. It was hard enough getting approval to acquire seismic in the bush without trying to acquire it in the suburbs of Perth so the board approved my recommendation to drop these blocks from the application. Area 5/98-9 was granted as EP-440 on 13 March, 2005.

During March, 2005, Emerald Gas Limited made an offer to farm into EP-104 and R1 in the Canning Basin by paying 25% of the cost of drilling Stokes Bay-1 to earn an 18.75% interest in these tenements. They also wanted an option on the West Kora application area L98-1 at the same terms.

The APPEA Conference was held in Perth from 10-13 April, 2005. At the PESA Deal Day, I gave a presentation on the Dune and other prospects in EP-435 in the onshore Carnarvon Basin. Follow up negotiations resulted in Rough Range Oil farming out a 25% interest in EP-435 to Australian Oil Company Limited by them paying 50% of the cost of drilling Dune-1. We also had a booth at the exhibition and following up contacts made at the conference, further discussions and presentations took place with Noble Energy, a large US independent oil company. They were interested in farming into Empire Oil's Perth Basin permits but further negotiations never eventuated as the opportunity was not approved by head office.

Mo and I decided to attend the 2005 ANZAC Day service in Albany. We drove down to Albany and stayed overnight at a motel. We got up before dawn and drove as close to Mt Charlotte as we could find parking. Everything was so well organised and we all got bused up the hill to the cenotaph. The service was very well attended and during the service I looked up to the west and saw Venus. It was incredibly bright, so bright it looked like a comet on a collision course with the Earth. Then the sun gradually rose on a clear blue day and the view of Prince George Sound was truely breathtaking. It made the ANZAC service twice as special. We then spent the next day touring Albany and enjoying its many attractions.

In July, 2005, Application Area 3/02-3 in the Perth Basin was finally granted to Empire Oil as Exploration Permit EP-426. Native title negotiations had been held up by one of the two claimants making unacceptable demands in the agreement. The grant was only settled by Empire Oil deciding to drop what I considered to be non-prospective blocks off that area claimed by the difficult group.

In August, 2005, I attended the Central Australian Basins Symposium which was held in Alice Springs. I was keen to catch up the latest studies on the Amadeus Basin and how that might relate to the Palaeozoic sections in the Canning Basins. My original theory, set out in my PhD thesis, that there was an Ordovician inland sea extending from the Amadeus Basin and connecting to the Canning Basin was now being rejected by evidence from new field mapping. After working on the Ordovician sediments in the Canning Basin, I was beginning to have the same doubts.

After the Central Australian Basins Symposium, Mo met up with me, we hired a car and drove down the Stuart Highway towards Adelaide. We turned west along the Lasseter Highway and headed for Kings Canyon with its spectacular outcrops of Mereenie Sandstone. We did the easy walk around the base of the canyon, watched the sunset and stayed overnight at the resort. The next day we drove back to the Lasseter Highway and headed for Uluru past the spectacular Mount Conner on the south side of the road. The Uluru resort was very busy with lots of coaches and cars unloading tourists for the rock. The next day we did the tour of the rock. We did not climb it but walked around thc base. There is something eerie about Uluru – the smooth erosion of the sides gives the rock a strange colour. The sides are also pockmarked by caves making the rock look almost alive. Really spooky! The next day, we left Uluru and headed back to the Stuart Highway and drove into South Australia.

Our first stop was Coober Pedy where we stayed in a motel buried into the sides of a hill. Mo said she felt like she was in a fairytale. Most of the accommodation in the town was similar due to the scorching temperatures in summer. The first thing we did was go noodling for opals. This was a bit like fishing – we had a lot of fun but didn't catch anything worthwhile. After Coober Pedy, we drove across to William Creek, down to Maree and Leigh Creek and detoured out to Arkaroola. The road was terrible but we had hired a 4WD vehicle. From there, we went down past the spectacular outcropping Cambrian limestones **(Plate 46)** to stay at the Prairie Hotel in Parachilna. The restaurant in this hotel was famous for serving every kind of native animals for dinner – excluding the endangered ones of course.

Figure 46
Outcrops of Cambrian limestones

Mo's 60th birthday had been in August but rather than celebrating it in the middle of winter, she decided to have it in October on one of the Unforgettable Houseboats on the Murray River. So we loaded up the station wagon with food and booze and headed off for Adelaide. We had a great crowd – there were twelve of us so we had ordered the six berth houseboat which was even more palacial than the ones we'd used before. There was even a spa bath on the top deck!

The Good Oil Conference was held at the Esplanade Hotel in Fremantle on 7-8 September, 2005. Empire Oil had a booth and I tailored our presentations to emphasise to brokers and investors the company's excellent acreage position in the Perth Basin and the potential for further oil and gas discoveries. At the same time, Empire Oil had a meeting with Andy Padman, then working for Arc Energy, to form a strategic alliance to explore the onshore Perth Basin. We agreed in principal but it was not going to be easy working out the fine detail as there was quite a difference in prospectivity across the various permits.

After nine and a half years, we decided to exchange our ocean view for city and river views. In October 2005, Mo sold the Observation Rise apartment in Scarborough and bought an apartment at 1 Henry Lawson Walk in East Perth. The old apartment was sold to an investor with the giant couch included and was immediately rented out to Danny Green. We saw the old couch on TV during an interview with him.

I was invited to join the Geological Survey of WA Exploration Technical Sub-committee as Empire Oil's representative and attended my first meeting in November 2005.

Rough Range Oilfield

As the oil price had been steadily rising from $US18 per barrel in 2001 to over $US40 per barrel at the end of 2004, I calculated the economics of developing the Rough Range oilfield based on an estimate of 250,000 barrels oil-in-place. The seismic time maps were useless due to velocity variations within the overlying Tertiary limestones. However, within a radius of 500 metres from the discovery well, there were 5 additional wells (Rough Range-4, -6, -10 & -11 and Central Rough Range-1) so I simply contoured the top Birdrong in each well to make a depth structure map.

The Rough Range-1B production well had already been drilled by Empire Oil so the only development costs were for the beam pump, storage tanks and loading facilities. The Rough Range oilfield was initially discovered by Wapet in 1953 with the first well intersecting an 8.6 metre oil column in the Early Cretaceous Birdrong Sandstone. It was also the first well drilled in Western Australia after WW2 and set off an exploration boom and stockmarket frenzy. Alas, between 1953 and 1955, Wapet proceeded to drill twelve appraisal wells on the structure, all of which were dry holes. Ampolex, one of the shareholders of Wapet, retained EP-41 after the other shareholders decided to relinquish the permit. In 1986 Ampolex undertook an extended production test of Rough Range-1A which produced a total of almost 9,000 barrels oil. Due to the then low oil prices and high transport costs of trucking the oil to the BP refinery at Kwinana, the test was abandonded. Rough Range Oil, a subsidiary of Empire Oil, took over the area and drilled Rough Range-1B in May, 2000. This well was a twin of the discovery well and encountered a 9 metre oil column in the Birdrong Sandstone.

Due to the high wax content of the oil, new storage tanks would have to be heated and special heated road tankers would also be required to truck the oil to the Kwinana refinery. These costs were estimated to be $750,000 including $150,000 for the special tankers so it was calculated that the field could be commercially developed and bring in much needed, if somewhat small, income. We had a meeting with the DoIR who recommended that Rough Range Oil apply for a Production Licence over the field in conjunction with an application for an Extended Well Test. In this way, Rough Range Oil could produce the field while the grant of the production licence was going through the government and native title approval processes.

Following successive relinquishments over the years (as required by the WA Petroleum Act), the old EP-41 permit consisted of three discrete and unconnected areas. An application to convert these areas into three separate permits was approved in November, 2005, and the new permits EP-433, EP-434 and EP-435 were granted by the DoIR. An application for a Location and Production Licence over the Rough Range oilfield was then immediately submitted to the department. Following settlement of the Native Title Agreement, Production licence L 16 was finally granted over the Rough Range oilfield in November, 2010.

The Rough Range oil field began production during July, 2005 as an Extended Well Test. Designs for the production facilities and the plans for the safety case were submitted to the DoIR in December, 2004. In November, 2004, Empire Oil had negotiated a $1 million loan to fund the construction of these facilities and the road tankers. Approval to construct the Rough Range facilities was received from the DoIR in February 2005 and final approval to produce the field was received in April 2005. The timing of the concept of the project to final approval had been about a year. An oil price was agreed with BP and was based on a basket of Widuri, Murban and Minas oils less a $3 discount (for being a nuisance). The total discount to the WTI price was ~ $5-7 per barrel. In April, a production licence PLA 1/04-5L over block 7089 was offered by the DoIR subject to negotiating a Native Title Agreement. The production facilities were completed in June with power being supplied by LPG tanks delivered to the site.

A Lufkin beam pump was purchased from China and arrived with the red star still painted on its side **(Fig 47)**. Phelps Transport was contracted to operate the five special road tankers. Approval to commence production was received in July and the first truckload of oil was delivered to BP at a price $US58 per barrel discounted down from the international oil price

Figure 47
Rough Range beam pump. Note the red star

of $US67 per barrel. By February, 2006, the Rough Range oilfield was producing 80-90 bopd and making $30,000 per month.

At the end of 2006, the Rough Range Oil Field was shut-in. Water cut had increased to over 90% and continued production was uneconomic. This did not compute with the oil-in-place reserve estimate which although small was still well in excess of what had been produced. I could only conclude that the well had been coned by production in the late 1980's and equilibrium had not been reached on the oil-water contact.

In December 2005, a Canadian company, Red Sky Resources agreed to farm into 13/00-1 in the Carnarvon Basin by acquiring an airborne geochemical survey subject to listing its Australian subsidiary on the ASX. Red Sky would then have the option to earn a 50% interest in the permit by drilling one exploration well. After six years, Application Area 13/00-1 in the onshore Carnarvon Basin was finally granted as EP-444 on 3 March, 2006. Red Sky Energy listed on the ASX and the survey was conducted in August 2006. The interpretation of this survey indicated a large number of anaomalies all over the permit but none of any significance. I was reminded of the Kandji-1 well in the inner Dampier Sub-basin and the gas seeps detected on the sea floor by the site survey. Perhaps the same process was occurring onshore. Red Sky never exercised their option and the agreement lapsed at the end of 2007.

In February, 2006, I had a meeting with Tom Fontaine and Keith Martens regarding an outstanding commitment Empire Oil had with Bounty Oil & Gas. Prior to my joining Empire Oil, the company had committed to drilling an exploration well in EP-412. Unless this dispute with Bounty was resolved, Empire Oil would be required to drill this well by 2008. As there were no drillable prospects in EP-412, I reported this to the board who gave me authority to settle the dispute as soon as possible. The dispute was finally settled in 2009 when it was agreed by both parties that Bounty would remove the drilling liability and that the equities in EP-412 would remain at Bounty 65% and Empire 35%.

In February 2006, Craig went to London for a meeting with Adrian Eyre, a director of Wharf Resources, to tie up farm-in deals in the Perth Basin. It turned out that Adrian Eyre was directly related to the famous Australian explorer Edward John Eyre. Soon after, Adrian Eyre informed Empire Oil that he had formed two new companies, Phoenix Resources Plc and

Jurassica Oil & Gas Plc. Phoenix farmed into EP-435 by contributing to the cost of drilling the Parrot Hill-2; and Jurassica farmed into EP-439 and EP-460 by contributing to the cost of drilling of Lake Macleod-1 and Star Finch-1. Empire Oil also farmed out the drilling of Dune-1 to Asia Pacific Energy and Bounty Oil & Gas, and together with the October farm-in by Australian Oil Company Limited, the cost of drilling Dune-1 was now fully funded. Dune-1 was drilled during May-June 2006 to a total depth of 1,400 metres but was a dry hole. The well was relatively trouble free and the cost over-run was only 8%.

In March, 2006 Arc Energy agreed to farm into Gulliver's interests in EP-104 and R1 in the Canning Basin to earn a 21.5% interest in each by paying 28.75% of Gulliver's share of the cost of drilling the Stokes Bay-1 well. Arc Energy had already bought out Kimberley Oil and Pelsoil and would contribute their combined interests of 12.6%. FAR Resources had previously agreed to pay their 8% share of the cost while Pancontinental agreed to increase their interest to 10% from 8% by farming into Gulliver's interest. This together with the Emerald Gas farm-in, meant that we had 85% of the well cost covered. This only left 15% of the well cost to be funded by Empire Oil and Indigo Oil. Phoenix Resources Plc then farmed into EP-104 and R1 by agreeing to pay $450,000 towards the cost of drilling Stokes Bay-1 to earn a 10% interest in each of these tenements. This completed the farmout deals and allowed for the drilling of Stokes Bay-1 and Valentine-1 to proceed. It took a while for the significance of this farmin to be appreciated by the stockmarket. Arc Energy was going to operate the drilling of two exploration wells with significant potential on Empire Oil's Canning Basin permits. In the end, the market overreacted and during August, Empire's shares increased in value from 0.8 cents/share to over 6 cents/share.

Empire Oil applied in May 2006 to renew EP-416 in the Perth Basin. Based on the available seismic control at the time, I was unable to farmout the Wellesley Prospect and the permit still had a commitment to drill one well. The permit was coming up for renewal so, with the approval of the DoIR, it was agreed to relinquish 43 blocks (instead of the required 28) in return for a waiver of the well commitment to keep the permit in good standing. Empire Oil applied to keep 14 blocks over the Wellesley and Leschenault prospects in the northern part of the permit. The permit was renewed by the government in August 2006.

The 2006 APPEA Conference was held on the Gold Coast in May. The most significant papers at this conference were on the progress of the coal seam gas boom in Queensland which I have discussed earlier. At the PESA Deal Day, I presented the Gingin West Prospect in the Perth Basin in the hope of attracting a farminee to acquire 3D seismic over the structure. Alas this was not successful. Following the conference, Mo and I hired a car, drove to Brisbane and caught the ferry to Stradbroke Island. We stayed at the Straddie Bungalows at Amity Point where we had a thatched roof bungalow with a verandah and barbeque. Down the side ran a road to a jetty where you could fish or watch the great sunsets in the evening. I didn't need to fish because up the road was a private house that sold fresh fish very cheap and very tasty when cooked on the barbeque that night.

An Operating Committee Meeting was held on EP-104 and R1 in September 2006. I reported that the Year Two well commitment in EP-104 would be satisfied by the drilling of Stokes Bay-1 while the Year Four well commitment in R1 would be satisfied by a Stokes Bay appraisal well or a well on the Valentine Prospect. Craig summarised the number of agreements which needed to be finalised and lodged with the government. These were the Sale Agreement between Arc Energy and European Gas (formerly Kimberley Oil); the Gulliver-Phoenix Farm-in Agreement; the Gulliver-Arc Energy Farm-in Agreement and the Gulliver-Pancontinental Farm-in Agreement. The joint venture also agreed that Arc Energy would become the operator on 31 December 2006 even though they legally had no interest in the tenements until the transfers were approved by the government.

After the September PESA technical luncheon, a few of us retired to the Adelphi Bar for a few more drinks. We carried on discussing the technical talk and other events happening or rumoured to about to happen in the industry, and before I knew it, the time was around 6pm. There were no taxis and so I hopped on a bus to East Perth and promptly fell asleep. I woke up way past my stop, got off and started walking back along Royal Street. By this stage, all the booze caught up with me and I was very drunk. I fell over and broke my glasses which ended up cutting my face. A store owner came out to assist me, called an ambulance and phoned Mo. I have no memory of what happened next. Mo took me to the hospital where they patched me up. I woke up the next morning with a terrible hangover, sore face covered in bandages, sore arms and legs from falling over and generally feeling very sorry for myself. A few days later, I had to go to OPSM to get new glasses. I made up some feeble excuse about falling over running for the bus but I think they thought that I had picked a fight with someone and lost! I also decided that for a 62 year old man to fall down drunk on the pavement was not a good look! Old age was definitely catching up with me. At all future PESA lunches, I resolved to leave all PESA lunches (including the Xmas Lunch) no later than 4pm and organised Mo to phone me on my mobile. She didn't have to because I scrupulously kept this resolution. As time progressed, PESA lunches became less social with everyone returning to the office after the talk.

Back in May, Mo had said she wanted to go to the AFL Grand Final. I told her that we didn't know which teams would be playing but she didn't care as she had never been to a grand final. I had been to the 1999 grand final with Ross Macleod so it was only fair that she should attend one too. So I went into the AFL website and booked a package through the Hawthorn Club. This was an excellent package and included tickets to the ball on Friday night, breakfast on Saturday morning and of course the grand final itself. As it turned out, the Eagles made it to the grand final to play against their old rivals Sydney – I couldn't believe it! Sydney had won the 2005 grand final by a few points so the Eagles were after some revenge and it looked like it was going to be a ripper match. So off we went to Melbourne and me with my facial scars looking like some Hollywood thug. The Hawthorn package turned out to be fantastic and even included a speech by the president Jeff Kennett. But better still, the grand final was another classic but this time the Eagles won even if it was only by one point.

In October 2006, Rough Range Oil undertook the Anagonda Geochemical Survey covered parts of EP-359, EP-412 and EP-435 in the Carnarvon Basin. Bob Moffit flew into Perth with all the equipment and we drove the long boring drive up to Exmouth. Prior to the survey, we drove out to Ningaloo Homestead to obtain access and permission. It was some 30 kilometres off the main Exmouth Road along an appalling, corrugated road to the homestead which was also the site of some very neat looking tourist accommodation. After Bob explained the survey method, they looked at us as if we were both totally insane - but they gave us access anyway. We ran the survey along the old seismic lines which were still visible as weak tracks. After the survey, we returned to Exmouth to analyse the samples which were not encouraging. The following day, it was the long boring drive back to Perth.

The Western Australian Basins Symposium was held at the Convention Centre in Perth on 5-8 November, 2006. It was held in conjunction with the AAPG International Conference and was a major event. Unfortunately, the technical papers presented were not published as a symposium journal as was the case for the last three conferences. Empire Oil had a booth at the conference to showcase the company and attempt to attract farminees for its many projects in the Perth Basin.

The main item for discussion at the November board meeting was the EP-104/R1 farmins where Arc Energy wanted to exchange Valentine-1 as the earning well with Stokes Bay-1 being heads up. Due to the exorbitant cost of building a causeway across the tidalflats to the Stokes Bay-1 location, Arc Energy wanted to drill the vertical Valentine-1 well first (located on the hard ground) and then pull back and deviate out to the Stokes Bay-1 location from the same well pad. The board approved this sensible proposal as did the other joint venture parties to the farmout.

The third match of the 2006-07 Ashes series was held at the WACA in Perth. Being members, we decided to go to the opening day.When we got to ground, the queue was a mile long and we were never going to get a decent seat. So we went home and watched it on television. We did go on Sunday afternoon and watched Adam Gilchrist's century, the second fastest after Viv Richards. But we had to alternate between sitting in the sun and cooling off in the bar. We heard a rumour later that the WACA sold more memberships than seats in the Members Stand. Due to the exchange rate, it was very cheap for the Poms to buy a membership and attend all five days. We had been members for over 20 years and were appalled that the WACA treated its long term members like something on the bottom of their shoes. We cancelled our memberships and never went back.

Following our cruise around New Zealand in 2005, Mo and I decided to take a fly-drive holiday package there. So on 3rd February, 2007, we flew to Christchurch, picked up the hire car and spent a day in the city. We then drove across to the western side of the island and continued south to the Franz Joseph Glacier and stayed overnight. There was a very interesting sign stating where the front of the glacier was in 1795 and it was then a very long drive up the mountain to see where it is now. It shows that even the southern hemisphere has warmed up considerably since the Little Ice Age. We continued south to Fox Glacier and then to Lake Wanaka where we had lunch overlooking this spectacular lake. We then drove on to Queenstown where we took the gondola up the mountain and had dinner in the cafe at the top. Again the views were spectacular overlooking yet another beautiful lake. The next day we drove out to Milford Sound for a cruise. It was a perfect day and it was so calm that the captain sailed out into Pacific Ocean. He said that it was very rare that conditions allowed him to do this and showed us the old smugglers beaches and camps.

We then drove on to Dunedin where we stayed overnight. The manager of the motel asked us where we were heading for and we told him north to Akaroa. He told us to take a detour inland to Lake Tekapo. The next day we took his advice and took the detour out to the lake. We couldn't believe our eyes. This wasn't just a case of "Oh no, not another lake"! It was huge, breathtakingly beautiful and totally unreal. It was a light turquoise colour due to the rock flour washing into the lake from the glaciers and had the Southern Alps and Mount Cook in the background.

We returned to the main highway and headed north to Akaroa. This was an eroded volcanic crater and involved a very twisting climb up to the crator edge and then the same tortuous route on the way down to the centre of the crater where the town was located. The southern side of the crater had completely eroded away and the town was located on a narrow channel emptying into the ocean. Akaroa was a small town, quaint and very French. It was originally a small French settlement called Port Louis-Philippe in 1840 just as New Zealand became a British colony. Fortunately for the French settlers, the Napoleonic wars had ended and they weren't all executed as spies.

The next day we headed north to Blenheim for the Marlborough Wine Festival. On the way, we stopped at the town of Kaikoura where there were large numbers of seals along the rock ledges. The noise was incredible as the males defended their territory and the mums kept their babies out of the way. We stayed overnight in Blenheim and headed to the festival in the morning. The festival was well organised and buses were available to take you there. It was staged in a large field with stalls arranged around the edge. The stalls alternated wine stall-food stall, wine stall-food stall, allowing you to eat as you tasted the various wines on display. We had a great time, had too much to drink and, as planned, stayed another night in Blenheim before heading for Nelson the following day. The next morning we were up at some ungodly hour of the morning to drive to Picton in time to catch the ferry to Wellington. We had to hand in our existing car and pick up a new one in Wellington. The crossing was mild (by Cook Strait standards) and from Wellington, we drove north to Palmerston North where we stayed overnight.

The next day, we drove to Lake Taupo and had lunch by the lake with all the ducks. Lake Taupo covers the remains of a crater that was formed by a super-volcanic eruption which occurred about 25,000 years ago. The explosion blasted three trillion tonnes of rock, dust and ash into the atmosphere and Lake Taupo now covers the great hole that was left behind. However it is only about a quarter of the size of the super-volcanic eruption at Lake Tabo in Sumatra. This occurred about 75,000 years ago and blew so much dust and ash into the atmosphere that it cooled the earth by about 3 degrees. As the earth was already in the middle of the last ice age, that was not very helpful!

We continued north along Highway 1 to Rotorua where we did a tour of Hell's Gate including a mud bath and spa. The next day we drove south to find the "Hidden Valley" of Orakei Korako. Peter Hall had told us it's a must to visit on the North Island but it's not promoted on the tourist maps. It was certainly off the beaten track but well worth the trouble. At the bottom of the valley is a small lake with a shop and wharf at the end of the road. On the other side of the lake, a thick pile of volcanic ash, mud and detritus flowed into the lake like a glacier. A ferry takes you across the lake where a track takes you past the volcanic mudflow and into an area of boiling mud pools, active getsers and pools coloured by sulphurous compounds. It was not as large as Rotorua but far more interesting and picturesque.

From the Hidden Valley, we travelled south heading for Wanganui on the southwest coast. We should have taken another route. It was a hair-raising drive with washouts and landslides narrowing the road down to one lane in many places. There also seems to be this natural law in New Zealand that whenever to come to a switchback in the road, there will be a logging truck coming in the other direction. Somehow we made it safely to Wanganui and the headed north to New Plymouth for our next overnight stay. New Plymouth is a busy port but the approach is truely spectacular with the Mount Egmont volcano rising majestically above the Taranaki Basin. From there we drove north to Hamilton, past the beautiful estuary at Mokau and then onto Aukland where we stayed before catching a plane back to Perth the next day.

Application Area 4/04-5 in the Perth Basin was granted 100% to Empire Oil as Exploration Permit EP-454 in February 2007. This permit adjoined EP-430 and was located on a terrace between the Dandaragan Trough and the Darling Fault. Amazingly, the permit covered an area of almost 1,000 sq kms but not a single exploration well had been drilled there. The seismic coverage wasn't much better - there was a small amount of Wapet 1970's seismic plus some Discovery Petroleum 1997 seismic. Empire Oil reprocessed the existing seismic and interpretation of this new data indicated a structure which we called the Garibaldi Prospect, where the Permian section was expected to be at drillable depths of around 2,000 metres. The prospect needed more seismic to mature it to drillable status.

While interpreting the seismic over the Lake Macleod Prospect in EP-439 in the Carnarvon Basin, Peter Kirk had delineated an amplitude anomaly in the Birdrong Sandstone on the crest of a compressional Miocene structure. My only explanation of how hydrocarbons could migrate into such a young structure was for these hydrocarbons to have leaked vertically up fractures from a deeper Devonian accumulation. Peter pointed out a serious problem - the structure was outside the area of EP-439 - Yikes! Fortunately this area was vacant so Rough Range Oil immediately applied for a Special Prospecting Authority ("SPA") over this amplitude anomaly (the Star Finch Prospect) to put its foot on the title. Once the work program geochemical surveys were completed and an exploration permit granted over the selected area, we could then distribute the various equities to the permit holders of EP-439. In addition, Rough Range Oil also applied for a Special Prospecting Authority over the western flank of the Lake Macleod Prospect which had been excised from the original application as it bordered the Ningaloo Marine Park.

During early 2007, EP-435 was farmed out to Phoenix Resources and Black Fire Minerals who funded the majority of Rough Range Oil's costs of drilling Parrot Hill-2. Bounty Oil & Gas and Rough Range Oil increased their equities to 10% and 38% respectively by both farming into AOC who reduced their equity to 17%. Asia Pacific withdrew from the joint venture. The new equities in EP-435 became Rough Range Oil 38%, Phoenix 25%, AOC 17%, Bounty 10% and Black Fire 10%. Parrot Hill-2 was drilled during March-April, 2007 but intersected the top of the Birdrong Sandstone 50 metres below the top in Parrot Hill-1 and was a dry hole.

In March 2007, the sale of the Maneroo shareholders interests in EP-438 and EP-448 to Empire Oil was finalised. An independent valuation of these interests was undertaken by Peter Cameron to confirm that the transactions were fair and reasonable. The sale was approved by the Empire Oil shareholders at a general meeting in September 2008 allowing Empire Oil to increase its interest in EP-438 from 20% to 80% and its interest in EP-448 from 25% to 70%.

The 2007 APPEA Conference was held in Adelaide during April. For me, the most interesting paper was by lead author Darren Ferdinando on the preservation of porosity in deep Lower Permian sandstones. One of the rules of exploring the Perth Basin back then was "Never drill down to the Lower Permian sandstones below 3000 metres" as these are always silicified and tight. The paper showed that coating of quartz grains by illite/smectite clays can prevent silicification and preserve porosity and permeability. Ongoing exploration was to lead to the discovery of the giant Waitsia Gas Field which will be discussed later. Empire Oil had a booth in the Exhibition area in order to promote our many prospects in the Perth Basin. Following the conference, we caught up with a group of our friends and hired another Unforgettable Houseboat for another adventure on the Murray River.

In July 2007, Alcoa reached agreement with Arc Energy to provide them with a $40 million loan towards their exploration programs in the Canning Basin. The loan was to be repaid from gas sales and Alcoa agreed to buy 500 PJ gas over 15 years. For an industrial company to take such a step was an indication of the tight domestic gas supply in Western Australia at that time.

In early August 2007, Mo and I flew to Exmouth to celebrate her birthday. She had found a great package deal with the Novotel Resort which included return airfares. The resort was very impressive with a great restaurant and a beautiful pool area. I booked a whale watching tour which included snorkling with sea turtles. I warned Mo that there was no guarantee that we would see any whales but things were to turn out better that I could ever have dreamed of. Not only did a couple of males swim around the boat to see if we were acceptable, but then they proceeded to show off by giving us an unbelievable aerial display for about half an hour. They flung their bodies almost completely out of the water, rolled over to expose their white bellies and flopped back into the ocean to produce the same effect as an exploding depth charge. The whole tour got their money's worth that day!

We caught up with Doug Swift for dinner that evening and the next day, we borrowed his vehicle for a couple of hours. We drove up to the top of Cape Range to see the old Cape Range-2 well location **(Fig 48)**. This was drilled by Wapet way back in 1956 and encountered gas bearing sands in the Middle Jurassic. Of course back then, there was no market for natural gas and now that there is, it's located in a national park! From there we drove up to Lighthouse Hill with its breathtaking views of Exmouth Gulf and Ningaloo Reef. We could even see the Floating Production Storage and Offtake vessels for the producing oil fields in the Exmouth Sub-basin. We then drove around to Turquoise Bay where I did a bit of snorkling on the reef. All in all, it was a very enjoyable and exciting holiday.

Figure 48
Wapet Cape Range-2 well location

In late August 2007, I flew up to Exmouth to carry out the Sandalwood and Whitlock Dam Geochemical Surveys in EP-412. I was accompanied by Doug Swift and we dropped into Bullara Homestead to get directions to Whitlock Dam where the tracks had been washed out by heavy erosion. Tim Shallcross gave us directions to enter the area from the south and we found this a lot easier than coming in from the north. We then followed this up with the Bullara and Midway Geochemical Surveys in EP-359. The previous operator had mapped a large Permian fault block called the Nabalgee Prospect part of which was located under the Learmonth Air Base. One of our survey lines ran along the road around the perimeter of the air base and I wondered if anybody on the base would notice us and be concerned about what the hell we were doing!

During October, 2007, Arc Energy proposed to merge with Anzon Energy to form a $1.1 billion energy company with producing assets in the Perth, Gippsland and Bass basins. In November, there were objections to this merger from Clive Palmer who requisitioned a shareholder meeting to remove the Arc Energy board. His company, Mineralogy, had a 5% shareholding in Arc Energy. At the meeting in December, the Arc Energy shareholders rejected Mineralogy's requisition and also approved the merger with Anzon Energy. However, in January, Nexus Energy launched a higher bid for Anzon Energy and Arc Energy withdrew. As it turned out, Nexus Energy withdrew its offer following the drilling of a dry hole at Basker-6 reduced Anzon's oil reserve estimates. Anzon Energy was eventually taken over by Roc Oil in September, 2008.

The Howard era ended on 24 November 2007 when the Labour Party was elected and Kevin Rudd became Prime Minister of Australia. This ushered in an incredible period of relative political instability for Australia. In June 2010, Kevin Rudd resigned and Julia Gillard was voted by the Labor Party to be the next (and first woman) prime minister. The next election in August was a hung parliament but Julia Gillard remained prime minister with the assistance of the independents Andrew Wilkie, Rob Oakeshott and Tony Winsor. In June 2013, Kevin Rudd became the prime minister after a spill motion within the Labor Party but at the next election in September, Labour lost and Tony Abbott became prime minister. However, he was removed in a Liberal Party spill in February 2015 and Malcolm Turnbull became the new prime minister. So in the seven year period from 2007-15, Australia had six prime ministers and all this occurred during the Global Financial

Crisis which began in 2008. Despite the lack of any effective government (or perhaps because of it), the Australian economy continued to grow and survived the GFC better than other developed countries. It appears to me that our economy might actually be better off if we shut down the Federal Government (that should gradually pay off the national debt) and let the technocrats run the country!

We spent Christmas 2007 on Rottnest Island with Chris and Alex. On Christmas Day the temperature got up to 41°C but outside, it was hot enough to melt the bitumen on the roads.

FARMOUT OF PERTH BASIN PERMITS

In December 2005, Empire Oil had received farm-in offers from a UK company Wharf Resources and Hermes Holdings, a company controlled by Tom Fontaine. The offers were to fund 100% of the costs of acquiring the Gingin West and Mullering 3D Seismic Surveys; plus to fund 35% of the costs of drilling Moriary-1. These companies also had options to increase their equities in EP-389 and EP-432 by drilling one exploration well in each permit. In January 2006, Hermes Holdings withdrew from the Perth Basin farm-in offer but in February, Craig went to London for a meeting with Adrian Eyre, a director of Wharf Resources, to finalise the deals. Wharf Resources would earn a 25% interest in EP-389 by paying a fixed sum of $700,000 towards the cost of the Gingin West 3D Seismic Survey; plus earn a 25% interest in EP-426 by paying a fixed sum of $750,000 towards the cost of drilling Moriary-1; and earn a 25% interest in EP-432 by paying a fixed sum of $900,000 towards the cost of the Mullering 3D Seismic Survey. Wharf Resources also had options to increase its equities in EP-389 and EP-432 to 60% and 50% respectively by drilling one exploration well in each permit. They also agreed to earn a 20% interest in EP-416 by contributing $500,000 to the cost of the Wellesley 2D Seismic Survey. A new Eyre Group company, Allied Oil & Gas Plc, would eventually take over the Wharf farm-in deals on EP-426, EP-432 and EP-416.

After lengthy negotiations, Empire Oil and ERM Gas finally signed the Farmin Agreement on our Perth Basin acreage in November, 2007. This was very timely as Arc Energy had recently withdrawn from negotiations on these permits. Based on my evaluation, Empire Oil agreed to give up 75% of its coal seam gas rights on all its Perth Basin permits in order to achieve its objectives in the conventional exploration projects. I considered the Perth Basin coals to be too thin and dirty to be candidates for commercial coal seam gas development. ERM Gas had agreed to:

(i) pay 32% of the costs of acquiring the Gingin West 3D Seismic Survey to earn a 12.5% interest in EP-389 with the option to pay 32% of the cost of drilling an exploration well to increase their equity to 30%;

(ii) pay one third of the costs of drilling Moriary-1 to earn a 25% interest in EP-426;

(iii) pay 30% of the costs of acquiring the Mullering 3D Seismic Survey to earn a 12.5% interest in EP-432 with the option to pay 30% of the cost of drilling an exploration well to increase their equity up to 25%;

(iv) pay $70,000 to the costs of acquiring the Wellesley 2D Seismic Survey to earn a 5% interest in EP-416;

(v) fund 100% of the costs of seismic reprocessing and interpretations to earn a 12.5% interest in EP-430, EP-440 and EP-454 with options to increase these equities to 50% by contributing 100% to the costs of acquiring 2D seismic surveys; and finally

(vi) ERM Gas had the option to conduct a separate $5.5 million exploration program to earn 75% of the Coal Seam Gas rights in all the above permits.

In my opinion, this deal saved Empire Oil from a slow death like West Oil, kickstarted serious exploration of our Perth Basin permits and ultimately led to the discoveries of the Gingin West and Red Gully gas/condensate discoveries in 2009-11.

MORIARY-1

The 2004 APPEA Conference was held in Canberra in March. At the PESA Deal Day, I presented the Moriary Prospect in EP-426 in the Perth Basin. This prospect was a fault controlled anticline similar in style to the Mount Horner oil field. The main objectives were the Jurassic Cattamarra Coal Measures.

In mid September, 2005, I drove to Dongara to arrange for a surveyor to survey the Moriary-1 well site in EP-426. The Moriary well had one of the best locations possible for an exploration well. It was at the intersection of the Mooriary and Ludlow roads where there was already a gravel pit to locate the drilling equipment and a swamp to the north for water. I also arranged to visit Stuart Smart on site to discuss access for the drilling. He had no major issues but wanted a biosecurity clause in the access agreement to protect his water aquifers. This was not a problem as the combination of the conductor pipe and the 9-5/8" intermediate casing ensured that his aquifers were well and truely protected.

Following on from its success at Hovea, during 2004-05, Arc Energy made further oil and gas discoveries at Xyris, Xyris South, Apium and Eremia in the Perth Basin. This continuing success in the basin plus the interest received from the 2004 PESA Deal Day presentation assisted in Empire Oil farming out EP-426 to ERM Gas, Allied Oil & Gas and Black Fire Energy and these companies contributed to most of the costs of drilling Moriary-1.

Empire Oil finally drilled the Moriary-1 well during February-March, 2008. The well was a dry hole and cost $3.15 million, some 10% over budget. In December 2008, Black Fire notified the operator that it was withdrawing from EP-426. The new equities in EP-426 were now Empire 50%, ERM Gas 25% and Allied 25%.

Arc Energy completed the aeromagnetic surveys over EP-438 and EP-448 in the Canning Basin thereby earning a 5% interest in each permit. In March 2008, they exercised their option to earn a 75% interest in EP-438 by drilling one exploration well but did not exercise their option on EP-448 and withdrew from that permit.

The 2008 APPEA Conference was held in Perth on 6-9 April. The 2008 Eastern Australia Basins Symposium was held in Sydney on 14-17 September. At the PESA Deal Days at both of these conferences, I presented the Bee-Eater Prospect in EP-359 and EP-435. This prospect was the old Parrot Hill East prospect of Ampolex, a tilted fault block formed prior to the Miocene compression and which straddled the border between the permits. It had a very significant geochemical anomaly over its crest.

After the conference, we drove up to Newcastle to visit Chris. He had just bought a house in Bruce Street, Cook's Hill and we helped him settle in. He sold his half of the house in Ostend Road Scarborough to Alex and used the money as a deposit for his new house – all arranged through the Bank of Dad!

BUDAPEST – AMSTERDAM CRUISE

In June 2008, Mo and I decided to go on a cruise from Budapest to Amsterdam travelling along the Danube, Maine and Rhine rivers. We flew from Perth to Dubai and stayed overnight to break up the long flight. That evening, we went out on a tour into the Arabian desert and then had dinner in specially set up tents. The next day, we flew to Vienna and stayed overnight. We caught up with Liz and Wolfgang Zimmer for dinner that evening. The next day, we caught the train to Budapest. The most remarkable aspect of this journey was the large number of wind farms in the Austrian countryside – there were thousands of them. Suddenly they all ended and I said to Mo "I think we've just crossed the Hungarian border."

Budapest is a beautiful city with the Danube River cutting it in two. The western side is very hilly due to outcrops of Triassic limestones and is the site of the old city of Buda with its historic castle and palace. Alas, the Germans used the high ground to defend the city during WW2 and the Soviet army blew it all to bits. Parts of the castle are still being restored. The Triassic limestones are the reservoirs for a number of oil fields in the Pannonian Basin to the south. Matra's Blue Topaz-9 drilled into them and tested salt water in permeable limestones (see Hungary p280).

The eastern part of the city is on a flat plain and is the site of the old city of Pest and today is the main commercial, government and shopping part of the city. Budapest also has an extensive underground transport system. The first time we used it was an experience. If we had been anywhere in Western Europe, we could have figured out which signs meant north, south, east or west. But Hungarian is a totally different language with no Latin base and we had no idea on which platform we should be on! Fortunately, we found a local who spoke some English and pointed us in the right direction.

After two days in Budapest, we joined the cruise and set off up the Danube. This was our first river cruise and we were very impressed with the vessel "Avalon Imagery" Although it was really just a glorified barge, it was very luxurious and our accommodation was more spacious than we had expected. We left in the late afternoon and had drinks on the upper deck as we glided down the Danube past the palace and farewelled the beautiful city of Budapest. The next morning, we had docked in Bratislava, the capital of Slovakia. This city was not as impressive as Budapest but we wandered around and visited Burg Castle and St Martin's Dome.

Our next stop was Vienna which we had visited in 2002 for Al Raam's 60th birthday. We hopped on the city tram which goes around the old city and found a place for lunch. The next stop after Vienna was Durnstein, a small town famous for its castle where Richard Lion-Heart was imprisoned from 1192-94 on his way home from the Crusades. Far more interesting was our next stop, the city of Melk. The famous monastery at Melk is a glorious building meticulously restored in the 18th century **(Fig 49)**. Equally impressive were the interior frescos.

Figure 49
The famous Melk Monastery

From Melk we cruised down to Lintz, an industrial city on the western side of Austria. Here we had a choice of two tours; one south to Saltzburg or one north to the town of Cesky Krumlov in the Czech Republic. We decided on the latter on the basis that we had never been to the Czech Republic. We were not disappointed. Cesky Krumlov was an amazing town set in an S bend in the Vitava River. There was a large castle complex on the eastern part of the "S"and the main town on the western part. There was a huge square in the main town and we were there on market day so it was very busy. We had a very pleasant lunch in a restaurant overlooking the river. I ordered a Czech beer which was excellent.

From Lintz, we continued up the Danube into Germany. Just before we arrived at Regensburg, we passed Walhalla, a very ostentaceous Greek Temple built in the early 19th century to honour the dead from the Napoleonic wars. Modelled on the Parthenon, it seemed completely out of place on the banks of the Danube. St Peter's Cathedral in Regensburg was very impressive but I was also intrigued by St Jacob's Church built in the 11th century by Irish missionaries. The main entrance to the church was decorated with strange creatures resembling crocodiles and winged hippopotami. Inside the church, the round columns were also decorated with more weird creatures very much in the style of the temples of ancient Egypt. It was well worth the visit.

After Regensburg, we motored on to the town of Kelheim where we turned off the Danube and headed north up the Main-Danube Canal. By this stage, we had been through 16 locks since leaving Budapest. The Main-Danube Canal is 170 kms long and was completed in 1992 at a cost of €2.4 billion. As you pass through the 16 locks along the canal, the locks raise you up 175 metres from the Danube and lower you down to the River Main. Half way along the canal, we stopped at the infamous city of Nurenberg.

At the end of the canal, we arrived at the historic and very pretty town of Bamberg and joined the River Main. From Bamberg, we continued along the Main winding its way through the countryside. Just before arriving in Wurtzburg, we passed the Kappele Church with its minaret-like structures. In Wurtzburg, we visited the impressive Marienburg Fortress converted to a Renaissance palace with beautiful gardens. After Wurtzberg, the River Main passes through many vineyards before eventually joining the Rhine at the city of Mainz. The vineyards rise up very steep slopes from the river and we watched a mechanical harvester collecting the grapes and wondered how the hell it didn't topple over as it wound its way through the vines.

After the leisurely cruise along the Danube and Main, the Rhine was a very busy working river with lots of traffic. Soon after Mainz, we stopped at the interesting town of Ruedesheim with its Niederwald Monument. This monument was built to commemorate the Franco-German War of 1870-71 and is dominated by its "Germania" statue, a rather dark and militaristic construction.

From Ruedesheim, we cruised down the Rhine past numerous castles and the famous Lorelei rock before arriving at the major city of Cologne. We visited the famous cathedral which somehow survived the Allied saturation bombing of Cologne during WW2. Being an original and not reconstructed cathedral, it still had that same amazing effect that I'd felt in the Barcelona Cathedral.

From Cologne, we continued down the Rhine and joined the Amsterdam-Rhine Canal. By the time we got to Amsterdam, the Avalon Imagery had passed through 68 locks. Amsterdam itself was a very overcrowded city and we were shocked how dirty it was. We went on a cruise around the famous canals but they were choked with rubbish. We even saw the skipper of a boat literally tip his personal rubbish into the canal! Can you imagine what would happen if you did that in Sydney Harbour? From Amsterdam, we flew to Cork, hired a car and drove to Bob's for a few days before returning to Perth.

In July 2008, John Doran suddenly died of a heart attack. I hadn't seen him since my SOCDET days when he was CEO of Command Petroleum. He was CEO of Roc Oil at the time and only 62.

In August 2008, Arc Energy was taken over by Australian Worldwide Exploration Limited ("AWE") in a friendly deal worth $550 million. Because AWE had no interest in the Canning Basin, all Arc Energy's Canning acreage was hived off into a new company called Buru Energy Limited with $75 million cash. Arc Energy shareholders received 19 cents cash plus 0.3 AWE shares plus 0.425 Buru shares for each of their shares. AWE also ended up with 15% of the shares in Buru Energy.

Empire Oil also farmed out EP-439, EP-460 and EP-461 to DVM International who contributed to the cost of drilling Star Finch-1 and

Lake Macleod-1. Together with the Jurassica farmins, these wells were mostly fully funded with Empire Oil retaining a 22.85% interest in EP-460; and a 30% interest in EP-439 and EP-461. Star Finch-1 was drilled in July 2008 but was a dry hole. There were absolutely no shows in the Windalia Radiolarite or Birdrong Sandstone so I cannot explain the seismic amplitude anomalies. But then I remembered – "the rocks are smiling at us". Lake Macleod-1 was drilled during October-November 2008 and was also a dry hole. Most of the Gneudna Formation had been eroded by the Base Cretaceous Unconformity and it was only 146 metres thick in the well. I concluded that the best prospectivity would be in the northern part of the permit where the Gneudna Formation source rocks would be thicker and more mature. Alas, there were no existing seismic lines in that part of the permit.

It was my 65th birthday in December 2008 so Chris flew over from Newcastle and we all went to the Vines Resort for a couple of days. Chris, Alex and I had a game of golf in the morning and we were all back at the resort for a very late lunch around the pool. The Vines is a very tough course and it took us a lot longer than expected to go round given the time spent in the numerous bunkers!

PERTH BASIN 2008 SEISMIC SURVEYS

By the middle of 2004, I had concluded that Empire Oil needed to acquire 3D seismic surveys over some of its prospects in the Perth Basin. Because 3D seismic surveys were very expensive, they had initially only be used over discoveries to optimise the appraisal drilling program. However, technology was improving rapidly and 3D seismic surveys had now been proved to reduce exploration risk on a world-wide basis. I had already proposed a 3D seismic survey over the Mullering anticline in the first year of EP-432 and now considered that it was now critical to be acquired over the Gingin gas discoveries in EP-389. It was difficult to fund seismic surveys via equity placements as investors wanted to see drilling (and hopefully discoveries) rather than "boring seismic"! The 3D seismic surveys simply had to be farmed out. This was not going to be easy either as most companies didn't want to commit to the expense of acquiring seismic knowing that there was a significant risk that the seismic wouldn't define any structures to drill.

I submitted the first application to acquire the Gingin West 3D Seismic Survey in August, 2004. The 3D seismic survey in EP-389 was also located in difficult environments and would have to cross the Brand Highway, two natural gas pipelines and enter the Boonanaring nature reserve. I therefore decided to start the approval process as soon as possible so that by the time the permits had been farmed out, the acquisition of the surveys would hopefully also have been approved by the government. In December, I also submitted applications to acquire the Mullering 3D Seismic Survey in EP-432 and the Wellesley 2D Seismic Survey in EP-416. I soon discovered that things had dramatically changed since my Barrack Energy days and the attached Environmental Managements Plans ("EMP") were unacceptable. It appeared that it would be very difficult to get permission to enter the Boonanaring reserve and that even the vacant crown land in EP-432 was also environmentally sensitive. It was also no longer possible to have the botanist surveying ahead of the bulldozers but a detailed botanical survey had to be acquired before the survey started. In addition, the use of bulldozers was no longer allowed to cut seismic lines in native bush so rollers and crushers had to be used. In January, 2005, I met up with Greg Woodman, a consultant botanist, to assist me in preparing revised EMP's and to start the ball rolling to obtain access into the Boonanaring nature reserve. This was followed up in April with a meeting with CALM and the Environmental Protection Authority ("EPA") regarding access for these 3D seismic surveys.

These new environment conditions and the need to conduct botanical surveys prior to acquiring seismic surveys was putting pressure on our work commitments. Empire Oil had to continuously apply for and be granted suspensions of the conditions to acquire seismic surveys in the EP-389, EP-416 and EP-432 permits. This was actually a great relief for us as these permits had not been farmed out and we weren't in a position to fund the acquisition of these surveys in the short term.

During March and April, 2006, a lot of my time was tied up with meetings with Greg Woodman and Enesar in the preparations of the Environmental Management Plans for the acquisition of our proposed seismic surveys. These meetings were followed up with presentations to CALM, the Conservation Council and the Wilderness Society.

On the 8th September, I attended a field trip with Greg Woodman, botanists from DEC and representatives from the DoIR over the area of our proposed Mullering 3D Seismic Survey. We all met up at the Cataby Roadhouse and drove out along Woolka Road to the survey area. I showed them one of the 1990 Barrack Energy seismic lines as an example of rehabilitation based on the methods used in that era. This of course included using bulldozers which cut the bush above the topsoil. The line was barely visible and from the incredulous looks on their faces, I suspect that some of them didn't believe that it was actually an old seismic line. I tried to explain that the seismic line was in fact 16 years old and the native bush simply wasn't that fragile. Provided the topsoil was not disturbed, as per the 1990 state government guidelines, regrowth would be natural and inevitable. Again, I'm not sure they believed this either. The area of the survey also contained the declared rare flora *Macarthuria keigheryi* and discussions were held as to managing the survey to minimise impact on these plants. Since the area of the seismic lines was such a low percentage (<2%) and the lines were to be rolled, it was decided that Empire Oil should apply for a Permit to Take. This prevented Empire Oil from being prosecuted and fined in the event that the seismic survey accidently disturbed a declared rare flora. The application was lodged in March, 2007.

In November 2006, I joined Terry Grocke and John Schulstad and travelled down to the little town of Benger south of Harvey, to scout the area of the Wellesley 2D Seismic Survey and begin the process of obtaining access from the private landowners. There were 28 private landowners in all plus Western Power, Landcorp and Kemerton Silica Sands. Fortunately there were a number east-west oriented existing roads in the area which we could use thereby reducing impacts on the environment and disturbance to private landowners. These were the Wellesley, Mitchell and Swamp roads and West Lane. In addition, Campbell Road ran roughly north-south over the crest of the structure and provided an excellent strike line to tie the survey. Swamp Road ended at the Benger Swamp but a CALM access road went through the swamp and we would need their approval to use it. The Benger Swamp had been infested with typha grass which dried out in the summer and became a fire hazard. CALM was progressively trying to eradicate it from the swamp so to gain access we would need strict biohazard protocols. Our east-west lines also needed to cross the wetlands of the Wellesley River which we would need to undershoot. So what originally looked like being a simple 2D seismic survey on private dairy farms was turning out to be consideraby more environmentally complex.

As part of the approval process for the Gingin West 3D Seismic Survey, I had to give a presentation to the board of the state EPA. I assumed that this would be to a couple of the board members but when I walked into their boardroom, there were at least a dozen board members present. I wondered to myself why all these top bureaucrats would be wasting their precious time being involved in a tiny 50 sq km seismic survey. Admittedly, it was in one of their reserves but given the number of billion dollar mining and petroleum developments being proposed in WA at that time, it did seem a bit of overkill. They were satisfied with my presentation and their only question was the potential impact on the nesting of the Carnaby Cockatoos. This occurs around November so when I informed them that our survey would be completed no later than May, they were completely satisfied.

During December 2007, I had a meeting with Colin Wedge who own the property to the west of the vacant crown land where the Mullering 3D Seismic Survey was to be acquired in EP-432. Part of the survey was going to extend onto his land. He had no objections to the survey and agreed the compensation but wanted an extra $5,000 to move his cattle during the survey. This was probably overkill but wasn't worth arguing over compared to the enormous cost of the botanical surveys in the vacant crown land to the east. The total cost of the environmental approval process was $230,000.

Early January was spent mostly tidying up the access agreements for the Wellesley seismic survey. This included visits to the offices of the Shire of Harvey and the Silica Plant and meetings with Land Corp and Western Power regarding access to the uncleared land.

After three and a half frustrating years since the original applications, Empire Oil finally received formal approval from the government to acquire the Wellesley 2D, the Mullering 3D and the Gingin West 3D seismic surveys. The contracts were signed with Terrex Seismic and the surveys commenced in late March and completed in early May 2008. During the recording of the Mullering 3D seismic, there was a bush fire which we believed was deliberately lit. The fire was quickly put out by Terrex but could have delayed the survey by continuing to smoulder. Fortunately, there was a light shower of rain which completely killed the fire and allowed the survey to continue. However this delay added an extra five days to the survey and caused the project to go over budget by $550,000 (35%). The Wellesley survey also went $150,000 (22%) over budget due to rain in the Harvey area slowing down the recording. The Gingin West survey was within budget.

CANNING BASIN DRILLING

In September 2007, the Valentine-1 well was drilling **(Fig 50)** and had penetrated an eight metre sand with good gas shows at 2,250 metres in lower part of the Anderson Formation. This sand was interpreted to be gas bearing but was tight. However, it did appear to correlate with similar tight, gas bearing sandstones in Point Torment-1, Kora-1 and West Kora-1.

Figure 50
Valentine-1 drilling on the edge of the Stokes Bay mud flats

As the well drilled on, even stronger gas shows were encountered during the drilling of the Early Carboniferous Laurel Formation and deeper marine shales overlying the Devonian. I interpreted these to be the Tournaisian source rocks and a potential shale gas/oil play. The objective Devonian sandstones turned out to be limestones and were not hydrocarbon bearing.

During November, the drilling of Stokes Bay-1 intersected two sands with good oil and gas shows in the lower part of the Anderson Formation. As in Valentine-1, these were interpreted to be gas bearing but tight. Based on a new seismic interpretation by Arc Energy, the R1 Joint Venture decided to deepen the well to the Devonian Nullara Group. Based on the tight limestones encountered in the Devonian in Kora-1, West Kora-1 and Point Torment-1, Empire, Indigo and Phoenix elected to go non-consent on the cost of this deepening. The well lost circulation at 2,755 metres and our decision to go non-consent was looking foolish. The zone was tested but recovered only 5,000 barrels of fluid interpreted to be mud filtrate. After several days swabbing, the test was abandoned due to the onset of the wet season. Based on pressures from the test, Arc Energy interpreted that there could be a 75-80 metre gas column in well.

During the drilling of Stokes Bay-1, Craig organised a broker trip to the site. This involved flying to Broome, driving to Derby, checking into the motel, and driving out to the wellsite on Point Torment.

After the safety inductions, we showed the brokers the various aspects of the drilling rig and then visited the mudlogging shack. The drilling and electric logs were being displayed in real time and Craig and I got to look at the logs over the Anderson sands. We returned to Derby and the next day took them all out on a fishing trip north to the Buccaneer Archipeligo. Fishing there was similar to Exmouth – throw your line in the water, count five seconds and you've got a fish. Most were undersized or not good eating and were returned to the ocean but I managed to catch a barracouta which we had for lunch **(Fig 51)**. The highlight for me was I caught a shark. It fought like hell and took me about ten

Figure 51
Look what I caught!

miniutes to reel it in so I thought I had "Jaws" on the other end. When I got it to the surface, it was only about a metre long and after all that effort it was cut loose to live another day. Now I know why sharks have survived for the last 350 million years!

The year 2008 was a busy year for takeovers in the Australian petroleum industry. In April, British Gas made a $13.6 billion bid for Origin Energy at $14-70 per share which was a 40% premium at the time. The offer was rejected by Origin which managed to hold off the takeover attempt and remain independent. In August, AWE completed its friendly takeover of Arc Energy in a $550 million deal. After the debacles with Arc Enery and Nexus Energy, Anzon Energy was eventually taken over by Roc Oil in September 2008. In October, British Gas made a $5.6 billion bid for the Queensland Gas Company by offering $5-75 per share. This was approved by the QGC board but it took another three months to mop up the remaining10% of the shares due to some shareholder dissent. Finally, in December, the board of Sydney Gas recommended acceptance of a takeover offer by AGL at 42.5 cents per share.

Adjoining EP-426 on its western side was Exploration Permit EP-368. The permit holders were AWE, Origin Energy and Norwest Energy, and based on the operator's mapping, there were a number of Jurassic leads in the permit. The permit also contained the western side of the North Erregulla structure. In late 2008, AWE and Origin Energy withdrew from the permit leaving Norwest Energy with 100%. Over the next few months, Empire negotiated an agreement with Norwest Energy whereby the EP-426 joint venture would take an 80% interest in EP-368 in exchange for Norwest Energy taking a 20% equity in EP-426. On notifying the EP-426 Joint Venture of this deal, ERM Gas and Allied decided not to take up their interest in EP-368 so Empire ended up with the entire 80%. The new equities in EP-426 became Empire 30%, ERM Gas 25%, Allied 25% and Norwest 20%.

Following the farmins by ERM Gas and an onshore gazettal by the state government, Empire and ERM Gas decided to apply for two blocks in the south Perth Basin. It was agreed that ERM Gas would be the operator and the interests would be split ERM 60% and Empire 40%. The northern block was east of Mandurah and had very little existing seismic data. The southern block adjoined EP-416 on its southern side and covered any possible extension of the Leschenault Prospect to the north of EP-416. ERM Gas submitted the two applications in October, 2008 and was notified by the DMP in June 2009 that the applications were successful and the permits offered subject to reaching native title agreements. ERM and Empire agreed to accept the offers and ERM commenced native title negotiations. The permits were granted as EP-479 and EP-480 in June 2012.

Following the funding of the reprocessing of existing seismic, ERM Gas earned a 12.5% interest in each of EP-430, EP-440 and EP-454. However, in July 2009, ERM Gas notified Empire that it was withdrawing from EP-430.

The 2009 APPEA Conference was held in June in Darwin. This was the last time the conference was held there as the number of delegates increased to exceed the capacity of their conference centre. The most interesting paper for me was co-authored by John Gorter on the Permo-Triassic boundary in the Perth and Bonaparte basins. There were also informative papers on the Hermes and Pluto fields in the Carnarvon Basin. The Hermes oil field was particularly frustrating as the discovery well Lambert-1 was drilled by BOCAL in 1973. However because the existing seismic at that time was of such poor quality, BOCAL's interpretation didn't show how big it actually turned out to be.

After the conference, Mo and I took the Ghan down to Adelaide. We were not impressed with the service as we had to load our own luggage onto the train. We were OK but an elderly couple had to be assisted by other younger passengers while the Ghan staff looked on. The best part of the trip was the stop at Katherine where we did a tour of the gorge which was just spectacular.

Canadian Holiday

In August 2009, Mo and I went on a holiday to western Canada. We flew from Perth to Hong Kong and connected with a Japan Airlines flight to Narita and then on to Vancouver. The service on Japan Airlines was surprisingly bad, even in business class and we had to virtually ask for everything once the meals had been served. In Vancouver, we joined up with the APT Tour and spent the next day sightseeing in the city. We were curious about the number of beggars in the city and our guide explained that these were "professional beggars" who were not poor but made their living this way.

The next day we caught the train to take us across the Rockies. The service on the "Rocky Mountaineer" was excellent and put to shame what we had received on "the Ghan". Our first stop was Kanloops located in a valley with a unique microclimate. You don't expect to see a desert in the middle of the Rocky Mountains but that's exactly what you get at Kanloops. On the day we arrived, it was incredibly hot – over 40°C so we didn't do much sightseeing. The Rocky Mountaineer continued on to Jasper where we stayed overnight and transferred to a coach the next day.

We headed south towards Banff passing the Athabasca Falls and various glaciers along the highway. We stopped at the Columbia Icefields and were driven onto the Athabasca Glacier on a special snowmobile. Our guide advised us that even though it was summer, the wind whistled down the glacier and it was very cold so we needed to be well rugged up.

Our next stop was Lake Louise, one of the most beautiful spots on the planet. We stayed at the Fairmont Hotel and the view from our room across the lake was spectacular with the cliffs of Devonian limestones rising up and forming a backdrop to the lake **(Fig 52)**. On the way to Banff, we stopped at Peyto Lake which was that light turquoise blue due to the glacial rock flour suspended in water. It reminded us of Lake Tekapu in New Zealand.

At Banff, we stayed at the Fairmont Hotel which was very luxurious on the inside but on the outside looked like a medieval castle from a horror movie. That evening, we walked around the town and had dinner at the Grizzly House Restaurant. We chose the hot rock menu which boasted buffalo, venison, crocodile, wild boar and rattlesnake as well as good old fashioned beef! The next day, we did a cruise of Lake Minniwanka – yes, you read it right. It's not a typo! Around the edge of the lake were huge cliffs of limestones and in one area, a spectacular outcrop of a reverse faulted anticline. After Banff, we headed back to Vancouver, firstly through Glacier National Park and then southwest to Kelowna and the Okanagan Valley. From there, we drove across to Whistler and then back to Vancouver. The next day, we took the ferry across to the pretty town of Victoria on Vancouver Island.

Figure 52
Lake Louise with Devonian limestones in background

From Vancouver, we took a cruise north up the coast of Alaska, through the Inside Passage to Juneau which is completely landlocked and the only way in is by sea or plane. It was only a small town and it was difficult to believe that it is actually the capital of Alaska. From there, we cruised up Glacier Bay and at the far northern end we reached the retreating wall of the glacier. Glacier Bay has an interesting history. Early French explorers reached the area around 1770 and recorded a giant glacier. Being at the end of the Little Ice Age, Glacier Bay had then been filled with the advancing glacier some 100 kilometres south of where our cruise ship was admiring its present position. The local Tlingit tribes told these explorers that their ancestors had enjoyed much warmer conditions and lived far to the north. Their villages had been destroyed by the advancing glacier and tribes had been forced to move south. This verbal history suggests that Alaska also enjoyed the Medieval Warming Period without invoking major industrialisation and the burning of fossil fuels.

The ship then cruised around to Skagway where we caught a train which took us up to the Canadian border along the old trail used during the 1890's Alaskan gold rush. The train crossed some very dodgy looking wooden bridges but the views were worth it. The cruise returned to Vancouver where the tour ended and we flew to London. In London, we got the train to Broxbourne and caught up with old neighbours for dinner. We then flew to Ireland and spent a week with Bob before returning to Perth.

Norwest Energy, our joint venture partner in EP-426, also operated EP-413 and adjoining TP/15 in the northern Perth Basin. They were preparing to acquire an aerogravity survey over these permits and suggested that while the equipment was in the area, we also acquire aerogravity over EP-368 and EP-426. This was agreed by each joint venture and the survey was acquired in October 2009. The results of this survey showed a large gravity high over the North Erregulla structure (as expected) but also delineated another gravity high to the southeast where there was very little seismic. If future exploration on the North Erregulla structure was successful, then new 2D seismic over this gravity high would be an obvious next step. The EP-368 and EP-426 joint ventures also jointly agreed to reprocess all the seismic over the Lockyer and North Erregulla structures in these permits. The seismic surveys in this area were acquired at different times and by different operators and it was hoped to provide and consistent data set. This was completed in June 2010 and I recommended that a 100 sq km 3D seismic survey be acquired over the North Erregulla structure and that Empire proceed immediately to attempt to farmout the prospect.

In December 2009, Allied Oil & Gas agreed with Empire to reduce its interests in EP-416, EP-426 and EP-432 in the Perth Basin to 10% with Empire taking back the balance. The new equities became:

In EP-416	Empire 85%, Allied 10% and ERM 5%;
In EP-426	Empire 45%, ERM 25%, Norwest Energy 20% and Allied 10%; and
In EP-432	Empire 77.5%, ERM 12.5% and Allied 10%.

ERM Gas still had an option to increase its equity in EP-432 but in February 2010, declined to exercise it. However, they did express interest in the shale gas potential of the Cadda Terrace where the Cadda-1, Woolmulla-1, Gairdner-1 and Peron-1 wells had encountered good gas shows in Permian shales and coals. Most of these shows were from coals in the Irwin River Coal Measures which also contained interbedded tight sandstones.

In December 2009, EP-466 in the Carnarvon Basin was granted 100% to Rough Range Oil. This permit covered the old Mardie area where excellent oil shows had been encountered in the Cretaceous Mardie

Greensand in several shallow wells. The Mardie Greensand was a low permeable reservoir but evidence offshore indicated that horizontal drilling could improve production rates as shown by Chevron on their Saladin Oil Field. My study of the area suggested that a two metre sand at the top of the Mardie Greensand may pinch out updip creating a large stratigraphic trap. I presented this prospect at the PESA Deal Day at the 2012 APPEA Conference in Adelaide.

The interpretation of the Mullering 3D Seismic Survey was carried out by Peter Kirk who delineated two main structures, the Cooljarloo and Woolka prospects. Both prospects were tilted fault blocks with significant fault throws so cross fault seal was still going to be the main risk. I prepared an interpretation report and began the farmout process to fund the drilling of one of these prospects to satisfy the Year Four permit obligations. Following an extension to Year Four, this well had to be drilled by April, 2009. These initial farmout efforts were unsuccessful, so the joint venture applied to move the Year Four well into Year Five ending April, 2010.

The EP-439 permit had a seismic survey commitment due in early 2010 and the joint venture decided to acquire another geochemical survey to assist in locating the proposed seismic lines. We needed to be north of Lake Macleod-1 so we ran a line up the road from the Blowholes to Gnaraloo Homestead and we ran lines north of the Boolathanna Homestead and west of Boologooro Homestead on the western side of Great Northern Highway. Bob Moffit and I drove up to Carnarvon in July 2009 to conduct the survey. The operation of the survey was routine except for one amusing incident north of Boolathanna Homestead. We were driving along a dirt track and came to a large goat with his back feet on the track and his front feet half way up a tree so he could reach the juicy leaves at the top. He had one eye on the leaves and one eye on the vehicle bearing down on him – "just one more leaf! I've got time for one more leaf!" I slowed the vehicle down and it wasn't until we were a few metres from him that he eventually took off. By this time, Bob and I were in fits of laughter. Back in Carnarvon, the results of the analyses were mostly disappointing except for a significant anomaly on the road to Gnaraloo Homestead just south of the northern boundary of the permit. We then drove up to Exmouth for some additional detail geochemical surveys in EP-359 and EP-435. By this stage, we'd had enough of the boring drive back to Perth, so I arranged for the vehicle to be trucked back to Perth while we flew back in comfort.

THE WORST 12 DAYS OF MY LIFE

During 2009, I received a National Bowel Cancer Kit which I completed and then sent back to them. The results were negative. However, in November, I received a second kit as the first was found to be faulty. I completed this and results came back positive. I was surprised by this as I had no symptoms of anything being wrong with my bowels other than a drop in my weight from 80kg to 76kg. After consultation with my doctor, in April, 2010, I went for a colonoscopy performed by Dr Tarquinio at Mercy Hospital in Mount Lawley. When I was admitted, I was asked the last time I was in hospital and I relied 1953 (when I had my tonsils out at age 9). The nurse asked the question again to make sure she heard me right. The colonoscopy itself was a very easy procedure and when I came round, I felt nothing to indicate I'd even had anything done. But I was horrified when the doctor told me that I had multiple polyps on my bowels. I'd been so fit and healthy all my life – how could this be!! He removed all but two during the colonoscopy but the remaining two were large and could not be removed.

I then had a second colonoscopy at Charles Gardner Hospital on 31 August. This was a more invasive procedure where they pump air between the colon wall layers and try to cut the polyps out. Unfortunately, this procedure failed. The next step was to have an operation to remove those parts of the bowel with the polyps. I was referred to a specialist surgeon Mr Tan who briefed me on the operation. Because the two polyps were a long way apart, the operation would have to remove over half my bowel. I thought this a bit extreme but Mr Tan assured me that most of the bowel was redundant (this really didn't reassure me) and the only change I would notice would be more trips to the toilet. He explained that I would be in hospital for 7-10 days as after the operation, the bowel shuts down and it takes a couple of days for it to start working again. Apparently bowels get very annoyed at being cut into bits and then stapled back together again! Can't say I blame them!

Just before the operation, Mo and I went to Sydney for the 50th anniversary of my class at North Sydney Boys High School. It started with a tour of the school which hadn't changed much in the past 50 years. I didn't recognise many of the old boys except John Shedrick and Paul Askins. For whatever reason, Ted Scribner did not turn up. After the tour, we went to the Narabeen golf club for the reunion dinner. It was all very well organised with lots of drinks, good food and the usual speeches.

I had the operation on Tuesday afternoon 26th October, 2010. The last meal I had had was on the previous Sunday. On the way to St John of God Hospital, Mo was far more nervous than I was and took a wrong turn so we had the scenic tour to the hospital. After admission, I checked into my private room – all very comfortable – TV, phone, view of the car park. Before the operation, they gave me an ECG and the nurse complimented me on having the heart of a 40 year old. After the operation, Mo came back to visit about 7-00pm and I vaguely remember coming to and waving at her. On the following day, I came to with two drips connected to my arm and another tube attached to my dick going to a bag at the bottom of the bed – not a pretty sight. I felt some pain but one of the drips was the pain killer and I had a button to control the amount. I had a look at my tummy and saw five key holes where various instruments went in and a large cut from my rib cage down to my belly button. Basically the surgeon pulls the bowel out, slices and dices the required bits which are removed and the two ends of the bowel are stapled together and stuffed back into your body, and the large cut is then stapled together. Easy to say but of course all this takes a skilled surgeon about 3-4 hours to perform.

On the Wednesday evening, I was mistakenly given some soup for dinner – I should have been on water only until the bowel restarted. Any food in the stomach cannot be removed until the bowel starts working so it stays there and becomes horrible green brown bile. I started throwing up on Thursday and that's when they inserted "the Tube". Now even the worst sadists during the Spanish Inquisition could not have devised such a heinous instrument with which to torture the innocent. They stick this plastic tube up one nostril and down into your stomach to siphon off the accumulating bile into another bag just like siphoning petrol out of the tank of your car. And now I had yet another tube coming out yet another orifice of my tortured body.

Although the Tube was efficient, it did not remove all the stomach contents so I threw up for five days. By Monday, I was in a pretty awful state and felt like crap. This was just as hard on Mo as it was on me and she was very distressed at my condition and could do nothing to help. Eventually the bowel started up, my stomach emptied completely and I immediately felt much better. I also figured out a way to cheer Mo up. After eight days without food and being sustained only by a saline drip, the lack of protein makes your balls swell up – and I mean really swell up. So when Mo visited me the next day, I got out of bed and said "have a look at these" and opened my gown. She let out a shriek that brought the nurses running to my room, and then she fell about laughing hysterically. It made her day and with me on the mend, I hoped she felt much better when she left the hospital.

The next day was Melbourne Cup, so when Mo came she had all the horse's names in a bag. We each put a dollar on each horse, divided the names into two bags, each picked a bag and watched the race. I ended up with the winner and most of the money but I was much more interested in getting out of hospital.

During recovery, it is very important to get some exercise. Mo would come to the hospital twice a day and she would take me for a walk up and down the hospital corridor. Getting me ready for the walk was rather complicated. The two drips were attached to a trolley. I had to get out of bed without twisting all the tubes, attach the pee bag and stomach bag to the trolley, make sure my gown was not open at the back (not a pretty sight!) and then away we went around the hospital corridors with me pushing the trolley.

As my bowel functions improved and I got better, the surgeon said I could be discharged on Saturday 6th November. I had spent 12 days in hospital! Before being discharged, a nutrition specialist visited me (fortunately Mo was also there at the time) and we discussed the foods I could and could not eat for the next two weeks. I had to avoid high fibre foods – the opposite of what you're supposed to eat. She gave me a comprehensive booklet which listed high and low fibre foods. Alcohol was not forbidden but beer was discouraged as it was gassy and makes you fart. In any event, I didn't feel like alcohol anyway.

When I got home I weighed myself and I was only 65 kilograms – I had lost 11 kilograms in hospital. I had lost all the muscles and weight from my shoulders, arms and upper body, and felt so weak. I also had diarrhoea when I got home but with the low fibre diet, this passed after a couple of days. With Mo's good cooking, I soon recovered and was back at work within a week. However, it took until May 2011 before I was back to 76 kilograms but I never got to 80 kilograms again. As a matter of fact, I can now pretty much eat whatever I like and my weight never gets above 76 kilograms. Having said all that, I don't recommend having half your bowel removed as a method of weight control!

After a month, I had another visit to Mr Tan to assess my rehabilitation. Everything was OK and he was satisfied by my progress. Before I left, he gave me a picture of my liver taken during the operation. He said my liver was in good shape and I had to agree with him – my liver looked terrific. I was much relieved because, as you have gathered from reading this book, I enjoy a drink or two. As I grow older, my capacity to drink alcohol is diminishing so there is now a good chance that my liver will outlive me.

In December 2010, Buru Energy tested the Nullara limestones in Stokes Bay-1 but the well recovered only salt water. It was a disappointing result that, following Empire's gas and condensate discovery at Gingin West-1, supported a developing strategy that Empire Oil should sell their Canning and Carnarvon assets and concentrate its limited funds for exploration in what I considered the more prospective onshore Perth Basin.

EP-389 and Red Gully Discoveries

The mid July 2008 Share Purchase Plan had raised $3.23 million and the placement of 25 million shares had raised a further $500,000. Empire Oil then had $9 million, $6.4 million in cash plus $2.6 million in the joint venture accounts allowing it to plan its drilling program in EP-389. Following the acquisition of the Gingin West seismic survey, the new beneficial equities in EP-389 became Empire 62.5%, Wharf 25% and ERM Gas 12.5%.

At the Operating Committee meetings in March 2009, I presented the results of the interpretation by Peter Kirk of the Gingin West 3D Seismic Survey. Peter had defined five prospects for the joint venture to consider:

1. The Gingin Gasfield Prospect was an elongate downthrown fault trap updip from the Gingin-1 gas discovery in the D Sands made by Wapet in 1965. It was not fully covered by the 3D seismic.
2. The Gingin Deep Prospect was a tilted fault updip from the Gingin-2 gas discovery in the Lower Cattamarra Coal Measures made by Wapet in 1966.

3. The Gingin West Prospect Block A (later renamed as the Red Gully Prospect) was a tilted fault updip from the Bootine-1 gas discovery in the D Sands made by Mesa in 1982.
4. The Gingin West Prospect Block B was a simple anticline updip from Bootine-1.
5. Finally, the Bootine Deep Prospect was a tilted fault updip from oil shows in the Lower Cattamarra Coal Measures in Bootine-1.

Empire was initially in favour of the Gingin Gasfield Prospect but ERM Gas preferred the Gingin West Prospect. I presented these prospects at the PESA Deal Day at the 2009 APPEA Conference in Darwin.

Following the interpretation of the Gingin West 3D seismic, in June 2009, ERM Gas exercised its option to pay 32% of the cost of drilling Gingin West-1 to earn an extra 17.5% in EP-389, taking it to a 30% interest. Including the farmout to Wharf Resources, the funding for the drilling of Gingin West-1 was now completed. The estimated cost of the well was \$6.67 million with Wharf paying 25%, ERM paying 32% and Empire paying 43%.

The original location of Gingin West-1 was to be on the Block A structure. However, this location was directly beneath the landowner's house. Due to the disposition of the roads, sheds and other infrastructure surrounding the house, the surface location would have to be moved a considerable distance to the east and the well deviated to the bottom hole location. Given the fact that we had to drill a deviated well anyway, the joint venture decided to drill the lower risk but smaller Block B structure first and then drill the Block A structure as a deviated well from the same drilling pad. Block B then became Gingin West-1 and Block A was renamed Red Gully-1.

Gingin West-1 spudded in November 2009. Incredibly, Wharf Resources notified Empire that it didn't have sufficient funds to pay for its 25% share of the well cost. It wanted to reduce its interest in EP-389 to 10%. Given the global financial crisis gripping the world, there was little choice except for Empire to take back the 15% interest in EP-389 from Wharf. Empire also had to find an extra \$1 million for the cost of drilling Gingin West-1 which Craig quickly organised via a Section708 placement to sophisticated investors.

Gingin West-1 encountered gas shows in the A Sand, the Coaly Unit and the D Sand but not in the sand that flowed gas in Bootine-1. Following log analysis, the only gas pay was at the top of the D Sand. The well intersected 4 metres of net gas pay overlying 10 metres of shale so we didn't have a good fix on the gas/water contact. At the end of December, the well was completed and suspended for testing. The well was tested in April at 7.66 MMcfg/d plus 300 b/d condensate - a very respectable result. A longer production test was carried out in July and the well flowed at a stabilised rate of 7.5 MMcfg/d plus 375 b/d condensate **(Fig 53)**.

Figure 53
Gas flare from Gingin West-1

At the EP-389 Operating Committee Meetings held in June 2010, I reported on the results of the Gingin West-1 well. It was possible that the gas accumulation in the D Sand may have extended to the Bootine-1 well where there was a potential one metre gas column based on logs. If this was the case, potential reserves could have been up to 4 BCF gas. In addition, the permit was up for renewal and the DMP had indicated that they wanted a well in the first year of the renewal. The joint venture initially didn't agree with this but as we were going to drill Red Gully-1 anyway, it seemed pointless to argue. In the end, the permit was renewed in April 2012 with one well in Year One, a 50 sq km 3D seismic survey in Year Two and one well in Year Four.

In July 2010, Wesfarmers announced that it had completed a gas sales agreement with Santos to purchase 60 PJ per year for 5 years at a price of \$7-80/per GJ. This was useful information for our potential negotiations to sell gas from our Gingin West-1 gas discovery.

Red Gully-1 was drilled in January-February, 2011 and encountered a 56 metres gross gas column in the B Sand and an 18 metres gas column in the D Sand. The thick gas column in the B Sand was trapped on the downthrown side of the fault by a very shaley Coaly Unit devoid of any interbedded sands. It was a very pleasant surprise and we were "smiling at the rocks". The well was completed and the D Sand was production tested in March and flowed at a stabilised rate of 12 MMcfg/d plus 832 b/d condensate.

SOUTH AMERICAN HOLIDAY

In February, 2010, following the success at Gingin West-1, Mo and I took a holiday in South America. Neither of us had been to South America and Scenic Tours was advertising an interesting itinerary including a cruise from Buenos Aries to Santiago via Terra del Fuego. So we flew to Sydney and connected to a Lan Chile flight to Santiago via Auckland. I had this vision that Santiago was about the same distance across the Pacific as Los Angeles but the South American continent is skewed anticlockwise and Santiago is in fact further east than New York! The flight from Auckland to Santiago was 14 hours but fortunately the Lan Chile service was excellent.

We spent a day sightseeing in Santiago before joining the tour and flying to Rio de Janeiro. Rio was an amazing city with beautiful beaches, lagoons and huge hills and cliffs of granite setting off the city. We stayed at the Sofitel Hotel and from our balcony, we had panoramic views of Copacabana Beach and the Atlantic Ocean, including the oil rigs on the horizon!

We did the usual sightseeing tours taking the train up Mount Corcovada to see the staue of Christ the Redeemer **(Fig 54)** and the cables cars to the top of Mount Sugarloaf. The views from the top of these mountains were breathtaking and revealed the natural beauty of this city. We also took a very relaxing tour of the botanical gardens located west of the city. At the end of the gardens, we were treated to spectacular views of Mount Corcovada, Rodrigo Lagoon and the city in the distance. We also watched the hangliders landing from their flights from one of the many granite hills in the city.

From Rio, we flew to the Iguassu Falls on the border with Argentina. We stayed at the very elegant Cataracus Hotel which lies on the western edge of the Iguassu National Park. The Iguassu Falls are on the Iguassu River which flows from the east across Cretaceous volcanic rocks, over the

Figure 54
Me and Mo in front of Christ the Redeemer

Figure 55
The huge Iguassu Falls

falls and then joins the Parana River on its way from northern Brazil south to Buenos Aires. There's no way to describe the Iguassu Falls to do them justice – you literally have to see them to understand their sheer size and magnificence. The walk around the falls seems to go on for ever and around each bend, there are yet more falls even more breathtaking than the last **(Fig 55)**. When you think how much fresh water is going over those falls every day, it's enough to make any Australian cry.

From Iguassu Falls, we flew to Buenos Aires which is an elegant, European style city of 12 million people. Our hotel was just off the Avenue del Liberator and just to the southwest was the Avenue 9 de Julio which is the main thouroughfare through the city. It was built to commemorate Argentina's independence on 9th July 1816. It is a broad avenue broken up by median strips of parks and gardens with many pink flowering trees which were in full bloom while we were there. Our first tour was to the Recoleta Cemetery to visit the grave of Eva Peron who is still revered in Argentina. In the cemetery, we were attacked by a plague of mosquitos. I found this very amusing because to get a visa for Brazil, we needed to get immunised for yellow fever but we never saw a mosquito the entire time we were there. Buenos Aries itself is located just south of the Parana River delta so mosquitos would not be unexpected. We were very glad of our yellow fever injections and hoped like hell that malaria had been eradicated from Argentina! The following day, we did a tour of the Parana River delta and the Rio Tigre where some 3,000 people live in houses on the many islands along these waterways. The most lasting impression we have of Buenos Aries was the dog walkers. Most people live in apartments and Argentinians love their dogs so there are people who make their living walking their dogs. You see them in parks some with up to ten dogs in tow. There has to be some skill in this job in order to avoid massive dog fights!

From Buenos Aries, we took a cruise on the Star Princess around the southern tip of South America. Our first stop was Montevideo where one of the guns of the Graf Spree forms a monument to the Battle of the River Plate in WW2. This battle took place in December 1939 when the Graf Spree which was raiding commercial shipping, was engaged by three Royal Navy cruisers off the coast of Uruguay. The Graf Spree was damaged and sought refuge in Montevideo as Uruguay was a neutral country. The captain decided to scuttle his ship rather than fight another battle against the larger British fleet.

From Montevideo, the ship headed out to the Falkland Islands and docked in the main town of Stanley. The Falkland Islands are very flat with few trees and reminded me very much of the Shetland Islands except for the scattered outcrops of Palaeozoic limestones. From Stanley, the ship cruised southwest to the scattered, bleak volcanic islands which form Cape Horn and then entered the Beagle Channel stopping at the town of Ushuaia in Argentina. It was a beautiful day and Ushuaia is a very pretty town and all the gardens were in full bloom. The town's postage stamps boast "Fin del Mondo" (the End of the World) which for the southern hemisphere, it most certainly is! From Ushuaia, the ship continued through the channel past many glaciers one of which had a hole in the base with a small waterfall **(Plate 56)**.

Figure 56
The Romanche Glacier

We eventually sailed into the Pacific Ocean, cruised up the coast of Chile, into the Gulf of Corcovado and docked in the city of Puerto Mont. It was Sunday and being a good Catholic country, everything was closed. So we sat on the foreshore next to the statue of General O'Higgins and enjoyed the scenery. General O'Higgins was a Chilean independence leader of mixed Spanish and Irish ancestry and he assisted Chile in achieving independence from Spain in 1818. There was a lot of low cloud in the gulf but as the morning warmed up, this eventually disappeared to reveal the Cabuco volcano whose peak was covered in snow **(Fig 57)**. It was such a shame that everything was closed as it would have been a great spot to enjoy a light lunch.

Figure 57
The impressive Cabuco volcano dominating the distant horizon

After Puerto Mont, the ship cruised north to the port of Valparaiso. On the morning before we arrived in Valparaiso, the captain made an announcement that there had been an 8.8 magnitude earthquake just off the city of Conception. Being in deep water, our ship was not affected but a large tidal wave hit the coast of Chile causing extensive damage and killing over 500 people. Although damaged, the port of Valparaiso managed to find a place for the ship to dock and for us all to disembark. We transferred to a coach which took us to Santiago which fortunately was not badly damaged. In our hotel room, we still felt the aftershakes which were up to magnitude 6 and very unnerving.

From Santiago, we flew to Cusco in Peru. Because Cusco is at an elevation of 3,400 metres, visitors feel out of breath and tired. You can drink the local tea or an energy drink to reduce the effects of altitude sickness. We drank lots of water and found that a berocca a day also helped. Cusco was the capital of the Inca Empire and is famous for its archaeological sites including a train trip to Machu Picchu. Alas there had been heavy rains in January and part of the train line had been washed away so we never got to see Machu Picchu. However, we made up for this by visiting many of the sites around Cusco.

Our first visit was to the ruins of the Temple of the Sun at Ollantaytambo. The stonework was quite remarkable, not rectangular like the Great Pyramids of Egypt but all different shapes and sizes and interlocking like a jigsaw puzzle. The blocks have no mortar and seemed to be fused together and it is believed that this is what makes these ancient structures earthquake proof. Even more remarkable were the remains of Sacsayhuaman just outside of Cusco. Some of the blocks are enormous, the one behind me measuring 4 metres high and 2.5 metres wide **(Fig 58)** and weighing an estimated 75 tonnes. According to our guide Manny, some of the blocks get up to 120 tonnes and one block that weighed over 200 tonnes was moved into the centre of Cusco itself.

Figure 58
Huge block forming part of the wall. Note interlocking stonework in background

Manny also told us that if you threw a floating object into the Urubamba River which runs just to the north of Cusco, it would not travel the 400 kilometres to the Pacific Ocean since the Andes Mountains are blocking the way. Instead it would travel north, join the Amazon River and after a journey of 6,000 kilometres, it would eventually enter the Atlantic Ocean. Check out a map of South America!

From Cusco, the tour took the train to Puno on the shore of Lake Titicaca, the highest lake in the world. The train travelled through the high plains of the Andes with spectacular outcrops of red beds and past a sign announcing an elevation of 4,000 metres. We stayed at the very fancy Hotel Liberatore in Puno and the next day visited the floating islands and canoes on Lake Titicaca. We then flew to Lima, spent a day sightseeing before flying back to Santiago and returning to Perth. It was a fabulous holiday but after five weeks away, we were glad to be back home.

In January 2011, Latent Energy merged with ASX listed Transerv Energy giving the combined company a 35% interest in both EP-321 and EP-407 including Warro gas field. Alcoa also committed to spend another $22 million to acquire a 3D seismic survey over the field and drill Warro-4.

In February 2011, Production Licence L 15 was granted over the West Kora application.

Back in the West Oil days, a number of oil industry friends would assemble at the Celtic Club in West Perth on Friday evenings for general discussions on the week's events. This was accompanied by numerous beers and for some of us, the odd cigar. When smoking in pubs and clubs was banned in 2006, we moved down to the Irish Club in Subiaco. Steve Walsh was a member and arranged for the club to make room for us downstairs in the storeroom with the beer barrels. It wasn't very comfortable and it was very cold in winter and prone to flooding. It occurred to me that the smokers of today are the equivalent of the lepers back in the 18th and 19th centuries so I named our gathering "the Leper Colony" **(Fig 59)** and the name stuck. I even had a sign made up but someone stole it.

The interpretation of the reprocessed seismic in EP-454 in the Perth Basin by Peter Kirk had defined a Permian structure we called the Garibaldi Prospect. I recommended that a 2D seismic survey be acquired over the Garibaldi structure to better define it for drilling. ERM Gas had the option to fund this seismic survey and increase its equity in EP-454 to 50%. They exercised this option in October. We immediately began planning the Garibaldi 2D Seismic Survey and Terry Grocke and I drove up to scout the

Figure 59
Some members of the Leper Colony

area. The survey was located on cleared agricultural land and did not need to enter any uncleared areas or reserves. We also visited the shire offices to notify them of the survey and any potential issues that may arise. The approval process was very straight forward and Empire received approval in time to acquire the survey in April 2011.

Following the withdrawal of ERM Gas from EP-430, Empire considered relinquishing the permit as there was an outstanding 2D seismic and one well commitment. Given the future activity planned in EP-454, Empire applied to the DMP to move the Year Five seismic commitment into Year Six, waive the Year Six well and grant a 12 month suspension of the permit to allow the seismic to be acquired at the same time as the Garibaldi survey in EP-454. In order to get this outrageous request approved, Empire agreed to retain only two blocks adjacent to EP-454 and relinquish the rest. This was approved by the DMP and the Garibaldi seismic survey was extended into EP-430 and total acquisition was 88 line kilometres.

Interpretation of the Garibaldi seismic survey by Peter Kirk showed that the Garibaldi Prospect was complexly faulted and may need a further 3D seismic survey to map the top Permian with any confidence. The interpretation also delineated the Launer Lead straddling the border of EP-454 and EP-430, and the Winchester Lead straddling the border of EP-430 and EP-321. I was coming to the conclusion that, outside the AWE-Origin tenements in the northwest, the Perth Basin was so underexplored that you could shoot seismic anywhere and find a new prospect!

One line of the Garibaldi survey was extended to the west to tie another seismic line and also ran over a shallow structure mapped by Arc Energy and which we called the Charger Prospect. This prospect was a simple, four way dip closed anticline with the main objectives being the fluvial sandstones of the Yarragadee Formation sealed by the Otorowiri Member at the base of the Parmelia Formation. The top of the Otorowiri was a strong seismic marker and could be tied back to the Warro wells and Waramia-1 where this unit consisted of a 50 metre thick lacustrine shale, an excellent seal. The Charger Prospect covered an area of 16 sq kms with about 60 metres of vertical closure and the potential resource was estimated at 10 million barrels oil.

The 2011 APPEA Conference was held in April in Perth. At the PESA Deal Day, I presented the North Erregulla Prospect, a large tilted fault block that straddles the border of EP-368 and EP-426, and covers an area of about 20 square kilometres. It was drilled by Wapet in 1967 but new interpretations of the reprocessed seismic indicate that the well was not drilled on the crest of the structure. Oil shows and log analysis indicated an oil-bearing sand in the Early Triassic Arranoo Sandstone and DST-1 over this sand recovered 20 gallons of 38° API oil. Oil shows were also encountered in the deeper Dongara Sandstone and DST-2 recovered 8 gallons of 38° API oil. Pressure data from this test indicated tight sandstone but DST-3 some 10 metres lower, recovered 22 barrels of formation water from porous sandstones. If a more crestal location could be found for the Dongara Sandstone updip from the old well, then these porous sandstones would be in the oil zone. However, the top of the Dongara Sandstone was a poor seismic event and the various seismic interpretations by Arc Energy, Norwest Energy and Empire Oil were all different. A 3D seismic survey was required over the structure to resolve these issues and delineate an optimum location for a new well. Based on our interpretation, the potential resource in North Erregulla was estimated to be 25 million barrels oil.

In October 2009, Empire Oil had been approached by a Russian group, Lion Energy, who wanted to farm into EP-439. They claimed they had access to Russian remote sensing technology and wanted to drill a well in EP-439. Their farmin offer was very tough – they wanted 100% of the permit with the existing joint venture parties taking a royalty. The joint venture eventually agreed on the basis that a small piece of something was better that a large piece of nothing! The farmin agreement took a long time to negotiate and in the end the contentious issues were resolved by Eric Tucker (representing Lion) and I and bypassing the lawyers. Sadly Eric was to die prematurely in 2013 from motor neurone disease. Eventually the farmin agreement was signed in June 2010 and Lion Petroleum gave us the location of the first well. This was in an area where there was no existing seismic whatsoever. Access to the location was easy as there were existing tracks on the pastoral lease. Fortunately, there were water bores in the area to give me the predicted section for at least the Tertiary-Cretaceous section. Below that I predicted Devonian Gneudna Formation beneath the unconformity (based on Geological Survey of WA maps) and put in some dolomites as primary objectives. I was really looking forward to the drilling of this well and gathering more information on what I thought was the prospective but underexplored Devonian sediments in the Gneudna Formation.

Rough Range Oil then lodged the application to drill Chapter-1 together with the Environmental Management Plan. The DMP notified us that we were required to conduct a botanical survey over the proposed access to the well site which was only a few kilometres from the existing tracks. I was stunned – the DMP actually wanted us to conduct a botanic survey over an area that had been grazed by sheep and goats for over a hundred years! Had they totally lost their minds? But they wouldn't budge and in the end Lion Energy spat the dummy and in April 2011, they cancelled the farmin agreement. They told us that they would never explore in Western Australia again and headed off to South Australia. There were now some serious problems developing within the DMP. Instead of promoting exploration in Western Australia, their complex and time consuming approval processes were now in danger of discouraging exploration in the state. And things would only get worse.

Following the processing and interpretation of the Wellesley 2D Seismic Survey, the Wellesly Prospect had been confirmed as a simple anticline on the downthrown side of the Darling Fault. There was only one fault on the flanks of the structure but the potential resource updip from this fault was still estimated at 35 BCF gas. Despite this, Empire received very little interest after the PESA Deal Day presentation at the 2010 APPEA Conference. The depth to the reservoir (4,000 metres) and the cost of the well were major deterents to farming out the prospect. I therefore recommended that the joint venture acquire a 2D seismic survey over the Leschenault Prospect. This prospect was a large horst block where the fault throws were so large that cross fault seal was an unacceptably high risk. However there were two small fault independent rollovers along the crest of the structure where the top Permian was at ~2,000 metres depth with considerably lower drilling costs. Although small in size, the prospects still had potential resources each estimated at 10 BCF gas. If the timing was right, the survey could be extended north into the application area L08-7 to satisfy work commitments there.

Based on Google Earth images, a 2D seismic survey could be acquired over the prospect with all the east-west dip lines running along existing roads and the north-south strike using existing CALM tracks running through the state forest. In January 2012, Empire submitted an application to acquire the Leschenault 2D Seismic Survey together with an Environmental Management Plan. Eventually I received a huge number of proposed amendments to the EMP from the DMP. For a simple 2D survey on existing roads and track and requiring no clearing of native vegetation, I considered these amendments totally unreasonable. An application to renew the permit had been made in August 2011 but had not yet been granted by the DMP. At this time, I was flat out dealing with the DMP on the approval processes for the more prospective and critical Wannamal 3D and Launer 2D surveys. I also didn't want to incur a lot of approval expenses until the permit was granted and title was secure. So the Leschenault Seismic Survey was put on hold and I gave serious consideration to recommending to the board that Empire not accept the renewal of EP-416.

Back in July 2010, Rough Range Oil had been contacted by UIL Energy who wanted to farm into EP-359 and EP-435 to explore for shale gas. This was fortuitous as we had exhausted all attempts to farmout the Bee-Eater Prospect. The deal was the same for each permit; that they would earn a 50% interest by funding the drilling of a well to 2500 metres to test the Birdrong Sandstone. They would then have options to drill an unconventional well in each permit to earn a 75% interest in the unconventional rights. The farmin agreement was subject to UIL Energy listing on the ASX and was signed in March 2011. EP-435 was renewed for a further five years in May 2011. Problems arose when UIL Energy decided it did not want to drill the Bee-Eater Prospect. They had selected a drilling location in the Paterson Trough which would initially test the pinchout of the Birdrong Sandstone and could also be deepened to test the Middle Jurassic sands that were gas bearing in Cape Range-2. The deal was renegotiated so that they would earn a 50% interest in EP-359 by funding the drilling of a well to test the Birdrong Sandstone pinchout. They would then have the option to deepen the well to the unconventional target to earn a 75% interest of the unconventional rights.

Red Gully Development

As compressor station 5 **(Fig 60)** for the Parmelia natural gas pipeline from Dongara to Perth was only two kilometres south of the Gingin West/ Red Gully well locations, we contacted APA, the owners of the pipeline. Our initial meeting was very strange – they wouldn't tell us what the tariff was and their first draft of the contract had very nasty penalties for not meeting contracted volumes. Given that we only had two producing wells, we needed some flexibility on these volumes to allow for maintenance and workovers.

Figure 60
Compresser Station 5

Over the first few months of 2011, we had a number of meetings with Alcoa and these were followed up with due diligence technical studies on Gingin West and Red Gully. As their long term contracts with the North West Shelf joint venture were gradually ending over the next few years, they were looking for new supplies of natural gas. Despite the fact that we didn't have anything like the volumes they needed, they were happy to take whatever they could get. These discussions ended in June with an offer from Alcoa to purchase 15 PJ (13.5 BCF) of gas from the project. The first 5 PJ would be pre-purchased at $5/GJ providing the joint venture with $25 million to build the gas plant required to process the natural gas prior to delivery into the pipeline. The next 10 PJ of gas sales would be at $5-50/GJ escalated at the CPI. After receiving this offer, I had a long meeting with Craig. This offer was a "Win-Win Deal". It saved us at least six months of delayed production and a lot of expense while we arranged the project financing for the construction of the gas plant. And the gas price of $5/GJ was below what we thought was the current pricing based on the Santos-Westfarmers deal, so Alcoa also got a good deal. For Empire, it was still a respectable $6-20 to $6-50 per Mcf after taking into account the conversion from GJ to Mcf and the the price being delived at the pipeline, not Perth. The discount on the price would also be more than made up for by not have to raise additional equity capital and further diluting our shareholders. My initial P2 reserve estimates for the fields were 24 BCF gas-in-place and 17 BCF recoverable gas and this was more than adequate to cover the Alcoa contract of 13.5 BCF gas. We made strong recommendations, firstly to the board and secondly to the EP-389 Joint Venture to accept this offer.

Regarding gas prices in Western Australia, these had historically been generally higher than in the eastern states where natural gas had to compete with cheaper coal fired power stations for electricity generation. The eastern states also had vast coal reserves compared to the small deposit at Collie in Western Australia. Back in 2006, the State Labor Government led by Alan Carpenter, introduced a Gas Reservation Policy which required producers to commit 15% of its production for domestic use. The policy was condemned by the industry as being a disincentive for further gas exploration. However it did not stop the development of existing fields such as Gorgon, Wheatstone, Macedon, Ichthys and Prelude and the potential development of Woodside's Browse Gas and Scarborough. Gas prices in Western Australia have fallen from highs in excess of $10 per GJ in 2008 to around $3 per GJ in 2019. As discussed earlier, the LNG projects in

Queensland have diverted signicant gas production to export at the expense of domestic gas requirements. Gas prices in the eastern states have now risen from the historic $3-4 per GJ to now over $10 per GJ and these high energy prices are impacting on the profitability of manufacturing in the eastern states. In my opinion, a Gas Reservation Policy in Queensland might have avoided this mess. Perhaps only two LNG projects would have been developed but the loss of those export earnings has to be compared with the current potential decline in manufacturing and loss of jobs in the eastern states. Of course, the banning of petroleum exploration and fraccing in NSW and onshore Victoria has not helped the situation and they certainly don't deserve any sympathy from the Queensland government! This situation could also get a lot worse. According to research by EnergyQuest summarised in the Australian Financial Review in May 2019, the Cooper Basin gas production is already in decline and by 2022, gas production from the Gippsland Basin is expected to start declining and the eastern states could have a severe natural gas supply shortage by 2025.

In June 2011, it was our 40th wedding anniversary. To celebrate, we took a cruise out of Sydney to New Caledonia and Fiji. Prior to the cruise, I bought a ruby ring for Mo as a surprise. It was an unusual design so I was somewhat apprehensive when I gave it to her but to my relief she loved it. On our return to Sydney, we caught up with Chris and Alex and had dinner on the Paddle steamer restaurant in Darling Harbour.

At the Good Oil Conference in September 2011, Empire Oil was awarded the Explorer of the Year Award. The following month was the 40th anniversary of the North Rankin gas discovery. Woodside put on a marvellous dinner for the old BOCAL employees and partners. It was great to catch up with many of the old crew some of whom we hadn't seen for years. On a sad note, Alan Burns died suddenly in December. I still have fond memories of my days at the Magnet Group, the listing of Stirling Petroleum and the drilling of Bruce-1.

I was concerned about the high condensate levels in the Red Gully gas fields. While this was a bonus for cash flow, it could also be a production problem. If you produce the gas too fast, the pressure around the well bore may drop to a level that allows the condensate to come out of the gas in the reservoir rather than the production casing, reducing permeability and hence the gas flow. Mark Pogson was commissioned to prepare a report on managing the production and his recommendation was to produce the wells at no more than 8 TJ/d (7.2 MMcfg/d) to maximise condensate recovery without damaging the reservoir.

Empire Oil now had to build a gas plant and the necessary infrastructure to produce the fields as well as explore the ten other permits it had in the Perth Basin and the fifteen permits in the Carnarvon and Canning basins. Craig and I had to divide our time with me looking after the exploration and Craig taking on the development. Managing twenty six permits and keeping them in good standing was simply not going to happen given our focus being firmly fixed on the Perth Basin. At a following board meeting, we decided to sell Gulliver Productions and Rough Range Oil so that we could concentrate our limited exploration funds on our more prospective Perth Basin permits.

SALE OF KEY PETROLEUM

In November 2011, Ted Ellyard, a director of Key Petroleum, organised a meeting with Craig and I plus a major shareholder of Key Petroleum. The major shareholders in Key Petroleum were unhappy with the performance of the company and wanted Key to be involved in new exploration ventures where there was more potential upside. For Empire, this was a great opportunity to sell Gulliver to Key Petroleum and remove the management and financial burden on Empire. In addition, the Canning Basin was having another boom. In September 2011, New Standard Energy had finalised a staged farm-in agreement with ConocoPhillips to spend US$109.5 million to earn a 75% interest in EP-443, EP-450, EP-451, EP-456 and three applications in the basin. In the following month, Buru Energy had discovered oil in Ungani-1 which flowed 1647 bopd from Early Carboniferous dolomites. The purchase of Empire's Canning Basin assets certainly appealed to the Key shareholders and they approved the purchase at the November AGM. It was also a good deal for Empire who ended up owning 55 million Key Petroleum shares and a 22.5% interest in EP-437 in the Perth Basin. The sale was finalised in December.

The Canning Basin boom continued in April 2012 with American independent Hess buying Kingsway Oil Pty Ltd who owned EP-429, EP-449 and four applications covering 60,000 sq kms in the basin. Then in

November 2013, Apache Energy farmed into Buru Energy's EP-390, EP-471 & EP-473 to earn a 50% interest in each by funding a $25 million of exploration program.

In November 2012, Buru Energy announced a State Agreement on five permits (EP-371, EP-391, EP-428, EP-431 & EP-436) covering 17,000 sq km in the Canning Basin. The agreement specified no relinquishments until 31 January 2024 and appraisal work could be credited against future commitments in other permits. In return, Buru and its joint venture partner Mitsubishi commited to supply 1500 PJ of gas over 25 years to the WA domestic market. This agreement raised eyebrows in the industry as it granted special consideration to only one of the several operators in the basin and was also rumoured to have been opposed by the DMP.

Soon after the PESA presentation at the 2011 APPEA Conference, Empire received a farmin offer from Origin Energy on the North Erregulla Prospect in the northern Perth Basin. This was a surprise as they had withdrawn from EP-368 in 2008. They were operating a large 3D seismic survey over L1, L2 and EP-320, and they proposed to continue this survey to the east to cover the North Erregulla structure. Empire had already had some discussions with Warrego Energy about sharing costs of mobilising a seismic crew for theit West Erregulla 3D survey but Origin had already contracted Western Geophysical to conduct their surveys. After further negotiations, Origin agreed to carry the two joint ventures for the cost of the 3D seismic to earn a 40% interest in each permit. Norwest Energy decided that they did not want to farmout but would pay their way for 20% of the cost of the 3D survey.

Origin Energy began the permitting of the landowners during 2012 starting with the L1 and L2 production licences and the Beharra Springs area. They encountered strong resistance from landowners in the area and obtaining access agreements took a lot longer (and a lot more expensive) than expected. There were only three landowners in the area of the North Erregulla 3D survey but their land was in prime wheatgrowing areas. In the end, Origin Energy couldn't reach agreement with the landowners and terminated the farmin agreements in November 2012.

So now I had to start all over again. My first priority was the landowners and their compensation demands. There were three landowners, Stuart

Smart, Daniel Michael and Peter Norwood but Stuart Smart was the largest landowner and led the negotiations. I met Stuart and Peter in March 2013 at Stuart's Erregulla Plains property. Their biggest problem was the compaction and wheel ruts caused by the vibrators as these caused damage to their seeding and harvesting machinery. I discussed with them the option of using a heliportable shot hole seismic survey similar to the Wannamal 3D Seismic Survey to be acquired in the Boonanarring Nature Reserve in EP-389. Unfortunately, this survey turned out to be prohibitively expensive and it would now be cheaper to use vibroseis and pay the required compensation to rehabilitate the land. I took Terry Grocke with me for another meeting with Stuart and Peter to explain the situation and continue the negotiations on rehabilitation. Soon after, Empire Oil was under attack by ERM Power who replaced the Empire Oil board in November 2013.

Following the purchase of Gulliver Productions by Key Petroleum, Empire Oil ended up with a 22.5% interest in EP-437 in the northwest part of the basin. Key Petroleum had a 22.5% interest with CalEnergy having the balance of 55%. CalEnergy was the operator and I attended a Technical Committee Meeting in March 2012. At this meeting, it was clear that CalEnergy had no further interest in EP-437 and soon after, they agreed to withdraw from the permit. So Empire Oil and Key Petroleum each ended up with 50% of the permit. Later in the year, Caracal Exploration Pty Ltd, a private company owned by Ted Ellyard, farmed into to earn a 10% interest in the permit from Key.

Empire Oil became the operator and my initial assessment was that there were two areas of interest; the Dunnart and Wye Knot prospects. The Dunnart prospect was a tilted fault block and was drilled by CalEnergy in December 2010. The well encountered good oil shows in the Basal Triassic Bookara Sandstone but these shows were not tested. The well needed to be redrilled and an open hole DST run in the Bookara Sandstone.

The Wye Knot prospect was a faulted anticline which was drilled by Discovery Petroleum in 1996. The well encountered gas in the Bookara Sandstone but potential reserves were very small. The gas did have significant hydrogen sulphide content indicating a possible biodegraded oil leg. The Bookara Sandstone was interpreted to overlie shales but due to the presence of radioactive sands in the Bookara, this interpretation was difficult to confirm.

Due to the corporate problems at Empire Oil, Key Petroleum took over the operatorship of the permit in mid 2013. In July 2014, they drilled Dunnart-2 which encountered excellent oil shows in the Bookara Sandstone. Due to hole conditions, no open hole DST could be run and when the electric logs were received, they were difficult to interpret. The well was cased for future production testing which took place in June 2015 but no commercial hydrocarbons were recovered.

By agreement with ERM Gas, it was decided to split EP-432 into two parts; a northern area (Area A) of eleven blocks covering the Cadda Shelf and a southern area (Area B) of five blocks covering the area of the Mullering 3D seismic. ERM Gas could then farm into Area A and increase its equity by drilling the Black Arrow prospect and still retain its 12.5% interest in Area B. As the permit was, following a number of extensions, due for renewal in October, 2012, this was the ideal time to split the permit and have two new permits granted.

The Black Arrow prospect was a follow up to strong oil shows in the Triassic Arranoo Sandstone in the Gairdner-1 well drilled by Barrack Energy in 1991 just to the north of EP-432. A large fault block extended from Gairdner-1 south into EP-432 and limited seismic indicated a structure at Black Arrow. When I visited the landowner to negotiate an access agreement, the track to the homestead crossed the outcropping fault block. I could clearly see the Lesueur Sandstone dipping back to the coastline at about 3°. My meeting with the landowner was interesting. He told me that the Greenies had been talking to the local community about the dangers of fraccing. Then he looked at me straight in the eye and said "Well – the Greenies are no friends of farmers so what's your side of the story" I told him that our first prospect was conventional and may require horizontal drilling but not fraccing. However I had come prepared and showed him a diagram of a fracced well **(Fig 61)**.

This clearly shows that surface aquifers are initially protected by the surface casing and then the intermediate casing. Production casing is then run down the well and through the target zone. After the well is stimulated, the hydrocarbons are produced through tubing sealed inside the production casing. The end result is that the surface aquifers are protected by four layers of casing inside the well bore and cannot be contaminated by the produced hydrocarbons. In addition, the aquifers cannot be contaminated from hydrocarbons in the fractures as these are normally two to three kilometres below any aquifers.

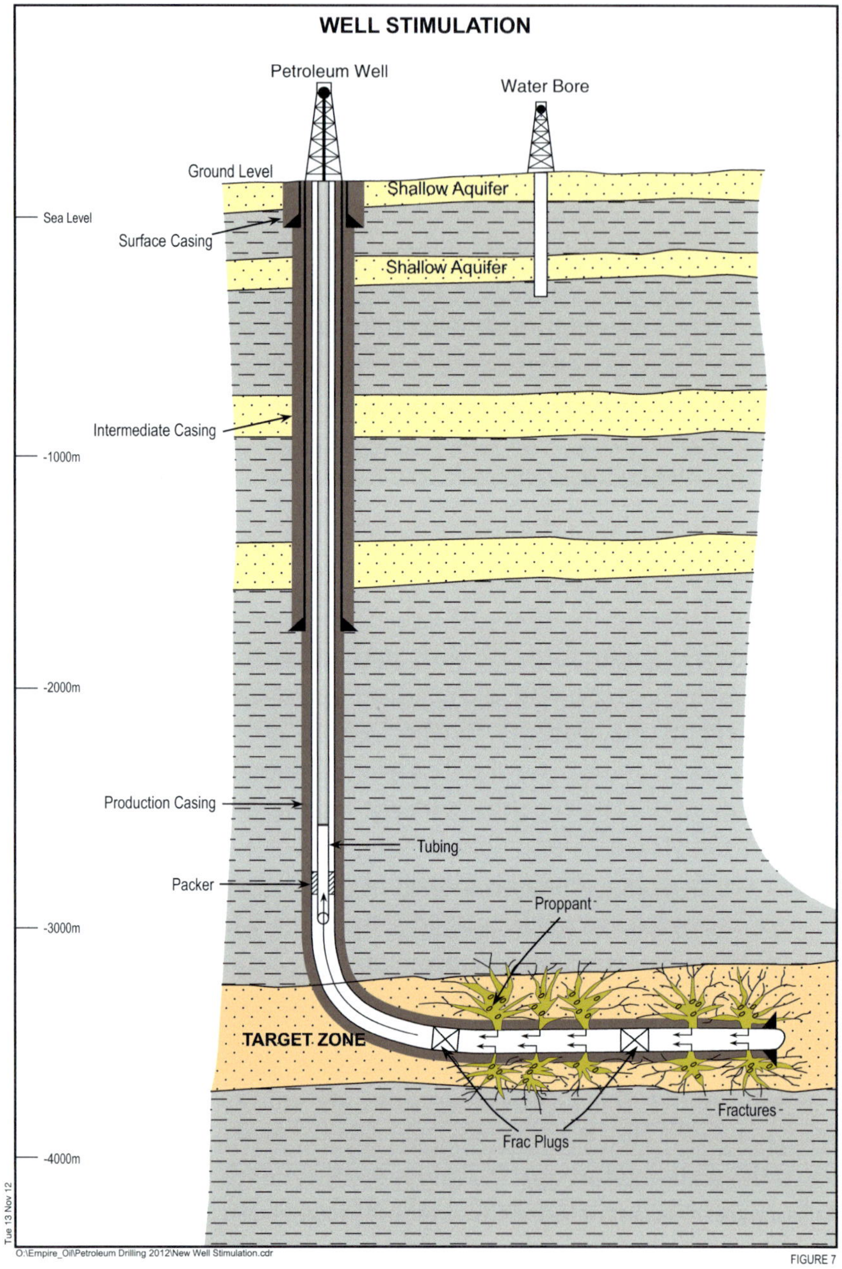

Figure 61
Diagram of a fracced well

The drilling of Black Arrow-1 was delayed by the slow DMP approval process and further extensions of the permit were applied for extending the permit to October 2013. By then, Empire Oil was under attack by ERM Power who replaced the Empire Oil board in November 2013.

Shale Gas

This is a good spot to discuss the Shale Gas Revolution in the USA. The first commercial natural gas production in the USA was from organic-rich Devonian shales in the Appalachian Basin. I referred to this under my Ashland Oil contract (p185). Prior to the development of horizontal drilling, production from each well was very low, even after fraccing, and the economics were tight. Historically, fraccing was first used in the USA on the giant Hugoton gas field in 1947. It was mostly used on conventional but tight reservoirs but was also applied to the shale gas wells in the Appalachians. The combination of horizontal drilling and multi-staged fraccing was pioneered by Mitchell Energy & Development in the Mississippian Barnett Shale from the Fort Worth Basin in Texas during the 1990's. The company was sold in 2002 for $US3.1 billion. The founder, George Mitchell, recently died in 2013.

Natural gas production in the USA rose from 20 TCF in 2005 to 27 TCF in 2012 driven by shale gas production. Over the same period, natural gas prices fell from a peak of $US13 per Mcf to a low of $US2 per Mcf in 2012. The Shale Gas Revolution was changing the world economic balance. The USA was saving an estimated $US100 billion per year by not having to import natural gas and their gas prices lowered energy costs and provided a boost to industrial activity in the USA.

The oil and condensate production associated with the Shale Gas Revolution also had an effect on the oil price. Since 2004, the oil price had been steadity rising from $US40 per barrel to a staggering $US145 per barrel in June 2008. Due to the panic associated with the Global Financial Crisis, the oil price fell to $US32 per barrel in December 2008. It reminded me of the oil price crash in 1986. In that crash, it took 14 years (1986-2000) for the oil price to rise to its previous high. Although the oil price hasn't yet risen back up to $US145 per barrel, it did get to $US115 per barrel in 2014. These high oil prices led to a Shale Oil Revolution in the

USA. With natural decline from conventional fields, US oil production had declined from almost 10 million bopd in 1970 to only 5 million bopd in 2005. Driven by shale oil drilling, US oil production increased to 9.4 million bopd in 2015. This increased oil production on the global market combined with low global growth following the Global Financial Crisis caused the oil price to fall to $US27 per barrel in February 2016. The oil price has now recovered to $US60-70 per barrel and the Shale Oil Revolution continues with US oil production reaching 11 million bopd in 2018 making it the largest oil producing country in the world – simply amazing!

In Australia, there was a concerted campaign by the Greenies to discredit fraccing. This was compounded by the anti oil industry film "Gasland" which was later totally discredited by further studies. Given that fraccing has been used in the USA for over seventy years and in Australia for over fifty years, this campaign has no scientific validity. In Australia fraccing was first used in the Northern Territory in 1967 and has been used in the Cooper Basin in South Australia, on Barrow Island in Western Australia (a Class A Reserve) and for coal seam gas in Queensland. There have been scientific inquiries into fraccing by the NSW, Victoria, Northern Territory and Western Australian governments. All four of these studies have concluded that "if fraccing is properly regulated, it is completely safe."

In June 2012, a community meeting was organised by the National Party and held in Gingin. Grant Woodhams, the Member for Moore gave the introductions and presentations were made by Bill Tinapple for DMP, Stedman Ellis for APPEA, Craig Marshall for Empire Oil, Peter Stone for CSIRO and Phil Gardiner for the National Party. Terry and I also attended. I had prepared a presentation on the fraccing technique and how the shallow aquifers were protected by three layers of casing but Craig gave a presentation on Empire's Red Gully project. Fortunately the video by APPEA showed how the fraccing technique was carried out and how shallow aquifers were protected. The main issues raised by the community were health hazards from the chemicals used, corrosion of casing protecting aquifers, the ability of the DMP to regulate fraccing, water consumption needed for fraccing and the general lack of community consultation. At the end of the meeting, the community was still divided with those opposing fraccing claiming that the DMP was not independent.

One of the recommendations was that an independent organisation, like the CSIRO, prepare a study on regulating fraccing and this might have more credibility with the local communities.

There was also a Shale Gas Conference to be held in Adelaide in August 2012. Although I had studied technical papers on the Barnett Shale, my knowledge of shale gas was meagre. I attended the conference but the papers were disappointing and demonstrated to me a worrying lack of understanding of the real shale gas potential of Australia's sedimentary basins. It appeared to me that the main reason why shale gas was so successful in the USA was their very low drilling costs, widespread oilfield intrastructure and access to large quantities of fresh water. It would be very difficult to make Shale Gas work in Australia with our high drilling costs and remote, isolated and often desolate basins.

Initially, the Red Gully development went reasonably smoothly. The Gas Sales Agreement with Alcoa was signed in October. Applications for the two production licences were lodged with the DMP in November and granted in April 2012 subject to negotiating native title agreements. The shipping agreement with the DBP was signed in February 2012. Application for the pipeline licence to take the gas from the gas plant to the Dampier to Bunbury Pipeline ("DBP") was approved by the DMP and granted as PL 96 in July 2012.

Soon after, Wharf Resources stopped paying their cash calls. Following the global financial crisis, raising equity funds in London for speculative companies had virtually dried up. Empire Oil tried to do the right thing by Wharf and offered to buy them out. Alas, they wanted an unrealistic price for their 10% interest in EP-389 and the Red Gully project. Empire Oil, in its capacity as operator for the EP-389 joint venture and according to the JOA, put them in default. After the 45 day default period, they still hadn't paid and, under the terms of the JOA, they were deemed to have withdrawn from the joint venture on 9 April, 2013.

To make matters worse, Empire Oil was then informed by BP that it had problems with the Red Gully condensate. The high content of LPG's in the condensate caused it to be highly volatile and unsafe to process through the Kwinana Refinery. Empire Oil had to modify the gas plant to remove the LPG's from the condensate and put them in the gas stream down the

DBP. This caused significant cost over-runs and delays to the building of the gas plant. The final cost ended up being $34 million and Empire Oil's share of the extra $9 million had to be raised on the equity market **(Fig 62)**.

Figure 62
Red Gully Processing Facility

All this had a number of consequences. Firstly Empire Oil needed to pay for most of Wharf's share of the ongoing costs and needed yet another capital raising. This Empire Oil did via a 1 for 9 rights issue. Secondly the DMP refused to register the transfers of Wharf's interest in EP-389 to Empire Oil and ERM Gas despite the fact that they had registered the JOA. Empire Oil and ERM Gas then had to take the DMP to court to prove that Wharf had withdrawn from the permit as set out in the JOA. Thirdly, BP didn't want to be involved in all this and was going to terminate negotiations on the condensate sales agreement. At my suggestion, Empire Oil set up a trust account to receive Wharf's share of the condensate until the outcome of the court case was resolved. Fortunately BP agreed to this and negotiations continued.

In September 2012, Empire Oil moved office from 154 Hampden Road, Nedlands to 229 Stirling Highway, Claremont. The new offices were larger so we could accommodate the extra staff for the development. There was also great area in the basement with lots of room to store all our technical reports and files.

In January 2013, Mo and I travelled to Newcastle for Chris and Tegan's wedding. Bob flew out from the UK to Sydney and met up with Alex. Mo and I arrived the following day and we all had dinner in Chinatown. Next day we all caught the train to Newcastle and stayed at the Honeysuckle Apartments. There we met up with the WA contingent Carol & Mark Jackson, Graham & Evie White, Andy & Helen Svalbe and Eugene & Joyce Urschell. Mike and Carol Wiltshire joined us later for the wedding. I hired a coach for the next day to tour the Hunter Valley and Graham kindly offered to do the driving. The next day turned out to be a stinker with the temperature rising to 42°C. Our tour consisted of rushing from the airconditioned coach to the airconditioned wineries but despite the heat we all had a great time. The coach had a GPS with a female voice which we knicknamed Lucy. Lucy must have also suffered from the heat as she kept trying to take us off the main freeway from Newcastle to Cessnock. That evening we all went to a nearby restaurant but alas the airconditioning had broken down so lots of beer and wine was drunk. Half way through the evening, a storm hit Newcastle and cooled everything down. Fortunately the next day was fine, sunny and cool and the wedding, reception and speeches all went off perfectly. The following day it poured with rain all day long and I told Chris and Tegan that their marriage had been truely blessed by the gods. After everyone had departed, Mo, Bob and I drove up to Bowraville to visit Mike Voelker on his farm. He and Bob got on very well and had many war stories to tell. It also gave Bob a chance to see the east coast of Australia and he was surprised at how green and lush it was.

LAUNER 2D SEISMIC SURVEY

As discussed earlier, the complex and time consuming approval processes of the DMP were now in danger of discouraging petroleum exploration in Western Australia. And things did get worse.

Empire Oil applied to conduct the Launer Seismic Survey in November 2011 but this turned out to be a lot more difficult than for the Garibaldi survey. The Launer survey, like the Garibaldi survey, was located on privately owned land used for cultivating wheat fields **(Fig 63)** and there should have been no major environmental issues. The survey was only

60 kilometres of 2D seismic to be acquired about 20 kilometres west of Coorow.

Figure 63
Launer Seismic Survey

The environmental management plan (EMP) was started in August 2011 and submitted to DMP in January 2012. In March-April 2012, Empire Oil received a series of questions from the DMP and a second version of the EMP was submitted in July 2012. More questions were received from the DMP in August and a third version of the EMP was submitted in September. Yet more questions were received in October and again in November. By this stage, Empire Oil's original application to conduct the survey had expired and we had to resubmit our application to conduct the survey. At the end of November, the DMP requested details of our negotiations with the various landowners. Empire Oil refused this request and I had a meeting with officers of the DMP to explain to them that their request would violate the landowners' rights to privacy and open Empire Oil up to potential serious law suits and damages. I also had to bite my tongue and not ask them "what the hell has this got to do with the environment". In January 2013, Empire Oil received the seventh set of questions from the DMP and submitted yet another version of the EMP. Eventually, after 14 months and $180,000 of consultant and management expenses, the DMP approved the conducting of the Launer Seismic Survey. The survey commenced on 28th March and was completed on 3rd April – after 14 months going through the DMP approval process the survey took exactly seven days to acquire – what a joke!!

One fascinating aspect of the Launer survey was obtaining landowner approval on one particular property. This was owned by Carbon Fund who grew trees to collect carbon credits which it then sold onto companies who needed them to offset their own emissions. To get the carbon credits they had to go through the most stringent conditions set down by the Federal government including the requirement of having to calculate all the carbon dioxide generated by their operations. This even extended to the managers travelling to and from the head office in Perth. To gain access to the property, Empire had to agree to estimate the emissions we would generate during the seismic survey. Following the survey, we gave them our estimated carbon dioxide emissions generated during the survey. The Carbon Fund managers must have been very patient people indeed to put up with all that tortuous bureaucracy!

Interpretation of the Launer survey confirmed the Launer and Winchester prospects. The Launer Prospect was a faulted anticline with its main objectives being the sandstones of the Jurassic Cattamarra Coal Measures. The Winchester Prospect was a tilted fault block at the top Permian but also had fault independent closure at the top of the Cattamarra Coal Measures. Empire Oil applied to renew EP-430 in September 2013.

After difficult and protracted negotiations, the EP-389 joint venture eventually bought the block of land covering the Red Gully and Gingin West fields for $1.325 million. A new company, Cattamarra Farms Pty Ltd was formed with 10,000 shares divided into 6,875 held by Empire Oil, 2,125 by ERMG and 1,000 by Wharf. The EP-389 joint venture signed an access agreement with Cattamarra Farms and we were now masters of our own destiny – or so we thought!

At the May board meeting, the company finances were discussed but thanks to the ponderous DMP approval process, all our exploration commitments had been deferred. Empire Oil was in good shape to start production and receive cash flow before the budget became too tight.

Also in May, the 2013 APPEA Conference was held in Brisbane. I did not present any prospects at the PESA Deal Day as by now, I had already presented all our drillable prospects. Empire Oil needed a basin-wide-farmout and I intended to promote this in our booth during the conference. Mo didn't attend but met me after the conference. We hired a car and made our way leisurely down to Newcastle to visit Chris and Tegan.

In July, I attended a presentation by the ASX on their new requirements for reserves reporting. Most of the presentation was by RISC who outlined the new definition of reserves, contingent resources and potential resources as per the SPE guidelines. At last agreement had been reached and from now on, hopefully there would now be consistency of reserves reporting on the ASX. Empire Oil needed an independent reserves report but in my opinion, we needed to perforate Red Gully-1 B Sand first. Most of our reserves were in the B Sand and Empire Oil needed to prove production from this sand before the 1P (proved) reserves could be independently verified.

In August, I had a very interesting meeting with the British Oxygen Company. They had heard of the high LPG content in the Red Gully gas field and were interested in purchasing the LPG. Subject to the economics, they even offered to build and pay for a special plant to remove the LPG's. After the meeting, I calculated the number of tonnes of LPG we were sending down the DBP but alas the volume wasn't sufficient to justify the cost of building a new plant. We needed to find more gas/condensate reserves in the vicinity of the Red Gully facilities. If we could, the LPG production could be a third source of significant revenue to the joint venture.

After a long wait of seven years, the next West Australian Basins Symposium was held in August at the Convention Centre. A lot had happened in those seven years and it was a terrific conference. There were excellent papers on the Macedon and Pyrenees members in the Exmouth Sub-basin – if only we had that quality seismic in the Perth Basin! There was another great paper on Triassic carbonate build-ups on the outer Carnarvon Basin – a potential new petroleum system. Onshore, there were a lot of interesting papers on the Canning Basin but none on the onshore Perth Basin. It certainly was the "black sheep" of producing basins in Australia. That was soon to change.

Rough Range Oil acquired the interests of AOC and Black Fire Energy in EP-435 thereby increasing its interest to 80%. In October, 2013, the DMP finally offered to grant the renewal of EP-359. It only took three years and 22 days from the date of our application to be offered the renewal. I sent an e-mail notifying the EP-359 joint venture companies of the offer of renewal and the acceptance forms but I also noted "hopefully you are all still alive"!

During September, I completed my study of the Red Gully B Sand. I divided the sand into five units based on environment of deposition and quality of reservoir. I interpreted the B Sand as a lake delta complex with the upper unit being a mouth bar environment; the middle units being turbidites and the lower units being a complex of two mouth bars.

Basin-wide Farmout

In mid 2012, I had several meetings with Hong Sit who was now with MI Energy, the largest privately owned petroleum company in China. I hadn't seen Hong for decades – not since my Pancontinental days. MI Energy wanted to get into Australia and was interested in the Perth Basin. I gave him a presentation on the various petroleum systems in the basin and a summary of the prospects in each of Empire's permits. MI Energy was also interested in the unconventional potential in the basin. The US Department of Energy had estimated that the unconventional potential of the Perth Basin was 45 TCF gas but this was a very preliminary assessment and in my opinion, rather optimistic. Due to the lack of core data and source rock analyses in most of the wells, I had to confess that the unconventional potential of the Jurassic was not well understood at all. Empire Oil had only recently commissioned RPS to evaluate the unconventional potential of the Jurassic sediments in EP-389 but they had only just started the study and were still gathering what data there was. I told Hong that currently, all of Empire Oil's prospects were conventional but that while drilling these prospects, the necessary data could be collected to evaluate the unconventional potential as a continuing and ongoing study. In addition, the basin had excellent infrastructure with the two natural gas pipelines running down its axis so that the economics would be superior to those in the more remote basins in northern Australia. He returned to China with my various reports on Empire Oil's Perth Basin permits for further evaluation.

Soon after the meeting with MI Energy, I had a visit from a company called Dyas BV Exploration and Production. Dyas was a large Dutch investment company and was looking for oil and gas opportunities in Australia. So again I gave them a presentation on Empire Oil's Perth Basin permits and left them with a number of technical reports for evaluation back in the Netherlands.

In July 2012, Empire Oil received a proposal from Argonaut investment brokers to raise funds for a basin-wide farmout of Empire's acreage in the Perth Basin without diluting its shareholders. This involved selling an interest in Empire Oil Company (WA) Limited, the company which held all the permit interests in the Perth Basin. This was a brilliant idea but would have been administratively cumbersome and costly. Because of Empire Oil's 68.75% interest in EP-389, I developed the concept that if a farminee was prepared to commit to a significant exploration program over Empire Oil's permits, then it would be prepared to allow that farminee to earn equity in the Red Gully Project. I hadn't worked out the numbers yet but I put the concept to the following board meeting.

The Basin-wide Farmout was discussed at the following board meetings and in March 2013, Macquarie Capital was appointed to manage the Basin-wide Farmout. I prepared a draft technical presentation and with the assistance of Macquarie, this was expanded into a formal Management Presentation. From their world wide contacts, Macquarie obtained expressions of interest from a number of companies. I was somewhat disappointed in this list as there were only a few overseas companies and only two Asian companies. From July to October, I gave technical presentations to all twelve companies on the Macquarie list.

On 24th October, bids closed for the Basin-wide Farmout. By this stage, Empire Oil was under attack by ERM Power who had lodged a 249D Notice to replace the board. As expected, there were no bids. No company was going to farm into a company whose senior management may soon be replaced.

WANNAMAL 3D SEISMIC SURVEY

After the gas condensate discoveries at Gingin West-1 and Red Gully-1, the EP-389 Joint Venture needed to extend the Gingin West 3D Seismic Survey to the east and north to cover the Gingin-1 and Gingin-2 gas discoveries made by Wapet in the mid 1960's. This meant that we had to acquire seismic in the Boonanarring Reserve where the Department of Environment and Conservation (DEC) wouldn't allow entry of seismic vibrator trucks. Our seismic consultant, Terry Grocke, recommended the use of helicopter supported shot hole seismic surveying as he had experience with this in New Zealand's Taranaki Basin where there were many small farms and national parks.

This involved no vehicles entering the reserve with the drill rigs to drill the shot holes being lowered by helicopter into the reserve. We knew this was going to be more expensive than vibroseis but we also knew from the existing 2D seismic that the Gingin-2 gas accumulation extended under the reserve. If we could acquire the 3D seismic and define the Gingin-2 structure, we didn't have to enter the reserve to drill the well. With modern drilling technology, we could locate the well on our own land and drill a deviated hole to the east under the reserve to intersect any gas bearing sands in the structure.

So in August, 2011, Empire Oil commenced the environmental management plan to acquire the Wannamal 3D Seismic Survey part of which was going to be in the Boonanarring Reserve. During November-December, Empire Oil carried out a Botanical Survey to clear each of the ~ 400 shot holes in the reserve to ensure that the drill rigs, when lowered into the reserve, didn't impact on any declared rare flora. During 2012, Terry and I and the DEC spent considerable time and many meetings to develop an EMP for the Wannamal 3D Seismic Survey which was acceptable to the DEC and the EPA. Then, during November-December, 2012, Empire Oil carried out a Dieback Survey in the Boonanarring Reserve so that the seismic survey could be conducted without the risk of spreading dieback around the reserve. Fortunately, no dieback was discovered in the reserve.

During July 2012, Empire Oil acquired the Wannamal Experimental Shot Hole Test on private land to the north of the Boonanarring Reserve. This survey was designed to determine the optimum charge size and depth of drilling to generate the best quality seismic. The technology of explosives had improved since the dynamite days and the new explosives burned faster and generated a sharper sound wave. I went up to vist the survey and observed that as each shot was fired, there was no visible expression at the surface and only a dull "thump" under your feet.

In January, 2013, Empire Oil submitted an application to the DMP to conduct the Wannamal 3D Seismic Survey together with an EMP which had already been approved by the DEC and EPA. The DEC approved EMP required Empire Oil to conduct the survey before June and the winter rains potentially increasing the spread of dieback. We also needed to avoid the hot summer months with the danger of bushfires in the reserve. Therefore, we had a very narrow time window of April-May to acquire the survey and stressed this point to the DMP in the hope of having a timely approval. Despite the need for the survey to be started in April, the DMP took a further four months to approve the EMP which had already been approved by the DEC and EPA. One of the reasons for this delay was that the officer in charge of our project went on a course for three weeks and then on leave for three weeks and, despite the obvious urgency, the project was not handed over to another officer to complete. I was completely disgusted by this total lack of urgency at the DMP and worst still that nobody in the state government seemed to care that this was discouraging exploration in the basin.

Eventually, on 1 May, 2013, Empire finally received approval to conduct the Wannamal 3D Seismic Survey in the private land only with the final unconditional approval received on 10 May. The delay in receiving these approvals from the DMP cost the joint venture ~ $150,000 in standby rates. And it could have been a lot worse! The survey wasn't completed until mid June **(Fig 64)** and fortunately for us it didn't rain otherwise the survey could have been suspended and delayed for a year at enormous cost to the joint venture. We all understood that the Boonanarring Reserve is an environmentally sensitive area and expected that the approval process for the Wannamal survey would be extensive. However, having spent an entire year getting the approval of the DEC and EPA for the survey, we did not expect the DMP to then take a further four months particularly as they were aware of the DEC time constraints for acquiring the survey.

From August, 2011 to May 2013, the joint venture spent an outrageous total of $800,000 to obtain the approval to acquire the Wannamal 3D Seismic Survey. This might be manageable for companies the size of BHP and Woodside, but for small companies struggling to bring much needed gas reserves to the domestic market it was money we could ill afford.

During April, 2012, it came to our attention that a few of Empire Oil's shareholders were saying some very nasty things on the Hot Copper website. This included a lot of misinformation about the company plus defamatory statements about the board members. Initially we ignored this uninformed chatter but by July, this activity may have been affecting

Figure 64
Helicopter supported seismic survey in the Boonanarring Reserve

Empire Oil's share price. Following the Red Gully discoveries, the share price rose to 4 cents per share and, as expected, had settled down to about 2 cents per share by early 2012. But from April 2012, the share price fell steadily to 1.1 cents by October. This was not due to the oil price which bounced around between $70 and $100 per barrel. Empire Oil had over 12,000 shareholders and we began to get phone calls from some of those who were concerned by the misinformation on the company and allegations against the board. Alcoa also expressed concerns and wanted to know what was going on. Empire Oil had received the first tranche of $5 million from Alcoa in January and the second tranche of $15 million was waiting for certain milestones to be met. These milestones were also set out in the Gas Sales Agreement ("GSA") and due to delays in regulatory approvals Empire Oil had requested an extension. Therefore at any time, Alcoa could cancel the GSA. The company had also announced a Share Purchase Plan to fund ongoing exploration commitments and this was being jeopardised by the falling share price. The action of these few shareholders was potentially threatening the funding of the gas plant and the very survival of the company. We had to do something. After receiving legal advice, we were informed that Empire Oil could not sue its own shareholders but we, as individuals, could sue those dissident shareholders who had defamed us.

In August, Empire Oil received a Section 249D Notice from Mr Edward Smith to replace the existing board with him, Mr Crawford and Mr Downey. This timing of this notice was a total disaster and had the potential to cause the cancellation of the GSA, the failure of the share purchase plan and the eventual demise of the company. It definitely did take up a lot of management time and expense to refute the accusations in the notice and contact all the shareholders to vote against Mr Smith and his associates. The General Meeting was held in October 2012 and the existing board members were returned with a huge majority of 90%. It was a relief to win but it was also very satisfying that we had received such an overwhelming vote of confidence from the vast majority of our shareholders. Pleasingly, Empire's share price also rose to 1.7 cents per share in October. We thought this was the end of the matter and went back to what we were being paid for – running the company!

RESIGNATION FROM EMPIRE OIL

In July 2013, ERM Power notified Empire Oil that it was a significant shareholder having a 10% interest in the company. On 27 September 2013, ERM Power served a Section 149D Notice on Empire Oil & Gas to remove all the board members except me. I was initially stunned and then very annoyed. The last notice by Mr Smith had taken up so much management time as well as being very expensive. This notice was far more serious than the previous one. We knew that ERM Power had acquired 10% of the shares in the company and also had a lot of support from major shareholders in Queensland. We held a board meeting on 1st October to discuss the notice and develop strategies to fight and defend the current board. As I was tied up with Macquaries for the Basin-wide Farmout and

Craig was also flat out with the production facilities and planning the shallow drilling program, it was decided to appoint a professional public relations consultant to manage the defence of the current board.

By early November it was clear from the proxies being received, that ERM Power would win and get control of the board. The board decided that, to avoid further expense to the company, they would resign in favour of the ERM Power nominees. Now I had another agonising decision to make. Do I stay on as a director of the new Empire and be disloyal to Craig and Neil who I had worked with for 10 years? Or do I resign as well and possibly end up being disloyal to Empire's 12,000 shareholders? As ERM Power had no experience in running a petroleum exploration company, I decided to stay on to provide continuity for the new board. I had no illusions that they would keep me on more than about six months but that would have been sufficient to hand over to a new technical manager and fulfill my director's duties to the shareholders.

At the board meeting on 22 November, the Deed of Settlement and Release was signed. I signed the deed twice, once on behalf of myself and once on behalf of Empire Oil. Craig, Neil and Jeff had effectively resigned from the board but Mr Ianello and Mr Heading had only been appointed subject to shareholder approval at the AGM on 26 November. So for three business days, I was the only legal director of Empire Oil. The company secretary countersigned the Deed of Settlement and Release for Empire Oil.

The Annual General Meeting was a very short affair. It was adjoirned to 2pm to coincide with the Extraordinary General Meeting. At the AGM, the shareholders ratified the appointment of Mr Ianello and Mr Heading as directors of the company. The Extraordinary General Meeting was then held following the appointment of the new board members. This meeting gave the dissident shareholders the opportunity to tirade on about the old board members, including me. One shareholder did however ask me where my loyalties laid now that the old board had gone. At the invitation of the Chairman, I got up and stated that my loyalties were with the old board but that I had decided to stay on to provide continuity of management as I considered this to be in the best interests of all the Empire Oil shareholders – and I meant every word of it. To my surprise, most of the shareholders clapped. It made me feel very good and confirmed in my mind that I had, at least at that time, made the correct decision.

The first board meeting with the new directors was held on 27 November at the offices of Clayton Utz. At this meeting, I advised the board of the necessity to urgently appoint a Managing Director to replace Craig Marshall. The new board appointed Kent Quinlan, an ERM Power manager, as interim CEO of Empire but didn't appoint an industy experienced CEO until May 2014, some five months later. The next board meeting was on 13 December and although there would be an independent review of Empire Oil's acreage, I took the opportunity to give the new board my assessment and rating of the various permits.

Mo, Alex, Shannon and I then took off on Saturday to Adelaide to meet up with Chris and Tegan for my 70th birthday. We all stayed at the Grand Chancellor and the following day, drove up to the Novotel Resort in the Barossa Valley. Chris, Alex and I had a game of golf at the resort on Sunday. The golf was crap but nobody cared - the day was fine, we were almost the only ones on the course and we were back at the resort in time for some beers and a late lunch. Perrrfect!!! On the Monday, I organised a tour of the vineyards in a van just for the six of us. Our driver really looked after us and arranged private tastings at two of the vineyards. The only downside was that Tegan was not well and our driver ran her back to the resort while we were having lunch. That night we had a fabulous dinner in the main restaurant for my birthday. The whole trip was absolutely fabulous.

Back in Perth, there was an Operating Committee Meeting on EP-437 on 18 December. This meeting was to finalise the offer to the landowner for access to drill Dunnart-2.

On 19 December I had a teleconference conference with Mr Ianello and Mr Heading in Brisbane. They told me that the Empire Oil board had decided to sue the past directors (except Jeff) for the costs of the defamation legal costs and gave me until the following morning to agree to pay my share. This was like a kick in the guts. Not only was it not in the spirit of the Settlement Deed but after sticking my neck out to provide continuity for the handover to the new board, this is how I was to be repaid. I immediately resigned as director of Empire Oil and all its subsidiaries. I truly did not understand. They had won – and won at very little cost. There really wasn't any need for them "to go around after the battle bayoneting the wounded". I left the office to go to the PESA Xmas Party as it was a good opportunity to get pissed. But I just wasn't in the mood so I went home early. I was utterly disgusted and so very depressed at the whole affair.

After resigning as a director of Empire Oil, I told the Chairman that if they no longer wanted my services, I would agree to the termination my contract. To my utter amazement, they wanted me to stay on and complete my contract. How can any company want to continue employing someone they are suing! Fortunately, my contract was expiring near the end of January. The Empire Oil Xmas Party that day was held in the office and was very subdued. I had a couple of beers, wished everyone a Merry Xmas and went home. After a very stressful four months and the prospect of having to work for a company that was suing me, I was really looking forward to the Xmas New Year break.

Monday 6 January was the first day back after the break. I had been conditioning myself to do my technical work and say nothing that could be used against me in my lawsuit with the new board. That afternoon, I accompanied Kent down to the DMP for a meeting with the new director of the Petroleum Division, Jeff Haworth. It felt very funny introducing the new CEO of a company that was suing me to the new director of the Petroleum Division. But the meeting went well and Kent flagged that Empire Oil would be applying for some extensions to its various permits while the new management got up to speed.

The 20th January was my last day at Empire Oil. I had been with the company for ten years, the longest time with any company during my career. I was feeling very low that morning but I got a big hug, kiss and a squeeze from Mo before I left for the office one last time. I was tidying up my office and filing the last of my projects when I got a phone call from Chris. He had just phoned his mum and was now phoning me with the great news that Mo and I were going to be grandparents at last. That certainly cheered me up and made what could have been a very depressing day into one of the better days of my life. We chatted for a bit and I told him that I couldn't have received better news on my last day at Empire Oil. I then had a visit from Geoff, Ali, Libby, Karen and Christina wishing me farewell and all the best for the future. I thanked them all and told them the good news I had just received from Chris. It turned a sad moment into a happy one and there were lots of hugs before I left the office for the last time.

When I got home, all the chaos of the past two years as a director of Empire Oil finally sunk in and it was as if a huge weight had suddenly been lifted off my shoulders. A new and much more relaxing life now awaited me. Mo and I took off for a well earned break on Rottnest for a week.

Rhone Trip

Prior to resigning from Empire Oil, we'd booked a holiday cruise up the Rhone River with APT for June 2014. We flew from Perth to Dubai at some ungodly hour of the morning arriving early afternoon. We stayed overnight at the Grand Hyatt Hotel near the airport to break up the journey and get a good night's rest. We had a very nice breakfast in a restaurant set in a beautiful and tranquil tropical garden.

After this, we headed for the airport and flew from Dubai to Nice where we stayed at the Hotel Suisse. Although the room was small, we had a great view from our balcony looking down the Promenade des Anglais and the sweeping bay and beach. Nice was a very picturesque city with lovely churches and parklands. It also has an interesting history being founded by the Greeks in 350 BC and becoming a busy port during Roman times. During the Middle Ages, it flip-flopped between being controlled by either the French or the Italians but finally became part of France in 1860 after the Treaty of Turin.

From Nice we returned to the airport, joined the APT Tour and drove to Monte Carlo. We stayed at the very palacial Fairmont Hotel overlooking the Mediterranian Sea. Monte Carlo covers a very small area sandwiched between the sea and huge cliffs of Cretaceous limestones. It is very crowded with constant traffic jams. Even the casino is small and before you lose your money, they actually charge you to go inside! For all its wealth, I'd rather live in Nice.

From Monte Carlo we toured the surrounding countryside including a visit to the beautiful Rothschild Gardens. They were constructed from 1905-1912 by Baronesse Beartrice de Rothschild and donated to France upon her death in 1934.

After Monte Carlo, we returned to Nice and after a walking tour, continued on to Arles where we joined the MS Amadagio for our cruise up the Rhone.

Arles is a very old town being founded by the Greeks in the 6th century BC and becoming part of the Roman Empire in the first century BC. The Romans built a large amphitheatre there in 90 AD and it is still standing today. From Arles we motored north to Avignon, another old town being founded by the Greeks and becoming an important town during the Roman Empire. The old Roman walls still partly surround the town. However its most significant claim to fame was being the capital for the popes during a period of unrest in Rome from 1309 to 1403. The Palace of the Popes was built from 1335-1364.

North of Avignon, we visited the Pont du Gard **(Fig 65)**, the remains of the Roman Nimes Aquiduct which was built during the first century AD and originally ran 50 kilometres from the Alzon River to Nimes. It continued to function some hundred years after the fall of the Roman Empire but eventually became clogged with debris due to lack of maintenance during the dark ages.

Figure 65
Pont du Gard was built to last

As we continued north, we visited the picturesque towns of Uzes, Viviers and Grignan. On the way we passed the Cruas Meysse nuclear power plant, a very impressive structure on the banks of the Rhone **(Fig 66)**. The French are very big on nuclear power and get 70% of their electricity generation from this source without any major issues. Further north, we passed another nuclear power plant at St Maurice l'Exil. Simply amazing!

Figure 66
Yes this is for real!

Our next stop was the tourist town of Tournon with its impressive castle. We visited the Eden Gardens which although it was a hard climb up lots of stairs, gave us panoramic views of the town. Eventually we arrived in Lyon, the second largest city in France. We spent a day touring the city but were most impressed by the paintings on the walls of various buildings in the city. These were so lifelike that tourists actually start talking to the figures! It has to be a great way to avoid graffiti.

From Lyon, we had planned to catch the train to Geneva where we planned to celebrate our anniversary on 12th June. When we arrived at the train station we were told that there was a train strike and our train had been cancelled. We had to get on a bus to take us from the airport to Bourg-en-Bresse to catch the "last train out of Paris" to Geneva. The train station at Bourg-en-Bresse was under construction as it was to be on the route for the Tour de France in July so getting our luggage onto the correct platform was a major operation.

By the time we got to Geneva it was very late but we managed to have a lovely meal and evening at the last restaurant open at the Hotel Beau Rivage where we were staying. We managed to celebrate our anniversary in style. The Beau Rivage was definitely the most expensive hotel we've ever stayed at - $1300 per night but it was a very beautiful hotel and we had a view of Lake Geneva from our balcony. It even had cherubs painted on the ceiling. Not only did we have a view of Lake Geneva but when the cloud cleared we could see Mont Blanc in the distance. We only stayed one night and checked out the next morning to do a bit of sightseeing. There was a park next to the hotel with a fine statue of Charles II and a fearsome looking gryphon. We then took a leasurely cruise on Lake Geneva taking in some of the most expensive real estate in Europe before flying out to London that evening.

The next morning, we hired a car for our planned holiday in Devon and Cornwall. I had ordered a Hyundai i30 but instead they gave us a brand new Mercedes 300. While this was a nice surprise, it was not the right car to go driving down the narrow lanes in Devon and Cornwall and even worse, it was totally computerised – you couldn't even release the hand brake – there wasn't one. Mo was doing all the driving so she had a steep learning curve to get the hang of the vehicle. I managed to find the manual but it was over 200 pages long. To make matters worse, it started to rain and there we couldn't figure out how to start the windscreen wipers. Fortunately they came on by themselves. Our first stop was to drive down to Bournmouth to visit Rosemary's grave and place some flowers and Australian flags there.

From there, we drove west to the "Jurassic Coast" of Dorset and Devon staying in the Eyre Court Hotel in Seaton. I wanted to stay in the next town called Beer but it was booked out for that night. I guess I'm not the only tourist who wants to stay in a town called Beer! After we checked in, there was a knock on the door and there were our old friends from Broxbourne Eve and Brian. On their last visit to Perth, we had told them that we were going on holiday in southwest England and were staying at Seaton. We had a very pleasant evening catching up with all the gossip and enjoying the English twilight. The next morning Eve and Brian returned home and we did a short tour of the town with its broad pebble beach and outcropping Triassic red beds and Cretaceous chalk in the background (**Fig 67**). From Seaton we drove west and just before Torquay we stopped at Babbacombe for lunch. This was a very beautiful spot with lovely gardens and a panoramic view of the bay with outcropping Permian red beds and Devonian limestones **(Fig 68)**. From there, we drove along the coast and at Dartmouth we crossed the River Dart by ferry.

Figure 67
The pebble beach at Seaton

We then headed for the little town of Hope Cove but once we left the main highway, the road into the town was very narrow and challenging. We had to pass a double decker bus and Mo's driving skills were severely tested but when we got to the Sand Pebbles Hotel, there wasn't a scratch on the Merc! Once we had checked in, we walked down the hill to the town where there was a small beach and a sign to the gallery announcing "Lots of lush stuff to see & buy". We had a very pleasant lunch at the local (and only) pub overlooking the cove.

The next day we headed back to the main highway and headed west into Cornwall. We visited the Eden Project near the town of St Austell. This

Figure 68
Great view of the bay and cliffs

project consisted of two main enclosures located in a reclaimed kaolinite quarry. The project houses thousands of plant species from all over the world and as you walk through the enclosures, the plants are divided into the main continents of Europe, Asia, North and South America, Africa and Australia. From there, we drove southwest to Lizard Point and the tiny town of Mullion where we stayed at the Old Inn. This inn was built in the 16th century and has been preserved as a typical Cornish village pub.

From Mullion, we drove west to Land's End. The day was absolutely beautiful and you don't see the Atlantic Ocean looking like a mill pond very often. The whole area is dominated by outcropping granites. From Land's End, we headed north past the St Ives and Newquay to Padstow.

In Padstow, we stayed at the Metropole Hotel which was a large, old hotel but not as comfortable as the small, quaint hotels at Seaton, Hope Cove and Mullion. It overlooked a large estuary which was fine when the tide was in but not so pretty when the tide was out. From Padstow, we followed the coast past the historic town of Bude and around to Barnstaple.

From there we drove north to Woolascombe on the coast of the Bristol

Channel. On the way, we detoured into the seaside town of Infracombe where we had lunch. It was a very busy town with lots of locals and tourists enjoying the fabulous weather so it took some time to find a parking spot. In Woolascome we spoilt ourselves and stayed at a fabulous resort, the Woolascombe Bay Hotel with a swimming pool and large grassy picnic area running down to the sea.

The next day, we drove west to Bridgewater where we joined the M6 and headed north to Lancashire to meet up with Jean and Jon for dinner at the Midland Hotel in Morecambe. The next morning we toured around Morecambe but the town had declined over the years as the tourist trade had moved to the preferred sunny Mediterranian resorts. From there we drove north to the Lake's District skirting around the south side through Ulverston before going north to Coniston where we stayed at the Watershed Hotel at Coniston. This hotel was very dark on the outside but was very comfortable inside and was located just outside of the town with fabulous views of Lake Coniston. In the late afternoon, we walked down to the edge of the lake with our drinks and nibbles and enjoyed the outstanding scenery.

From the Lake's District, we drove back to Carnforth where Jean and Jon lived. The Lake's District is very popular and we encountered a massive traffic jam as we drove through Windermere. In Carnforth, we stayed at the County Hotel which was very central and quite comfortable. Jean's son Rennie lived in Kendal with his wife Sarah and two children Noah and Maya. It was Noah's birthday on 22nd June and they put on a party for him. So we all drove up to Kendal to join the celebration. Again it was a beautiful day and Rennie told us that Kendal has one of the highest rainfalls in Britain and he was very greatful that we brought the Perth sunshine with us!

After Carnforth, we drove down to Manchester airport and dropped the hire car off at the Thrifty depot. There was a lot of construction at the airport and Thrifty hadn't bothered to tell us that the depot had been moved. We had to go into the Europcar depot to find out where Thrifty was. They had their new phone number and they had to send a car for us to follow back to their new depot. We were not impressed! Fortunately we had allowed plenty of time to get to the airport otherwise we would have missed our flight to Dublin. In Dublin, we stayed at the Beresford Hotel and spent the next day touring the city. We were disappointed in Dublin – it was very dirty and run down. I enjoyed a visit to Trinity College with its magnificent library and a stroll through St Stephens Green.

The next day, we picked up our hire car from Europcar and this time got what we ordered – a Hyundai Getz – just perfect for the narrow roads in Ireland. We had planned to visit the archaeological site of Newgrange which was built circa 3200 BC and was designed so that the entrance is aligned to the rising sun during the winter solstice and sunlight illuminates the inner chamber. When we got there we were told that all the buses were full and the next tour wasn't until the afternoon. Given the fact that their website didn't allow you to book online, I was not impressed and very disappointed. So we continued on our drive across Ireland to the coastal town of Kilkee where I had been on that field trip way back in 1973. The countryside was very green but it was cloudy with the occasional shower of rain. In Kilkee, we stayed at a local pub but the next day, it was pouring with rain so I didn't get the chance to see the Carboniferous outcrops again.

From Kilkee, we drove down to Kilrush and caught the ferry across the Mouth of the Shannon to Tarbert. From there, we drove south through Killarney and Kenmare to Bob's place at Glengarriff.

We spent four pleasant days at his place which, although the accommodation was very basic, was still very picturesque. From there, it was a drive to Cork airport, drop of the car and flights to Heathrow, Dubai and finally back to good old Perth and home.

Empire Oil Writ

The Empire Oil writ was served on me in late December 2013 and in February 2014 we all attended the Supreme Court for the beginning of the proceedings. In January and February we had our first meetings with the lawyers appointed by our director's liability insurance company.

In March 2014, Warrego Energy farmed out their West Erregulla project in EP-469 to Dyas and Mazarine Energy in $40 million deal to earn 80%. The farm-in program included the acquisition of an 80 sq km 3D seismic survey and the drilling of one well in 2015. I had been talking to Dyas during Empire's basin-wide farmout efforts but that ended with the ERM notice to spill the board. But good luck to Warrego – timing is everything and they thoroughly deserved this initial success. After some three years of approval process negotiations with the DMP, Warrego's West Erregulla 3D Seismic Survey was finally acquired in October 2014.

In April, Empire Oil appointed Ken Aitkin as their Chief Executive Officer and in May, they sold their 52 million shares in Key Petroleum for $676,000.

Also in May, the Supreme Court of Western Australia ruled in favour of Empire Oil and ERM Gas in respect of Wharf Resources deemed withdrawal from EP-389. The new interests in EP-389 became Empire Oil 76.39% and ERM Gas 23.61%. Justice was finally done and the Supreme Court confirmed that Wharf Resources had legally defaulted on their joint venture obligations.

In June 2014, Empire Oil successfully perforated and tested the B Sand in Red Gully-1 allowing these reserves to be formally added to those of the D Sand.

We had our first mediation meeting with Empire Oil and their lawyers in July. This did not result in any settlement. So our lawyers started the tedious process of preparing our witness statements. For me, this took fom December 2014 until March 2015 but for Craig, he spent over six months preparing his.

In September 2014, Empire Oil announced that it had purchased ERM Gas' interests in EP-389 including the Red Gully gas condensate project, EP-426, EP-432, EP-416, EP-440 and EP-467 for $16.34 million. After being debt free during the 15 years since listing, Empire Oil now had $15 million debt to ERM Power.

In March 2016, we had our second mediation meeting but again no agreement was reached as the Empire board rejected our revised offer of settlement. The legal process dragged on but eventually the Supreme Court set a trial date for May 2017. I had my annual medical checkup in December 2016 and to my surprise, my blood pressure was higher than normal. I didn't feel stressed but with the impending trial, I obviously was. Eventually, in March 2017, we had our third mediation meeting where a settlement sum was finally agreed. The lawyers drew up the various settlement deeds for execution and they were then submitted to the Supreme Court. In April, the

Supreme Court terminated the proceedings and it was finally over. After three years and four months, I could now get on with my life. I had my next annual medical checkup in November 2017 and to my great relief, my blood pressure had returned to normal.

Consulting

Although now retired, I undertook a number of small consulting jobs after leaving Empire Oil. I really enjoyed these projects as they kept me in the industry and I would like to think that my clients benefitted from my experience in the Perth Basin.

Norwest Energy

In March 2014, I was contacted by Norwest Energy to undertake a quick evaluation of the lower Kockatea Shale in Arrowsmith-2 drilled in EP-413 in the northern Perth Basin. Arrowsmith-2 was drilled by Norwest Energy NL in May-June 2011 as a followup to Arrowsmith-1 drilled by French Petroleum Company (Australia) Pty Ltd in April-June 1965. This well flowed gas at an average rate of 4.0 MMcfg/d from the Permian Carynginia Formation but in 1965, there was no market for natural gas. Arrowsmith-2 encountered good oil and gas shows in the lower part of the Kockatea Shale. Gas shows with peaks up to 200 units continued down throughout the Permian Carynginia Formation, Irwin River Coal Measures and High Cliff Sandstone. The well was logged, cased and suspended for future stimulation. Operations to stimulate and flow back Arrowsmith-2 were conducted from July 2012 to February, 2013. The Fifth (and last) Stage was over the lower part of the Kockatea Shale which flowed at rates up to 100,000 cfg/d with associated oil and condensate. My evaluation was completed in April 2014 and concluded that the oil and gas shows come from a unit of interbedded limestones and shales at the base of the Saprolpelic Interval of the Hovea Member. I recommended that in any future wells drilled in the Arrowsmith area, a core be taken in these interbedded limestones for special analyses to determine their suitability for stimulation.

The APPEA Conference was held in Perth in April 2014. I paid for a one day attendance on the Tuesday and caught the best of what technical papers there were.

Our first grandchild, Georgia Courtney Warris, was born to Chris and Tegan on 22 August 2014. We booked a villa on Rottnest from 16-30 December and they all flew over to Perth and the entire family spent a very pleasant time on the island over the Xmas break. From our villa, we also had a great view of the manta ray which appeared to have taken up residence in Geordie Bay. It did its cruise around the bay about 5pm every evening just as we were having first drinks!

WAITSIA GAS FIELD

Remember the interesting paper at the 2007 APPEA Conference by lead author Darren Ferdinando on the preservation of porosity in deep Lower Permian sandstones. This paper showed that coating of quartz grains by illite/smectite clays can prevent silicification and preserve porosity and permeability. Well in September 2014, AWE drilled Senecio-3 as an appraisal well on their gas discovery in the tight Dongara Sandstone. As well as evaluating the Dongara Sandstone, the well was deepened down to evaluate the Lower Permian Kingia and High Cliff sandstones which turned out to be also gas bearing. The well was tested in March 2015 and flowed at the respectable rate of 12.5 MMcfg/d.

This discovery was followed up with Waitsia-1 in June 2015 which flowed gas from the Kingia Sandstone at 24.7 MMcfg/d and from the High Cliff Sandstone at 25.7 MMcfg/d. Waitsia-2 was drilled in July 2015 and flowed gas from the Kingia Sandstone at 38.5 MMcfg/d. And things only got better! In 2017, Waitsia-3 and Waitsia-4 encountered 150 metres of gas pay in the Kingia Sandstone and which were tested at the staggering rates of 49.5 and 89.6 MMcfg/d respectively.

Recoverable 2P reserves are now estimated to be of the order of 800 PJ and this excludes the Dongara Sandstone gas reserves in the Senecio Field. The field commenced Stage 1 production in 2016 from Waitsia-2 through the Xyris Production Facility with sales of about 10 MMcfg/d to Alinta Energy. Stage 2 of the development was planned to build three additional plants and increase production to 100 TJ/d.

The PESA 50th Anniversary was held at the Parmelia Hotel in July 2016. I was sitting next to Eric Streitberg and he asked me "Bevan, how did we miss Waitsia – the biggest gas field in the basin?" Because, I replied, it defies the three main rules for the basin. Firstly, it's a downthrown fault trap; secondly, it's in Lower Permian sandstones and thirdly it's below 3,000 metres. I didn't have the heart to tell him that "the rocks had been smiling at us" for over 30 years. So Waitsia is a classical lesson for budding petroleum explorers – never accept existing theories for any basin. Do your own evaluations and remember "if you don't drill, you won't find anything!"

Eneabba Gas

Back in May 2014, I had my first meeting with Eneabba Gas now controlled by Ventnor Capital. They asked me to undertake a risk analysis of three projects they were evaluating. The first project was EP-447, EP-488 & EP-489 in the onshore Perth Basin and owned by UIL Energy; the second was EP-432, EP-430 & EP-454 also in the onshore Perth Basin but owned by Empire Oil; and the third was EP-481 and EP-482 in the Merlinleigh Sub-basin and owned by New Standard Energy. I submitted my report in February 2015. It was clear that the Merlinleigh Sub-basin permits had significantly higher risk than the Perth Basin permits and was rejected. Eneabba Gas had preliminary discussions with Empire Oil regarding their permits but they were apparently not interested in farming out their permits at that stage. My recommendations were to farm into UIL Energy's permits and continue negotiating with Empire Oil on their permits.

While I was preparing my report, Eneabba Gas reached agreement with Black Rock Mining Limited (formerly Green Rock Energy Limited) in October 2014 to purchase their 100% interest in application area EPA 90. This area contained the Ocean Hill gas discovery which ironically was an old Barrack Energy prospect that was drilled by Arrow Energy (formerly Barrack Energy) and Sagasco (now Origin Energy) in 1991. This agreement was subject to completion of Native Title Agreements with the Yued and Amangu and which were completed in November 2015.

In April 2015, I prepared a technical report on EP-447 based on a seismic interpretation by Peter Kirk on the existing seismic plus the Badgingarra

Seismic Survey acquired by UIL Energy in 2013. In June, Eneabba Gas purchased GCC Methane Pty Ltd which owned 50% of EP-447. Eneabba Gas was now on its way to acquiring a very respectable acreage position in the onshore Perth Basin.

On a personal note, in April, I attended the funeral of Doug Swift who used to be Empire Oil's operations manager in Exmouth. It was a very sad occasion as he was only 53. He was born in Wittenoon and contracted mesothelioma from the deadly blue asbestos in the town. On a happier note, our second grandchild Samuel James was born to Alex and Shannon on 17 May 2015.

In July, I was awarded the Meritorious Service Award from the Petroleum Exploration Society of Australia. The timing of this award was perfect because as of April, I had been been working in the Australian oil industry for fifty years. I took the opportunity to remind the audience that in 1965 the oil price was only $US1-80 per barrel which is equivalent to about $US15 per barrel in today's adjusted dollars. So be of good cheer, the oil price is actually going up!

Also in July, I had another visit to the hospital – nothing drastic - just overnight. For several years I had been getting nose bleeds, always from the right nostril. These progressively got worse until one evening, I was dining out with Mo when suddenly my nose suddenly gushed blood all over one of my best shirts and I had to rush to the restroom to clean up – all very embarrassing. Soon after, I went to my doctor and got a referral to see a nose specialist Dr Hee whose office was in Hampden Road immediately opposite the old Empire Oil office. After a few visits, he concluded that I needed an operation on my nose to stop the bleeding. It was a quick procedure and I was out of hospital the next morning. Dr Hee did a great job and I haven't had any problems since – touch wood! When I received copies of the bills that had been paid by HBF, I noticed that the one night stay at the hospital was $1500. Strewth – it was cheaper to stay at the Beau Rivage hotel overlooking Lake Geneva!

In August, we flew over east to visit Georgia for her first birthday. We first flew to Melbourne and watched an AFT match between the Cats and Hawks. The next day, we hired a car and set off on the long drive to Newcastle. First stop was Tocumwal a pretty town on the banks of the Murray River. From there we drove north to Forbes and then Dubbo where we visited their open range zoo which was very impressive. Next day we drove across to Newcastle in time for Georgia's first birthday.

Back in Perth, the Good Oil Conference was held in September and Eneabba Gas gave a presentation and also hired a booth in the exhibition area. I assisted in preparing the posters for the booth and manning it during the conference. The oil price was in free fall down to $US40 per barrel from a high of $US115 per barrel in June 2014 so the conference was somewhat subdued. Nevertheless it was good to catch up with many of my friends in the industry.

In February 2016, Eneabba Gas reached agreement with UIL Energy to sell them all the shares in Oceanhill Pty Ltd, the beneficial owner of 100% of EPA 90, and GCC Methane Pty Ltd, the owner of a 50% interest in EP-447. The deal was for 90 million convertable redeemable preference shares and was subject to the grant of EPA 90 and completion of the sale agreement between Black Rock Mining and Oceanhill. UIL Energy now had 100% of EP-447, EP-488 and EP-489 plus 100% of EPA 90 (Ocean Hill) and Application Areas EPA 82, 98 & 99 all in all a very impressive acreage position in the onshore Perth Basin

The Ocean Hill application EPA 90 was awarded as Exploration Permit EP-495 in May 2016. The permit was then transferred in August to Oceanhill Pty Ltd so that it could be sold to UIL Energy.

In June, ERM Power sold its 18.72% shareholding in Empire Oil to Mineral Resources bringing that company's shareholding to 19.36%. Mineral Resources also agreed to sign a new $15.1 million facility with Empire Oil allowing them to refinance the debt obligation to ERM Power maturing in August. Now that ERM Power had no further involvement with Empire Oil, my hopes rose slightly that we may now be able to get a settlement with Empire Oil.

In July, I attended the funeral of Geoff Towner. He was only 66. I had a coffee with him a few months earlier after his first heart attack and he looked fine and on the way to recovery. So it was a big shock when I heard that he was gone. There was a huge rollup at his funeral as testament to his gregarious personality and many friends.

It was Georgia's second birthday in August so we decided to drive over for the celebration. So we loaded up the car and set off first to Coolgardie and then across the Nullabour to Port Augusta. From there we wound our way across to Mildura and onto Newcastle. Being the end of winter, the countryside was very green with large areas of wheat and canola fields. For her birthday, we gave Georgia (amongst other things) a hand made pram which Mo bought in Guildford. She loved it and it was a real hit. On the return journey we went via Broken Hill. I hadn't been there since my PhD days and it was just as busy as ever. It was hard to find any accommodation but we eventually found a nice motel and just got the last room. From there we drove across to Port Augusta and down around the Eyre Peninsula to Streaky Bay, one of our favourite holiday spots. After crossing the Nullabour, we went south to beautiful Esperence for two days and then headed home.

On our return, we decided to sell our apartment and purchase a house in the suburbs. After looking at a number of properties, we eventually found what we were looking for. Most of the new houses were on small blocks of land and had no garden or outside entertaining areas. We bought a house in Doubleview which was large enough for all our stuff and had a large courtyard facing north. The house wasn't perfect and required quite a bit of renovating but the price was heavily discounted compared to other properties in the area. While all this was going on, our third grandchild, Mia Isabell was born to Alex and Shannon on 24 January 2016. The Warris dynasty was really expanding! Settlement on the house occurred at the end of February and on the day we moved in, it was 42°C so it was a very long and exhausting day.

UIL ENERGY

At the end of November 2016, I was contacted by Barnaby Egerton Wharburton of UIL Energy to act as their technical consultant on their various permits in the Perth Basin. The consulting agreement was signed in December. During January and February, I attended a number of meetings with UIL and Aztech, a drilling and testing consulting company. These meetings were to determine the best way to drill a second well on the Ocean Hill-1 gas discovery.

Ocean Hill-1 was drilled by Sagasco Resources in 1991 on a large north-south trending anticlinal structure. The prospect was defined by Barrack Energy but Sagasco became operator when they acquired their 67% interest in EP-320. The well encountered significant gas shows over an 800 metre gross section of the Jurassic Cadda Formation and Cattamarra Coal Measures. A number of open and cased hole tests were undertaken but the best test was only 0.7 MMcfg/d from sands in the Cadda Formation. Analyses of these tests results indicated that the sands were tight but possibly also formation damaged. An independent report by De Golyer and MacNaughton in 2013 estimated a C2 contingent resource of 360 BCF of recoverable gas.

My assessment of the reservoirs in Ocean Hill-1 is that the sands are both silicified and clayey but that the sands in the Cadda Formation have significantly less clay than those in the underlying Cattamarra Coal Measures. I thought that horizontal drilling through these sands may result in better flow rates without the high cost of fraccing. I was keen to see that an Ocean Hill-2 well cut cores in the Cadda gas sands and that the vertical well be suspended for later re-entry. The cores would then be sent to experts in the US for special analyses to see if their permeability could be improved by horizontal drilling. Aztech did not agree with this assessment and recommended that the well be vertically drilled and then immediately stimulated. They did not agree that cores were necessary to design an effective stimulation and test. After the meeting, I discussed my concerns with Barnaby but not being an expert on unconventional reservoirs, I suggested a second opinion. As it turned out, one of the independent directors of UIL was Keith Skipper whom I knew from his days in Australia with Bridge Oil. Barnaby arranged a conference call with Keith and I was delighted that Keith agreed with me that, based on North American experience, cores are essential for the planning of both horizontal and stimulated wells.

UIL appointed Searcher Seismic to reprocess the existing seismic in EP-495. I had several meetings with Mark Ballesteros who was overseeing the reprocessing and interpreting the final seismic data.

The reprocessed seismic was a marked improvement on the old data with better definition of the main horizons and the fault planes. The Ocean Hill structure was similar to previous interpretations and contingent resources did not change significantly. The Ocean Hill South Prospect was more complexly faulted and a 3D seismic survey was therefore recommended prior to selecting a drilling location.

The APPEA Conference was held in Perth from 15-17 May. I didn't attend the conference but gave a presentation on Ocean Hill as the PESA Deal Day.

In August I attended the funeral of Lauchie MacLean who I had worked with in the old BOCAL days. He was 90 so he had a good innings. I had a few beers with Ian Whiteley afterwards.

The WA State election had been held on 11 March and the Labour Party won in a landslide. Mark McGowan became the new premier. In September, he announced the imposition of a statewide moratorium on fraccing until the completion of an independent inquiry chaired by Tom Hatton, Chairman of the EPA. It seemed like every state government had or was going to have an independent inquiry into fraccing. They all could have saved their taxpayers a lot of money by pooling their funds and having just one inquiry! After this announcement, UIL Energy put all exploration on hold until the results of the inquiry were known.

END OF EMPIRE

The Good Oil Conference was held in September 2017 at the Hyatt Hotel, a change from the Esplanade Hotel in Fremantle. On the first day of the conference, Empire Oil announced that they had applied for an extension of the voluntary suspension of their shares. Back in August, Empire Oil had undertaken a static gradient test of Red Gully-1 to provide data for the 2017 reserves assessment. On completion of the test, the well failed to produce at its prior rate. Empire undertook a nitrogen lift to bring the well back on stream but it was unsuccessful. On 14 September, the second day of the conference, Empire received a Notice of Default and Demand and Appointment of Receivers and Managers from Mineral Resources. On 15 September Empire appointed Ferrier Hodgson as administrators of the company. By November Mineral Resources announced that it had acquired all the assets of Empire including the Red Gully Processing Facility and all their exploration tenements. By February 2018 the creditors of Empire had agreed to a 1 for 20 recapitalisation and a placement of 50.75 million shares to Trident Capital Pty Ltd. And that was the end of Empire Oil! I didn't know whether to laugh or cry but I sure as hell hoped that most of our original loyal shareholders got out in time. Of course all the old Empire Oil prospects (North Erregulla, Lockyer Deep, Cooljarloo, Woolka, Black Arrow, Charger, Garibaldi, Launer and Winchester) now owned by Mineral Resources, still remain to be drilled or matured for drilling.

Also in September, Beach Energy announced that it had acquired Lattice Energy, Origin Energy's conventional upstream subsidiary, for $1.6 billion making Beach a $3 billion company. This deal also included a 50% interest in L1 and L2 in the onshore Perth Basin and which contained the giant Waitsia Gas Field. In Decemner 2017, AWE (who owned and operated the other 50% of L1 and L2) announced that a takeover offer had been received from China Energy Reserves and Chemical Group for AWE at 73 cents per share. This was immediately followed by another takeover offer by Mineral Resources at 80 cents per share. In January 2018, Mitsui made a third takeover for AWE at 95 cents per share. This deal was accepted by the board and shareholders and AWE became a wholly owned subsidiary of Mitsui and was delisted in May 2018.

I sold the apartment in 128 Adelaide Terrace that was held in the Warris Family Trust. Although I thought that this was probably the bottom of the real estate market in Perth, I no longer wanted the responsibility of owning a rental property. The apartment was managed by LJ Hooker and I watched the rents go all the way up during the mining boom and then all the way down after the boom. Unfortunately the last tenants did some damage to the unit and left owing a month's rent. Most of this was covered by the rental bond and Landlord Insurance but it was still a stressful time. As it turned out, it wasn't quite the bottom of the real estate market.

On 7 October 2017, our fourth grandchild, Harriet Shay was born to Tegan and Chris. We went over in November flying to Brisbane and then catching the train to Surfers Paradise. After a couple of nights there, we hired a car and drove down the coast and stayed the first night at Yamba one of the best kept secrets on the coast of NSW. From there we headed down to Crescent Head and then on to Newcastle. We spent a couple of days with Chris and Tegan, played with Georgia and Hattie and unloaded (in secret) their Christmas presents. After Newcastle, we got the train down to Sydney, had dinner with some old friends before flying out to Perth the following day.

The PESA Xmas Party was a real blast with a large crowd, lots of beer and wine and at around 5pm I shared a cab home with John Gorter. But there were still lots of people at the bar when we left – it's good to know that the oil industry is still in such good hands.

We spent Christmas morning with Alex, Shannon, Sam & Mia opening presents and sharing the joy of our grandchildren. On Boxing Day we had an open day at our place – the weather was perfect and Mo excelled herself with the catering – as usual.

At the beginning of 2018, the oil exploration industry in Australia was not in good shape. Only one offshore exploration well had been drilled in 2017 despite the fact that offshore drilling rig rates had fallen to one third what they were in 2013. During 2016 and 2017, the oil price had increased from $US27 to $US57 per barrel. Despite this, there were no exploration wells drilling anywhere in Australia for the first three months of 2018. It wasn't until April that Beach Energy drilled Lady Bay-1 in the Cooper Basin and Woodside began drilling Ferrand-1 on the Northwest Shelf. Oil prices continued their rise reaching $US77 per barrel (WTI) in October.

In March, Strike Energy announced that it had farmed into Warrego Energy's EP-469 in the onshore Perth Basin. It would earn a 50% interest in the permit by drilling West Erregulla-2 to a total depth of 4925m. This well will appraise the existing gas pay in the tight Dongara Sandstone and test the deeper Kingia and High Cliff sandstones productive in the Waitsia gas field to the west.

In July, Quadrant Energy and Carnarvon Petroleum announced the largest oil discovery in Australia for 30 years in their Dorado-1 well in the Bedout Sub-basin. The well intersected 80 metres of net oil pay in the Early Triassic Caley Sandstone plus another 40 metres of net oil pay in deeper sandstones. The potential P2 reserves for the structure were estimated to be 170 million barrels oil. This was on top of their gas discoveries at Phoenix South and Roc, the latter of which tested at 51 MMcfg/d and didn't include the enormous exploration potential in the rest of their permits in the basin. There had only been four wells drilled in this basin (and two had gas in tight sands) since exploration began so it truely was a frontier basin. However it was still a staggering result – how could such an area be overlooked for so long? Those sneaky rocks had been "smiling at us" for as long as they had been at Waitsia! It didn't take long for the penny to drop so in August, Santos agreed to purchase Quadrant for $US2.15 billion plus a contingency payment for the Dorado potential. Santos will bring its offshore operating experience to compliment Quadrant's technical staff.

The Good Oil Conference was held in Perth in September and I received a free pass for each day. All the talk was on the Dorado oil discovery and how much additional potential there was in their four permits covering the basin. I also looked in on the Beach Energy booth. They were very keen on drilling Beharra Springs Deep to evaluate the Kingia/High Cliff play in this area. They were also planning to acquire 3D seismic in the area to the southeast of Beharra Springs where there were other tilted fault blocks. If the Kingia/High Cliff play extended this far from Waitsia, then this would upgrade the Eneabba West Prospect mapped by Peter Kirk in the old Empire Oil days and now in EPA 82 owned by UIL Energy (now Strike Oil). It would also upgrade the Lockyer Deep and North Erregulla prospects now owned by Mineral Resources.

In October, Strike Energy announced that, subject to shareholder approval, it had agreed to a takeover offer to acquire UIL Energy. Once this goes through, I will be fully retired and this is a good point to close off this book. But before I close off, I am delighted to report that West Erregulla-2 was drilled in June-August 2019 and not only encountered gas in the Wagina Sandstone but also intersected 58m net gas pay in the Kingia Sandstone and 10m net gas pay in the High Cliff Sandstone. The well was tested at a staggering rate of 69 MMcfg/d. The drilling rig then moved to Beach Energy who drilled Beharra Springs Deep-1 which encountered 36m net gas pay in the Kingia Sandstone and tested at 25-31 MMcfg/d. The Perth Basin is no longer the black sheep of the Australian oil industry! It appears to me that over the next decades, every structure in the northern Perth Basin that was drilled but terminated in the Dongara/Wagina sandstones has to be redrilled and deepened to the High Cliff Sandstone. And one day, some of these may even find significant oil fields.

I've had a fabulous life and there's nothing I'd change if I had to live it over again – even the bad times. I've got a fabulous wife who has stuck by me through thick and thin; I have two great and successful sons and four gorgeous grandchildren. And of course, my fifty plus years in the Australian oil industry have been equally as rewarding. So, following the recent discoveries at Waitsia and Dorado, I go back to the theme of this book and give the following advice to all those future oil explorers – "the rocks won't smile at you" unless you drill that crazy exploration well!

I wrote this book to provide some history of what the old oil industry was like prior the oil price crash in 1986 and the subsequent transformation of the industry by computers. I also hope this book leaves something for future generations of oil finders to pursue.

Glossary Of Terms

Acidisation – a process of pumping acid into a hydrocarbon reservoir (usually a carbonate) to increase flow rates.

Aeolian – a sedimentary rock formed by wind blown sand.

Alluvial – sediments deposited in continental and flood plain environments.

Anticline – sediments folded into an upward convex geometry.

API gravity – the American Petroleum Institute standard method for specifying the density of crude oil – the higher the number the lighter the oil.

Appraisal well – a well drilled into a known hydrocarbon accumulation for the purpose of determining the extent of the discovery.

Aquifer – subsurface sediments capable of producing water from a well.

Basement – metamorphic or igneous rocks unconformably overlain by sediments.

Barrel of oil – 42 US gallons – 35 imperial gallons – 159 litres.

Bopd – Barrels of oil per day.

Boepd – Barrels of oil equivalent per day on the basis that one barrel of oil has approximately the same energy value as 6,000 cubic feet of gas.

BCF – billion cubic feet of gas.

Breccia – sedimentary rock composed of angular rock fragments cemented by fine grained matrix.

Calcarenite – a limestone made up of sand sized calcium carbonate grains.

Carbonaceous – containing carbon rich debris such as plants.

Carbonate – a sedimentary deposit containing mostly calcium carbonate.

Casing – steel pipe used to line the borehole to prevent wall collapse and the entry of formation fluids.

Clastics – sedimentary deposits containing mostly sandstones, siltstones and shales.

Coal Measures – a sedimentary deposit containing clastics interbedded with coal deposits.

Completion – an operation to complete a well for production including setting casing, perforating, stimulating and testing the well.

Conglomerate – a very coarse sediment deposited under very high energy environments.

Condensate – hydrocarbons associated with natural gas which are liquid at the surface but are gaseous in the subsurface.

Coquina – a sedimentary rock composed mostly of transported fragments of shells.

Centipoise - a measure of vicosity. One centipoises is the viscosity of water at 20°C.

Darcy – a measurement of permeability equal to one thousand millidarcies.

Deltaic – sediments deposited at the mouth of a river.

Development well – a well drilled specifically for production from a field which has been declared commercial.

Diagenetic – physical and chemical changes to sediments after deposition and burial.

Dry hole – an exploration well that did not discover significant hydrocarbons.

DST (Drill Stem Test) – testing of an interval in a borehole by isolating that interval with packers and allowing formation fluids to flow into the drill string and up to the surface.

Electric logging – physical measurements of rock properties such as natural radioactivity, density, resistivity and acoustic velocities made by wireline tools recorded continuously as they are winched to the surface.

Evaporites - are sedimentary rocks deposited from brines in basins where the amount of water lost by evaporation exceeds the amount of water from rainfall and influx from rivers.

Exploration well – the first well that is drilled into a potential hydrocarbon trap. Also called a Wildcat.

Facies – the features of a sedimentary rock which characterises its environment of deposition.

Farmin/Farmout – where a company (the farminee) acquires an interest in a tenement from another company (the farmor) by funding an agreed exploration program.

Fault – a fracture surface in a rock. Where the rocks are under tension, the fault is normal and when the rocks are under compression the fault is reverse.

Fluvial – sediments deposited in river systems.

Fracture stimulation – the process of artificially inducing fractures in a hydrocarbon reservoir to increase flow rates (see also Fig 61).

G-I-P (Gas in Place) – means the total volume of gas in a hydrocarbon reservoir but does not indicate the volume of gas that is economically recoverable.

GJ (Gigajoule) – is a measurement of energy (one billion joules) which is approximately equal to the energy value of one thousand cubic feet of gas.

GOR (Gas Oil Ratio) – is the volume of gas produced with oil and expressed as a ratio per barrel.

Graben – a faulted downthrown block.

Horst – a faulted upthrown block.

Inlier – an older rock formation outcropping and surrounded by younger rocks.

Karstified – topography formed by the erosion and dissolution of limestones and often associated with vuggy porosity.

Kerogen – the solid and insoluble organic matter in sedimentary rocks.

Lacustrine – sediments deposited in lakes.

Limestone –a sedimentary rock consisting mostly of calcium carbonate.

Metamorphic – sedimentary rocks that have been deeply buried, subjected to heat and pressure and deformed by compression.

Migration – the movement of oil generated from a source rock either laterally or vertically into a reservoir.

Millidarcies – a measurement of permeability.

MMcfg/d – the flow of gas from a well measured as millions of cubic feet of gas per day.

Multiples – seismic energy that has been reflected more than once thereby masking the real reflections.

NPV (Net Present Value) – the sum of all cash flows (positive and negative) of a project discounted to time zero at a given discount rate.

Operator - the company engaged by the joint venture to manage the exploration work program and liase with all government authorities as required.

Outlier – a younger rock formation outcropping and surrounded by older rocks.

Over-pressure – in sediments is caused by the inability of pore fluids to escape as they are compacted by the pressure of the overlying sediments. It can cause difficulties while drilling.

Palaeogeography – the interpreted geography and environments in a given area at a particular geological time.

Permeability – the ability of a fluid to flow through a porous rock.

Petrology – the study of rocks dealing with their composition, texture and structure using thin sections and microscopes.

Petrophysical – the physical measurements of the properties of rocks and their fluid or gas contents.

PJ (Petajoule) – is a million gigajoules and is approximately equivalent to the energy value of one billion cubic feet of gas.

Plugged and abandoned – the process of sealing a dry hole with cement to avoid contamination of the environment.

Porosity - the space in a rock not occupied by solid minerals that can be filled with hydrocarbons or water.

Prospect – a potential trap for hydrocarbons which is sufficiently defined for drilling.

Quartzite – a sandstone whose original porosity is now filled with silica cement and is therefore tight.

Red Beds – red coloured sediments generally deposited in oxidising environments.

Reef – a buildup of carbonate mounds formed by organisms such as corals, shells and crinoids.

Reserves – the volume of economically recoverable hydrocarbons in a trap.

Reservoir – porous sediments capable of containing hydrocarbons.

Resource – estimated volumes of hydrocarbons in a trap where the certainty of recovery and commerciality are insufficient to classify them as reserves.

Sandstone – a sedimentary rock made up mostly of sand grains.

Saturation – the percentage of water, oil and/or gas in the pore spaces in a reservoir.

Seal – an impenetrable bed of rock through which oil and/or gas cannot pass.

Seismic – method of determining the subsurface features by sending sound waves down into the various rock layers where they are reflected back and the time it takes for them to reach the surface is measured.

Shale – a fine grained sedimentary rock made up mostly of clay.

Siltstone – a sedimentary rock made up mostly of silt sized grains.

Source Rock – a fine grained sediment rich in organic matter which when buried deeply will under heat and pressure generate hydrocarbons.

Stratigraphy – the study of the subdivision, composition, age and correlation of sedimentary rocks.

Structure – a geometric configuration of sediments deformed into features such as anticlines and fault blocks capable of trapping hydrocarbons.

Subduction – a plate tectonic term for the sliding of one plate over another.

Syncline – sediments folded into an upwardly concave geometry.

TCF – a trillion cubic feet of gas.

Tenement - means any permit, licence or lease issued by state or federal governments and conferring rights to explore for and produce hydrocarbons in accordance with the relevant petroleum act.

Thermal Maturity – the level of heating that a source rock has been subjected to by depth of burial. It grades from no generation to oil generation then gas generation and finally no further generation (overcooked) as it is progressively buried deeper.

Tillite – a sedimentary rock deposited under glacial environments.

Tight – a term used to describe a reservoir with little or no permeability.

TJ (Terrajoule) - is a thousand gigajoules and is approximately equivalent to the energy value of a million cubic feet of gas.

Transgression – the advance of the sea across the land.

Trap – a subsurface reservoir rock with a sealing caprock in a configuration capable of containing migrating hydrocarbons.

Turbidite – a sedimentary rock formed by sediment gravity flow responsible for depositing clastic sediments in deep oceans or lakes.

Unconformity – a surface of erosion or non-deposition separating younger from older beds.

Updip – the location of an exploration well that is interpreted to be structurally higher than a nearby well that encountered reservoirs capable of trapping hydrocarbons.

Vuggy porosity – cavities, voids and fractures in carbonates formed by dissolution of calcium carbonate by migrating fluids.

Water drive – the ability of water to sweep and replace produced hydrocarbons.

WTI – West Texas Intermediate –light sweet Texas crude oil used as a benchmark for oil pricing.

ERA	PERIOD	EPOCH	AGE millions of years (not to scale)	MAJOR EVENTS
CENOZOIC	QUATERNARY	HOLOCENE		Modern Humans
			0.01	
		PLEISTOCENE		First Homo sapiens
			2	
	TERTIARY	PLIOCENE		First hominids
			5	
		MIOCENE		
			23	Mammals dominant
		OLIGOCENE		
			34	First elephants
		EOCENE		
			55	First horses
		PALEOCENE		
			66	Major Extinction Event
MESOZOIC	CRETACEOUS			First flowering plants
			145	
	JURASSIC			First birds
			200	Dinosaurs dominant
	TRIASSIC			First mammals
			250	Major Extinction Event
PALAEOZOIC	PERMIAN			
			300	Major Ice Age
	CARBONIFEROUS			
			360	First reptiles
	DEVONIAN			
			420	First fishes
	SILURIAN			
			445	First plants
	ORDOVICIAN			
			490	First corals
	CAMBRIAN			
			540	First abundant invertebrates
PRECAMBRIAN	PROTEROZOIC			Soft bodied animals First algae
			2500	First stromatolites
	ARCHAEN			
			4000	
	HADEAN			
			4600	Age of Earth

APPENDIX 1

Made in the USA
Las Vegas, NV
03 November 2026

57424226R00250